Composites and Advanced Materials for Industrial Applications

K. Kumar
Birla Institute of Technology, India

J. Paulo Davim
University of Aveiro, Portugal

A volume in the Advances in Chemical and Materials Engineering (ACME) Book Series

IGI Global
DISSEMINATOR OF KNOWLEDGE

Published in the United States of America by
IGI Global
Engineering Science Reference (an imprint of IGI Global)
701 E. Chocolate Avenue
Hershey PA, USA 17033
Tel: 717-533-8845
Fax: 717-533-8661
E-mail: cust@igi-global.com
Web site: http://www.igi-global.com

Library of Congress Cataloging-in-Publication Data

Names: Kumar, K. (Kaushik), 1968- editor. | Davim, J. Paulo, editor.
Title: Composites and advanced materials for industrial applications / K. Kumar and J. Paulo Davim, editors.
Description: Hershey, PA : Engineering Science Reference, [2018] | Includes bibliographical references
Identifiers: LCCN 2017037490| ISBN 9781522552161 (h/c) | ISBN 9781522552178 (eISBN)
Subjects: LCSH: Composite materials--Industrial applications.
Classification: LCC TA418.9.C6 C63134 2018 | DDC 620.1/18--dc23 LC record available at https://lccn.loc.gov/2017037490

This book is published in the IGI Global book series Advances in Chemical and Materials Engineering (ACME) (ISSN: 2327-5448; eISSN: 2327-5456)

British Cataloguing in Publication Data
A Cataloguing in Publication record for this book is available from the British Library.

Advances in Chemical and Materials Engineering (ACME) Book Series

ISSN:2327-5448
EISSN:2327-5456

Editor-in-Chief: J. Paulo Davim, University of Aveiro, Portugal

MISSION

The cross disciplinary approach of chemical and materials engineering is rapidly growing as it applies to the study of educational, scientific and industrial research activities by solving complex chemical problems using computational techniques and statistical methods.

The **Advances in Chemical and Materials Engineering (ACME) Book Series** provides research on the recent advances throughout computational and statistical methods of analysis and modeling. This series brings together collaboration between chemists, engineers, statisticians, and computer scientists and offers a wealth of knowledge and useful tools to academics, practitioners, and professionals through high quality publications.

COVERAGE

- Fatigue and Creep
- Ceramics and glasses
- Statistical methods
- Composites
- Polymers Engineering
- Thermo-Chemical Treatments
- Industrial chemistry
- Chemical Engineering
- Metallic Alloys
- Sustainable Materials

The Advances in Chemical and Materials Engineering (ACME) Book Series (ISSN 2327-5448) is published by IGI Global, 701 E. Chocolate Avenue, Hershey, PA 17033-1240, USA, www.igi-global.com. This series is composed of titles available for purchase individually; each title is edited to be contextually exclusive from any other title within the series. For pricing and ordering information please visit http://www.igi-global.com/book-series/advances-chemical-materials-engineering/73687. Postmaster: Send all address changes to above address.

Titles in this Series

For a list of additional titles in this series, please visit:
https://www.igi-global.com/book-series/advances-chemical-materials-engineering/73687

Energetic Materials Research, Applications, and New Technologies
Rene Francisco Boschi Goncalves (Federal University of Para, Brazil) José Atilio Fritz Fidel Rocco (Aeronautics Institute of Technology, Brazil) and Koshun Iha (Aeronautics Institute of Technology Brazil)
Engineering Science Reference • ©2018 • 367pp • H/C (ISBN: 9781522529033) • US $225.00

Chemical Compound Structures and the Higher Dimension of Molecules Emerging Research...
Gennadiy Vladimirovich Zhizhin (Russian Academy of Natural Sciences, Russia)
Engineering Science Reference • ©2018 • 221pp • H/C (ISBN: 9781522541080) • US $175.00

Emergent Research on Polymeric and Composite Materials
R. Somashekar (University of Mysore, India) and Thejas Urs G. (University of Mysore, India)
Engineering Science Reference • ©2018 • 341pp • H/C (ISBN: 9781522530237) • US $235.00

3D Printing and Its Impact on the Production of Fully Functional Components Emerging...
Petar Kocovic (Union – Nikola Tesla University, Serbia)
Engineering Science Reference • ©2017 • 115pp • H/C (ISBN: 9781522522898) • US $150.00

Advanced Applications of Supercritical Fluids in Energy Systems
Lin Chen (Tohoku University, Japan & Japan Society for the Promotion of Science (JSPS), Japan) and Yuhiro Iwamoto (Nagoya Institute of Technology, Japan)
Engineering Science Reference • ©2017 • 682pp • H/C (ISBN: 9781522520474) • US $235.00

For an entire list of titles in this series, please visit:
https://www.igi-global.com/book-series/advances-chemical-materials-engineering/73687

IGI Global
DISSEMINATOR OF KNOWLEDGE

701 East Chocolate Avenue, Hershey, PA 17033, USA
Tel: 717-533-8845 x100 • Fax: 717-533-8661
E-Mail: cust@igi-global.com • www.igi-global.com

Table of Contents

Detailed Table of Contents

Polymeric composites are being used widely for industrial applications. This is because of a number of advantages offered by the composite materials. Some of the advantages offered are low weight, enhanced mechanical properties, excellent durability, etc. Therefore, this chapter provides an overview of different industrial applications of polymeric composites. The different industrial sectors considered are automotive, military, aerospace, sports and leisure, and industrial construction. The authors are of the opinion that the importance of polymeric composites could be realized with the applications presented in this review.

The development of hybrid-fiber-reinforced composites has increased in recent decades because of its abundance, low cost, low weight, high strength, stiffness, and bio-degradability, thereby increasing its engineering applications. However, the major drawbacks of natural fibers in composites are their high moisture absorption and poor compatibility between fiber and matrix. Hence, chemical treatments are primarily considered to modify the fiber surfaces with the objective to improve interfacial bonding between fiber and matrix. This chapter addresses an overview of chemical treatments and their effects on natural fibers-based hybrid composites are reviewed. The chemical treatments include alkali, silane, maleated, and others, focused mainly on hybrid natural fiber composites. The significance of chemical treatment of natural fibers aimed to improve adhesion between fiber surface and

Quality products with low cost manufacturing routes are the major objectives for the product development in any application. The current statement is evident for polymer-matrix composites, particularly in high end applications such as aerospace and mass transit structures. These applications require advanced composite materials tailored to meet the property demands posted by dynamic load conditions, and hence, the use of wide spectrum of constituents and architectures are vital to cater the needs. Consequently, the development of novel composite materials with the permutations of ingredients leads to the innovative processing techniques. To address the gap in the manufacturing with economical processing routes of thick sectioned advanced composite parts showing superior properties at different wall sections, an innovative composite manufacturing technology coupling resin transfer moulding (RTM) processing and vacuum applications, namely vacuum enhanced resin transfer moulding technology (VERTMTy), is conceptualized, proposed, and developed.

Selection of material is a crucial part of the voyage of obtaining the optimal product. The task is challenging for the designer who has to select the optimal material from a list of thousands of materials that satisfy the tangible and intangible criteria. The problem of identification of the optimum material can be solved using different MCDM (multi-criteria decision-making) tools. This chapter proposes to select the optimal material for polymeric composite bumper beam using TODIM (tomada de decisao interativa multicriterio, which means in Portuguese "interactive and multi-criteria decision-making") method. Six different composite materials were considered. Glass fiber epoxy turns out to be the optimal material for the bumper beam. Further, finite element analysis was carried out to study the behavior of the candidate materials under specific loading conditions.

In the last decades, automotive industry has especially focused on developing and applying new materials and technologies for enhancing the comfort and security levels in the vehicles, but on the other hand for reducing the weight of the means of transportation in order to decrease the energy consumption. In this concept, textile-reinforced composite structures come in possession of one of the most favored materials in the automotive industry by satisfying these demands. In this chapter, usage of textile-reinforced composites in the automotive industry has been elucidated under three main sections: textile reinforced composites for 1) automobiles, 2) mass transportation vehicles, and 3) trucks. The aim of this chapter is to discuss the subject in detail by giving technical information about particular vehicle parts and composite structures utilized in the automotive industry and academia.

The concern towards environmental issues and the need for more polymer-based composites has extended its collection towards polymer composites strengthened with natural fillers. The main aim of the chapter is to make the side metallic sticks (called braces, where aluminum alloys are being used) of orthotic calipers with epoxy-based composite reinforced with coir dust as filler. Its mechanical behavior together with density, tensile and three-point bending are observed and compared with the existing aluminum-based calipers components after validating the experimental as well as virtual results. Virtual testing is done using 3D software CREO and analysis is done with the help of ANSYS workbench. It is seen that the strength and stiffness of epoxy-based composite with natural fillers is more than that of presently used aluminum alloy. The microstructure of the composites is considered to summarize the general filler distribution in the matrix using scanning electron microscope.

This chapter reports the recent advances in the fabrication methods, properties, and microelectronics packaging applications of various inorganic fillers and reinforced-polymer composites. Recently, inorganic particles, including ceramics and carbon-based material reinforced polymeric matrices, have attracted both academic and industrial interest because they exhibit good thermal and mechanical properties. The low dielectric constant and dielectric loss, the low thermal expansion coefficient, and high thermal conductivity make these kinds of composites suitable for microelectronics packaging. The filler ratio, surface modification, and preparation methods of these composites have a marked effect on the final properties of these materials. Herein, the preparation methods, thermal and dielectric properties, shortcomings, and microelectronics applications of polymers/inorganic composites are summarized and discussed along with detailed examples collected from the extensive scientific literature.

The development of advanced composites not only enhances strength, ductility, durability of materials, and endows materials with the multifunctional property, but also reduces the construction cost and promotes civil engineering infrastructure to make sustainable development. In this chapter, several representative advanced composites with abundant research achievements and wide applications are systematically introduced with regard to cementitious composites, fiber-reinforced polymer composites, novel thermally functional composites, and 3D printing composites in terms of their definitions, properties, research progress, and applications in civil engineering infrastructures.

This chapter presents a broad perspective of the recent research done on laminated composite hypar shells used as roofing units. Different types of analysis including bending, vibration, buckling, impact, and failure are included. The chapter is collated and categorized based on various aspect of research. The first aspect concentrates on typical analyses including problems in which various boundary conditions and laminations are considered. Then it focuses on the structural complexities which include stiffened shells, shells with cutouts, etc. The basic results of theoretical and experimental investigations of stress-strain state, vibration, buckling, and failure are summarized in this chapter incorporating the review of materials published in scientific and technical journals and proceedings in recent times.

This chapter explores the concept of visible light active nanocomposites for the enhanced photocatalytic hydrogen generation and dye degradation. Since the late 1960s, A. Fujishima has been involved in unfolding the fascinating characteristics of titanium dioxide (TiO2) as semiconductor oxide. The increased growth in population and industrial development has tremendously increased the generation of waste products and consumption of energy worldwide. This situation creates an immense need of clean and sustainable alternative sources of energy. Hydrogen, having a high energy capacity, is considered as a reliable fuel for the future energy requirements. In addition to that, due to the rapid industrialisation, our water is being contaminated with various harmful industrial effluents. This chapter illustrates the significance of visible light nanocomposites for the photocatalytic application of hydrogen generation for future energy security and dye degradation for the effective effluent treatment of textile industries.

This chapter aims to discuss the evolution of electroless nickel coatings with respect to their tribological behavior with special emphasis on their applicability at high-temperature-based applications. Electroless nickel coatings have tremendous potential as anti-wear and anti-friction coatings under ambient condition. The investigation of their tribological properties at high temperatures is relatively new. At demanding conditions, most conventional lubricants lose their properties and hence the use of self-lubricating coatings is unavoidable. Due to high melting point of nickel and oxidation resistance, electroless nickel-based coatings may prove to be suitable candidate at high temperatures. A review and analysis of the tribological characteristics of electroless nickel coatings especially at high temperatures is therefore necessary and has been reported in this chapter. Future research directions for the improvement of coating properties at high temperature are also identified from the review.

Preface

The editors are pleased to present the book *Composites and Advanced Materials for Industrial Applications* under the book series *Advances in Chemical and Materials Engineering (ACME)*. Book title was chosen as it converges upcoming technologies in composite materials for the next decade. In present time, "Composite" is the buzzword for major discipline and many scholars are working in these areas. This book provides insight for all researchers, academicians, post graduate or senior undergraduate students working in the area.

Materials are always a gift to the mankind although it always remains a conflict that whether a material gives birth to a new technology or vice versa. One method of producing new material with combination of properties is by combining the material in solid state which is referred as composite materials. Composite materials are the class of advanced material made up of one or more material in solid state with distinct physical and chemical properties. They are lighter in weight and have high Mechanical Properties. Today, composites are in use in a variety of structures, ranging from spacecraft and aircraft to buildings and bridges. This wide use of composites has been facilitated by the introduction of new materials, improvements in manufacturing processes and developments of new analytical and testing methods.

Unfortunately, information on composites and their applications are scattered in journal articles, in conference and symposium proceedings, in workshop notes, and in government and company reports. This proliferation of the source material, coupled with the fact that some of the relevant publications are hard to find or are restricted, makes it difficult to identify and obtain the up-to-date knowledge needed to utilize composites to their full advantage. This book intends to overcome these difficulties by presenting, in a single volume, many of the recent advances in the field of application of composite materials. The main focus of this book is on polymeric matrix, metal matrix, and ceramic matrix composites along with inorganic, organic and natural reinforcements in fibre and particulate form.

Industrial applications of composites are multifaceted and are visible in almost all industries. Few examples can be:

- Aircraft/Military (Commercial, pleasure and military aircrafts, including components for aerospace and related applications)
- Appliance/Business (Composite applications for the household and office including appliances, power tools, business equipment, etc.)
- Automotive/Transportation (The largest of the markets, products include parts for automobiles, trucks, rail and farm applications.)
- Civil Infrastructure (A relatively new market for composites, these applications include the repair and replacement of civil infrastructure including buildings, roads, bridges, piling, etc.)
- Construction (Includes materials for the building of homes, offices, and architectural components. Products include swimming pools, bathroom fixtures, wall panels, roofing, architectural cladding.)
- Consumer (Products include sports and recreational equipment such as golf clubs, tennis rackets, snowmobiles, mobile campers, furniture, microwave cookware.)
- Corrosion-Resistant Equipment (Products for chemical-resistant service such as tanks, ducts and hoods, pumps, fans, grating, chemical processing, pulp and paper, oil and gas, and water/wastewater treatment markets.)
- Electrical (This encompassing market includes components for both electrical and electronic applications such as pole line hardware, substation equipment, microwave antennas, printed wiring boards, etc.)
- Marine (Products for commercial, pleasure and naval boats and ships). etc. etc.

The chapters in the book have been provided by Researchers and Academicians working in the field and have gained considerable success in the field.

Chapter 1 introduces the readers to an overview of the wide industrial applications of polymeric composites. The different industrial sectors considered are automotive, military, aerospace, sports and leisure and industrial construction. It is felt by the authors that the importance of polymeric composites could be realized with the applications presented in the chapter.

Chapter 2 indicates a review of development of hybrid fiber reinforced composites. The class of composites has got an increased interest in recent decades because of its abundance, low cost, low weight, high strength, and stiffness, and bio-degradable and thereby increasing its engineering applications. However, the author feels that major drawbacks of natural fibers in composites are their high moisture absorption and poor compatibility between fiber and matrix. Hence considers chemical treatments to be of utmost importance to modify the fiber surfaces with the objective to improve interfacial bonding between fiber and matrix. So the chapter addresses an overview of chemical treatments and their effects on natural fibers based hybrid composites.

The chemical treatments include alkali, silane, maleated and others, focused mainly on hybrid natural fiber composites were studied. The significance of chemical treatment of natural fibers was aimed to improve adhesion between fiber surface and matrix along with reduction in water absorption property to improve physical and mechanical properties as compared with untreated fibers for use in components of engineering applications.

From here the book starts with the specific application to various sectors. Chapters 3, 4, 5, 6, and 7 discuss the application towards Aerospace and Automobile Industies.

Chapter 3 discusses about growing applications of Nitinol in aerospace industries and also towards MEMS and bio-medical industries. This is due to its unique properties of pseudo-elasticity, bio-compatibility and shape-memory effect. The authors feel that behaviour of NiTi alloy i.e. Nitinol can be changed by altering the composition, modifying the porosity and applying external thermal and mechanical treatment. The authors in there presentation has fabricated porous NiTi alloy with powder metallurgy by varying the composition of polypropylene as an organic binders from 0% to 15% and Young's Modulus and porosity of porous alloy were evaluated. The effect of process parameters viz.: compaction pressure, sintering temperature and sintering time were optimized for optimal output. It was observed that the particles initially act as a binder but with the increase of temperature, they evaporate and create pores. So with the increase of organic particle percentage, the porosity increases and Strength obviously decreases.

Chapter 4 illustrates the usage of silica gel reinforced aluminium composite in piston of IC engine, one of the most complex parts among all engine components. As during the operation, the pistons of IC engines are typically subjected to high loading and wearing, so to withstand these, it is essential to have the component made up with a material possessing high mechanical and excellent tribological properties. The authors have compared mechanical as well as tribological properties of the suggested composite with aluminium alloy which is presently being used in manufacturing of piston of IC engine and concluded that it can enhance the performance to a great extent.

In Chapter 5, an innovative composite manufacturing technology coupling resin transfer moulding (RTM) processing and vacuum applications namely "Vacuum Enhanced Resin Transfer Moulding Technology (VERTMTy)" is conceptualized, proposed and developed. The same has been developed for polymer-matrix composites towards high end applications such as aerospace & mass transit structures. These applications require advanced composite materials tailored to meet the property demands posted by dynamic load conditions and hence, the use of wide spectrum of constituents and architectures are vital to cater the needs. The chapter proposes to develop quality products with low cost manufacturing routes as it is the major objectives for any product development in any application.

The task is really challenging for any designer who has to select the optimal material from a list of thousands of material satisfying the tangible and intangible criteria of the product. Chapter 6 proposes an optimal method towards material selection in obtaining the optimal product. The authors have proposed TODIM (TOmada de Decisao Interativa Multicriterio, which means in Portuguese 'interactive and multi-criteria decision making') method, flagged under MCDM (Multi-Criteria Decision Making) tools. The chapter considers selection of the optimal material for polymeric composite bumper beam using six different composite materials and have concluded that Glass fiber reinforced epoxy composite is the optimal material for the bumper beam within the considered materials. Further, finite element analysis was also carried out to study the behaviour of the chosen materials under specific loading conditions.

Chapter 7 of the book proposes usage of textile reinforced composites in the automotive industry. In the last decades, automotive industry has especially focused on developing and applying new materials and technologies for enhancing the comfort and security levels in the vehicles, but on the other hand for reducing the weight of the means of transportation in order to decrease the energy consumption. In this concept, the authors feel that textile reinforced composite structures come in possession as one of the most favored materials by satisfying these demands. In this chapter, usage of the same has been elucidated under three main sections, i.e. passenger automobiles, mass transportation vehicles and trucks. The chapter focuses on discussion in detail by giving technical information about particular vehicle parts and the composite structures utilized in the same.

Chapters 8, 9, 10, 11, and 12 discuss the application of composites in various industries like Bio-Mechanical, Packaging, Civil etc.

Chapter 8 of the book proposes usage of green composite (epoxy based composite reinforced with coir dust as filler) to make the side metallic sticks (called Braces, where aluminum alloys are being used). The proposal serves two basic necessity of the society. First of all the concern towards environmental issues and secondly usage of a low weight member for better locomotion of Polio infected Patients. It is seen that the proposed composite offers more strength and stiffness than that of presently used aluminum alloy.

Chapter 9 illustrates recent advances in the fabrication methods, properties, and microelectronics packaging applications of various inorganic fillers and reinforced-polymer composites. Currently, inorganic particles, including ceramics and carbon-based material reinforced polymeric matrices have attracted both academic and industrial interest because they exhibit good thermal and mechanical properties. The low dielectric constant and dielectric loss, the low thermal expansion coefficient, and high thermal conductivity make these kinds of composites suitable for microelectronics packaging. The filler ratio, surface-modification, and preparation methods of these composites have a marked effect on the final properties of these materials. So in this chapter preparation methods, thermal and dielectric properties, shortcomings, and microelectronics applications of polymers/inorganic composites are summarized and discussed along with detailed examples collected from the extensive scientific literature.

In Chapter 10, development of advanced composites towards civil engineering applications has been elaborated. The usage not only enhances strength, ductility, durability of composites and endows composites with the multifunctional property, but also reduces the construction cost and promotes civil engineering infrastructure to make sustainable development. In this chapter, several representative advanced composites with abundant research achievements and wide applications are systematically introduced with regard to cement based composites, fiber reinforced polymer composites, novel thermally functional composites and 3D printing composites, in terms of their definitions, properties, research progress and applications in civil engineering infrastructures.

Chapter 11 also aims towards civil construction i.e. roofing unit. The chapter presents a broad perspective of the recent research done on laminated composite hypar shells used as roofing units. Different types of analysis including bending, vibration, buckling, impact, failure have been discussed. The article is collated and categorized based on various aspect of research. The first aspect concentrates on typical analyses including problems in which various boundary conditions and laminations are considered. Then it focuses on the structural complexities which include stiffened shells, shells with cutouts etc. The basic results of theoretical and experimental investigations of stress-strain state, vibration, buckling and failure are summarized in this article incorporating the review of materials published in scientific and technical journals and proceedings in recent times.

Chapter 12 aims towards study to explore the concept of visible light active nanocomposites for the enhanced photocatalytic hydrogen generation and dye degradation. Since the late 1960s, A. Fujishima has been involved in unfolding the fascinating characteristics of titanium dioxide (TiO2) as semiconductor oxide. The increased growth in population and industrial development has tremendously increased the generation of waste products and consumption of energy worldwide. This situation creates an immense need of clean and sustainable alternative source of energy. Hydrogen, having a high energy capacity is considered as a reliable fuel for the future energy requirements. In addition to that, due to the rapid industrialisation, water conserves are also getting contaminated with various harmful industrial effluents. This chapter illustrates the significance of visible light nanocomposites for the photocatalytic application such as hydrogen generation for future energy security and dye degradation for the effective effluent treatment of textile industries.

Chapter 13 aims to discuss the evolution of the electroless nickel coatings with respect to their tribological behaviour and special emphasis on their applicability at high temperature based applications. Electroless nickel coatings have tremendous potential as anti-wear and anti-friction coatings under ambient condition. The investigation of their tribological properties at high temperatures is relatively new. At demanding conditions, most conventional lubricants lose their properties and hence the use of self-lubricating coatings is unavoidable. Due to high melting point of nickel and oxidation resistance, electroless nickel based coatings may prove to be suitable candidate at high temperatures. A review and analysis of the tribological characteristics of electroless nickel coatings especially at high temperatures is therefore necessary and has been reported in this chapter. Future research directions for the improvement of coating properties at high temperature are also identified from the review.

First and foremost, the editors would like to thank God. In the process of putting this book together it was realized how true this gift of writing is for anyone. You have given the power to believe in passion, hard work and pursue dreams. This could never have been done without the faith in You, the Almighty. The Editors would also like to thank all the Chapter Contributors, the Reviewers, the Editorial Advisory Board Members, Book Development Editor and the team of Publisher IGI Global for their availability for work on this editorial project.

K. Kumar
Birla Institute of Technology, India

J. Paulo Davim
University of Aveiro, Portugal

Chapter 1
Industrial Applications of Polymer Composite Materials

Divya Zindani
National Institute of Technology Silchar, India

K. Kumar
Birla Institute of Technology, India

ABSTRACT

Polymeric composites are being used widely for industrial applications. This is because of a number of advantages offered by the composite materials. Some of the advantages offered are low weight, enhanced mechanical properties, excellent durability, etc. Therefore, this chapter provides an overview of different industrial applications of polymeric composites. The different industrial sectors considered are automotive, military, aerospace, sports and leisure, and industrial construction. The authors are of the opinion that the importance of polymeric composites could be realized with the applications presented in this review.

INTRODUCTION

Composites are materials consisting of two or more phases that are physically and chemically distinct. The different phases constituting a composite are separated by distinct interfaces. The combination results in a material with enhanced structural and functional properties (Mayer et al., 1998). Composite materials have a number of advantages such as high resistance to corrosion, high fatigue strength and low weight. The advantages associated with the composites have led to their widespread application. A wide array of industrial such as aerospace, automotive, sports leisure

DOI: 10.4018/978-1-5225-5216-1.ch001

goods, military and biomedical are using composite materials to produce different products (Shaw et al., 2010). Umpteen examples can be found of composites in nature, for instance the connective tissues in mammals.

Composite materials consist of matrix phase and a dispersed phase. Matrix phase has continuous character and is the primary phase. The matrix phase serves to hold the dispersed phase and shares the load with it. Dispersed phase on the other hand is discontinuous in nature and is embedded in the matrix. The dispersed phase is known as the reinforcement and is the secondary phase. The dispersing phase typically has higher strength than the matrix phase. Examples include that of wood wherein the matrix is lignin and the dispersing phase are the cellulose fibres. Reinforced concrete is another example of composite material wherein the steel is the dispersing phase and concrete is the matrix.

With the advancements, more advanced composites were invented with their roots expanding to industrial applications. These advanced composites have a higher performing reinforcement in polymeric matrix such as graphite/epoxy, Kevlar/epoxy and boron/epoxy composites. Further, the flexibility to obtain the desired mechanical properties suiting the structural needs and aspects, the formability, dimensional stability, thermal and acoustical characteristics etc. have rendered the polymeric composites for a wide range of industrial applications such as transportation, industrial and sports and leisure.

The construction industries have considered the use of fibre reinforced polymeric composites such as glass, aramid fibres in matrices of epoxy, polyester, vynilester resins (Bunshell and Renard, 2005). Basalt fibres have been considered and used by the construction industries to suit the high temperature situations (Şerbescu et al., 2006). The use of composites has made it easier to fabricate and assemble different building structures with enhanced load bearing capabilities (Taranu, 2009).

Similarly in the automotive industry, the traditionally used metals are being replaced by the polymeric composites which has not only become a trend but also a necessity. The reason can be attributed to the increasing legal restrictions by the different regulating bodies across the globe. The use of polymeric composites has increased the fuel economy of the different class of vehicles.

Thus polymeric composites (PMC) have become an integral part of a wide range of industries. As such their applications in the industries should be explored systematically. The present chapter is an attempt to explore the industrial applications of the polymeric composites. The chapter is organized to include the following sections: Polymeric matrix composites, discussing the advantages, disadvantages, factors affecting properties of PMCs and fabrication techniques of PMCs; Applications of PMC, discussing industrial applications of the PMCs. The chapter finally concludes with the concluding remarks.

Polymer Matrix Composites

The PMCs uses a matrix material which is typically a resin solution. A wide range of polymeric materials are available such as polyester, epoxy, vinyl ester, polyamide, polypropylene, vinyl ester etc. Generally, fibres are the reinforcement material, however in certain cases it can be common ground materials (Mkaddem et al., 2008). The strength of the final material depends upon the ratio of the two phases i.e., matrix and the reinforcement. Different production technique leads to different ratios. For instance, the hand lay-up technique produces PMCs with 60% resin and 40% fibre whereas it is 40% resin and 60% fibre when produced using vacuum infusion process.

The reinforcement leads to the fabrication of composites that has the following characteristics: high specific stiffness, specific strength, resistance to fracture; low cost and good impact, abrasion, fatigue and corrosion resistance. Some of the disadvantages are low thermal resistance and high coefficient of thermal expansion. The properties of the PMCs are affected by a number of factors. These factors are discussed next:

Factors Affecting Properties of PMCs

- **Properties of Matrix:** The application of the PMC is determined by the type of polymer used for the fabrication. Some of the advantages possessed by the polymeric matrix are easy to process, lower cost and specific gravity and good chemical resistance. However with advantages come disadvantages such as low modulus, low strength and low operating temperature (Huang and Talreja, 2006). A wide range of polymeric materials are used for the fabrication of PMCs such as thermoplastic polymers, elastomers and thermosetting polymers.
- **Thermosetting Polymers**: These are the polymeric material with covalent bonds between the molecules of the constituting particles. The thermosetting polymers therefore possess network structure. They do not soften on heating but decomposes. They cannot be reshaped once solidified by cross-linking process. Some of the examples of thermosetting polymers are epoxies, phenolics, polyesters, melamine, polyimides and silicone.
- **Thermoplastic Polymers**: The thermoplastic polymers are made of strong intra-molecular bonds. However, they possess weaker intermolecular bonds. These can be reshaped on application of heat and pressure. Examples of thermoplastic polymers are polyethylene, polystyrene, polycarbonates, nylons, polysulfone, polyether imide etc.

- **Elastomers**: The elastomers are viscoelastic polymeric materials that have comparatively lower modulus of elasticity while higher yield strain in comparison to the other materials. Some of the examples of elastomers are natural rubber, butyl rubber, synthetic polyisoprene, silicone rubber, thermoplastic elastomers, fluroelastomers etc.
- **Dispersed Phase-Shape and Orientation:** There is no preferred direction for the particles that enhances the mechanical properties and lowers the cost of the material (Bednarcyk, 2003). The particles used as a dispersing phase can have wide range of shapes such as spherical, platelet, cubic or some other irregular or regular geometrical shape. The particulates used as reinforcements have typically same dimensions in all the directions.
- **Interfacial Adhesion:** The behaviour of the composite material depends upon the cumulative behaviour of the reinforcement, polymeric matrix and the interfacial adhesion between the reinforcement and the particle. The interfacial adhesion should be strong to achieve superior mechanical properties. The extent of interfacial adhesion of the matrix molecules with the fibre is determined by the chemical reaction or adsorption taking place at the surface of the fibre. The investigation of the interfacial surface can be done using the techniques such as atomic force microscopy.

Fabrication of Composites

The important processes used for the fabrication of PMCs are discussed as follows:

- **Hand Lay-Up:** Hand lay-up technique is one of the oldest and simplest methods of manufacturing small and large reinforced products. The technique of hand lay-up can be used for flat surface, cavity or positive shaped mould made from metal, wood, plastics or their combination. There are different techniques to use the hand lay-up method: one handed guiding, two handed guiding, manual folding, hoop shearing, double tension shearing, tension-secured shearing, tension and sticking and mould interaction shearing.
- **Bag-Moulding Process:** It is a versatile process of fabricating composite parts. The mould is laid with the lamina which in turn is spread with the resin. The entire mould consisting of the lamina and the resin is covered with the diaphragm or a bag. This is then cured with the application of heat and pressure. The final product with the shape of the mould is obtained after achieving the required curing cycle (Tay et al., 2006).
- **Pultrusion:** The composite materials are automatically manufactured into constant and continuous profiles. The final product is pulled out from the die instead of using pressure to force the product out. Appropriate dies can

be used to produce a wide range of structural shapes such as rods, tubes and other such profiles.

- **Filament Winding:** The solid of revolutions such as tubes, pipes, spheres and cylinders can be fabricated using the technique of filament winding. The technique of filament winding is particularly used for producing large tanks and pipes for the chemical industry. The basis of the filament winding technique is the high-speed laying down of the reinforcements in predefined pattern.
- **Preformed Moulding Compounds:** Matched die moulding processes such as injection moulding, transfer moulding and compression moulding can be used for the fabrication of reinforced thermosetting resin products. In the matched die process the ingredients are added to the premix i.e., preformed moulding compound (Friedrich et al., 2005). Faster rates of production can be achieved using this technique.
- **Resin Transfer Moulding:** The process produces large and integrated products with high performance. In this process, a preform which is dry reinforcement material cut into desired shape is placed inside the mould cavity. To take care of the entrapping air, the resin is injected at the lowest point and fills the mould upward. To minimize the loss of resin clamping of the tube is done on leakage of the resin into the resin trap. The flow of resin is stopped once the resin starts to flow from the vent areas of the mould. The curing process of the mould component then begins. On sufficient curing, the composite gains sufficient green strength and then can be removed for further post-curing treatment.
- **Injection Moulding:** The method of injection moulding is used for both the thermosetting and thermoplastic materials. The composite is fed into a barrel that is heated to the desired temperature. The mixing is followed next and is then forced out into the mould cavity. On cooling, hardening takes place and the final product takes the shape of mould cavity. Injection moulding is used for the production of a wide range of products. Injection moulding has a number of advantages such as high production rates, low labour cost, less time to produce the product and minimized scrap. However, the expensive equipment and high running costs are the major disadvantages of the process.
- **Reaction Injection Moulding:** The process is used for thermosetting polymers and is similar to the injection moulding process. In this process the two parts of polymers are mixed together. Then using impinging mixer the mixture is injected under high pressure into the mould. The reaction injection moulding process is used for the production of a variety of automotive components such as fenders, spoilers and bumpers.

- **Reinforced Reaction Injection Moulding**: The process uses a reinforcing agent added to the mixture used for reaction injection moulding. Examples of reinforcements include mica and fibres. Rigid automotive panels are used via this process. In case the fibres are used as reinforcing agents, a mesh of fibre is arranged in the mould and then polymer mixture is injected into the mould.
- **Spray-Up:** Two different sprays are used for liquid resin matrix and reinforcing fibres respectively. The fibres are chopped into fibres of 25-50 mm length and are then sprayed simultaneously with resin spray. The main advantages of the method include: low cost, less expensive tooling and widely used for many years. Main disadvantages include: laminates tend to be very rich in resin content. Resin needs to be of low viscosity.

Applications of PMCs

Aerospace Industry

There has been constant research by the aerospace industry to improve the fuel efficiency, load carrying capacity and enhance the manoeuvrability of the airplanes. In the constant desire to achieve the above objectives, materials possessing low weight and good mechanical properties are at the forefront of the research community. With this option of low weight material, companies such as American Airlines could manage to save around 11 gallons of fuel everyday with the total fleet size of about 600 planes (Moniruzzaman, 2012). This has been regarded as a breakthrough for the other aircraft manufacturing and operating companies.

Aircrafts such as Boeing B-787 are designed and manufactured using composite materials to a great extent. The total mass of the composite material used reaches to around 50% of the total mass of the aircraft. The other materials used in its construction include aluminium alloys (20%), titanium alloys (15%), steel (10%) and the remaining 5% is made up by the other materials. Important parts of the aircraft such as wings, fuselage and tail are made by the composite materials (Merkisz and Bajerlein, 2011).

Laminates of sandwich types composite materials are also being used extensively in designing the Boeing B-787 Dreamliner. The structure is made from honeycomb core embedded in the two thin external layers. The sandwich structures provides for high stiffness, high impact strength, high resistance to corrosion, lower operating costs, minimized components and hence reduced assembly time.

The carbon-epoxy HexMC composite is used in the fabrication of aircraft components with reduced weight and good mechanical properties. The composite is manufactured by Hexcel (USA) and has rendered for the possibility to manufacture complex shapes sandwiches for the 787 Dreamliner. Window frames, fittings, brackets, etc., are some of the parts of the Dreamliner manufactured using the HexMC composite material (www.jeccomposites.com).

AeroComposit of Russia (Solano, 2012) was the first in Russia to develop the components strengthened with carbon fibres for the civil aircrafts. The AeroComposit company has not only used the carbon reinforced composites but also has replaced the traditional processes of manufacturing with the out-of-autoclave production. Fabrication and design of the MS-21s wing with the low weight carbon reinforced composite then become the next objective of the OAK AeroComposit which was done in tandem with the FACC of Austria. The company researched to obtain carbon reinforced composite with the newly developed cost-optimized resin.

The material structure used for A380 wing spoilers was manufactured by Patria Aerostructures Oy (Patria) of Finland for the airbus using composite sandwich box. The composite material was fabricated using T700 carbon fibre reinforcement joined together with the M21 prepreg resin systems in a honeycomb structure (Mustakangas et al., 2012).

A wide range of composite materials have been fabricated by Victrex Europa GmbH (Padey and Butz, 2012) for the aerospace application. The composites were manufactured using the polyetheretherketone (PEEK) polymeric resin which are characterised by enhanced mechanical strength and dimensional stability. The composites manufactured had the potential to withstand high temperatures and aggressive chemical environments. The reinforcements used for the composite were chopped glass fibres and carbon fibres.

The wings and the tail components of the SARAS aircraft were designed using carbon-epoxy composites. This was done by the CSIR-NAL of India. The stringers, ribs and spars used at the bottom of the plane's skin was too manufactured using composite materials heat treated in autoclave at a temperature of 175°C and a pressure of 7 bar. The final aircraft had 25% reduced mass in comparison to that produced using metal alloys.

The aircrafts food tray arms have also been developed using the high-performance thermoplastic compounds. These were developed by SABIC (Saudi Basic Industries Corporation) of Saudi Arabia. The resin used was polyetherimide and carbon fibres were the reinforcements (Moniruzzaman, 2012). The manufactured arm trays using the fabricated composite lead to saving of the weight of the aircraft. According to the estimate, around 0.5 gallons of the fuel was saved with 600 planes each with 200 passenger carrying capacity (Moniruzzaman, 2012).

The three most important types of composites used for aerospace industry are the materials that are reinforced with carbon fibre, fibreglass and aramid fibres. The primary benefits that are offered by the composites for the aircraft industry are reduced weight and simplification of assembly.

Automotive Industry

The use of composite materials has been very popular in the automotive industries. Automotive industries have been using composite materials for fabrication of important elements of a vehicle.

There are automotive companies that are combining exceptional design with innovative technologies while focussing on saving of energy and weight. Companies such as BASF and SMART (www.smartforvision.basf.com) have been using composite materials to meet their global commitments of delivering electric cars. BASF has therefore used their special epoxy systems with carbon fibres to produce components such as chassis and external framework. Mass saving of up to 150 kg per chassis was achievable with the use of composite materials while maintaining the strength equivalent to that of the light metal alloys. Another innovative technology is that of polymeric wheel rim made from another high performing material made by using polyamide matrix with long glass fibres as reinforcing material. This gave a weight saving of 12 kg per car.

Three South Korean giants namely LG Hansys, Hyundai Motor and Shinhan Mold have together designed a battery pack module carrier. They have used carbon fibres with PA6 polyamide for forming the compact elements. The use of fabricated composite led to number of advantages such as weight saving, elimination of anti-corrosion paint, improved recyclability and enhanced driving performance.

For the production of front end of AUDI A8, Bond-Laminates GmbH produced from polymer under the trade name Durethan BKV 30 H2.0 EF, have been used. The material is made from 30 vol% of glass fibres with PA 6 matrix as the polymeric matrix.

An energy efficient material, Bayflex RIM was introduced for the application in automotive industry by the Bayer Material Science group (Meyer, 2012). Bayflex RIM is a polyurethane material having remarkable heat stability, low coefficient of thermal expansion, low moisture absorbing tendency etc. Reaction injection moulding technique is used for the manufacturing of Bayflex RIM. The developed material was able to meet the industrial need of energy efficient vehicles with light weight construction.

Monocoque chassis was constructed using GMS EP270 epoxy prepreg with carbon fibres as reinforcements by Autohorizon a group of young researchers from Australian charitable foundation (Weller, 2012). The constructed chassis was analysed for strength and stiffness properties. The results were found to be satisfactory. The monocoque chassis was used for Ferrari 355 sports car. The carbon based chassis was used for Roding Roadster which is another two seat sports car (Kollmannsberger et al., 2012).

A prototype vehicle known as BioMobile (Perraudin and Hummer, 2012) was constructed by The University of Applied Sciences, Switzerland. The upper part of the frame of the constructed vehicle was made using the banana fibres whereas the lower part was manufactured using Greenlite consisting of cellulosic fibres. Bio-composites were also used for the frame of the same vehicle.

The thermoplastics were also used for the construction of trucks bumper. Magna Exteriors and Interiors Inc. have produced bumpers for the MAN TGX and TGS trucks while Werkzeugbau Karl Krumpholz GmbH & Co. Produced it for TGL and TGM models. Polyamide 6 with 60% glass fibres as reinforcements was used for the bumper. The material used ensured high strength and stiffness.

A material by the name Thermylene has been produced by Asahi Kasei Plastics, North America Inc. for bumpers, elements of brakes, clutch and luggage racks. The developed material is equivalent to Long-Glass-Fibre-Reinforced-Polypropylene and is characterised by very good mechanical properties. The stiffness possessed in the longitudinal and transverse directions have been observed to be excellent for a wide range of temperature (Iyer R, 2013).

The front end carrier of a vehicle is thin-walled element that should have high stiffness and mechanical properties with excellent surface quality. As such high strength polyamide has been used by Volkswagen for the production of front end carrier. The polymeric composite under the trade name Ultramid has also been used for fabrication of front end carrier of VW Golf 7. This was developed in collaboration between BASF and Volkswagen.

The application of polymeric composites for the Chevrolet Corvette Z06, a sports car from General Motors, USA. The power to mass ratio was found to be much better than its other counterparts such as Aston Martin DBS, Nissan GT-R or Porsche 911. Components such as fenders, bonnet etc., have been produced using lightweight polymeric composite with carbon fibres as reinforcements.

Today, the leading automotive makers are using polymeric composites on a very wide scale. The intake system, airframe and other engine components are now manufactured using polymeric composites in the new models from Porsche. The oil pans of cars from BMW are manufactured using polymeric composites (Gibas, 2013).

Polymeric composites based on carbon fibres and Kevlar have been used in the construction of a wide range of elements for Citroen DS3 a racing car. The composite material was developed and used by Galway Carbon Motorsports. The panels were made using acrylic resin as resin. The bonnet, nearside door pannels and the boot were fabricated using the developed polymeric composite. A substantial mass reduction of around 85% was obtained with enhanced stiffness and mechanical strength (Grealish and Gray, 2013).

Marine Industry

The modern ships are constructed using the different polymeric composites. The use of polymeric composites produces ships with lower fuel consumption and good manoeuvrability. The polymeric composites are corrosion and water resistant and therfore are widely used by the maritime industries for the construction of ships.

The main hulk and deck of speedboat Revolver 42 have been cnstructed using polymeric composite comprising of vinyl ester as resin and M-foam core reinforced with fibres. It was the use of polymeric composites that hekped the speedboat to reach speeds of 68 knots (Bell, S., & Arcuri, 2012).

The wing of the Vestas Sailrocket 2 sailboat was manufactured using preimpregnated carbon fibres filled with honecomb core from Nomex. The construction made the sailboat to break the speed record at a competion organised at Walvis Bay in Namibia.

For large elements of submarines, thermoplastic-matrix composites have been used by the marine industries over the world. The main goal of these companies has been to replace the metals and alloys with glass/epoxy or carbon/epoxy composites (Choqueuse et al., 2012).

In India the houseboats for Kerla tourism have been constructed using polymeric composites. Glass reinforced polymers have replaced the heavy wood and steel structures. This has not only led to lowering of production costs but also has shortened the production time for the houseboats. The parts of the houseboat that have been constructed using Glass reinforced polymeric composites includes inner ceiling module, supporting pannels for window opening and pagoda-style roof (Munshi, 2012).

An environment friendly polymeric composite has been developed by NavEcoMat for the construction of yachts, boats and small ships. The eco friendly polymeric composites are manufactured using polylactic acid with flax fibres as reinforcements. Naskapi Indians canoe is the recent application of the developed material.

Ifremer of France and Kairos Inc. have jointly designed boats using bio-sourced polymeric composite materials. The two have considered hemp, flax, jute and cellulosic fibres in their different forms such as woven fibres, unidirectional fibres and mats (Bosser et al., 2012). Tests were conducted on the boats manufactured and the results were found to be satisfactory.

La Rochelle of France has constructed electric powered passenger boat. The parts of the boat made from biosourced composites include cabin, bulkwark and roof.

Military Industry

The satisfactory level of reliability is of utmost importance for the military industry. As such the industry requires materials that are trustworthy, strong and safe.

The polymeric composites with phenolic resins and layers of S-glass fibres with aramid fibres as reinforcements have been manufactyred using vacuum assisted resin transfer molding tecnique by Astilleros Santa Pola S.L, a spanish shipyard. The developed material was able to provide for excellent anti-ballistic properties, corrosion resistance and resiatnce against UV.

Composites materials were used for the construction of Zumwalt, the most advanced destroyer of the US navy. The radar of the ship was one of the key elements manufactured using the composites. Carbon-fibre reinforced vinyl ester resin was used for building the ship's deckhouse. The fibreglass phenolic laminates were used for manufacturing of the ship's roof.

Vertical stabilizer, fuselage, horizontal stabilizer and wings of Locheed Martin F-35 Lightning II were made from the CFRP (Bielawski, 2014). Further, the structural parts of AV8B Harrier and Bell-Boeing V-22 Osprey were made from polymeric composites.

Polymer based composites are used for the construction of unmanned aerial vehicles (Ghilai, 2012). The aircrafts are required to operate for a relatively long period of time once refuelled. As such the major requirement is the lower mass. The wing of Triton UAV of the US Navy has been manufactured using polymeric composites with epoxy resin reinforced with carbon fibres.

The stealth qualities of B-2 are met because of the composite materials that provides the ability to be moulded into complex shapes to the B-2 fighter planes. It is estimated that around 40000 to 50000 pounds of advanced composite materials make up the body of the B-2 fighter planes.

Sport and Leisure

Composites have been used widely for the production of sports and leisure goods. Structures for bicycles have been manufactured using polymeric composite materials by AX-Lightness GmbH, Germany. The wheels for their mountain bikes are made using carbon fibres with epoxy resin prepregs. Good impact resistance and strength are assured using the prepregs.

A new design of the bicycle frame has been proposed by Braid, a German company. The company has proposed to use their Braiding technique instead of making tube using two half shells. In the Braiding technique the frame is filled with PU foam cores ensuring lighter frame with enhanced stiffness and mechanical properties. Further, the company was able to compete with their Chinese counterpart in terms of cost of production. Carbon fibre reinforced material were used by Werner Paddles Inc., for the production of paddles with enhanced aesthetics and performance and reduced cost.

Ultra-high performing surfboards have been manufactured by Australian manufacturer Samsara Surfboards. The surfboards are made from flax fibres which have least effect on the environment (Smith, 2012).

The skis snowboard are made from wood, metal and fibre composites. However, wood gets damp on absorption of moisture while on the other hand metal such as aluminum increases the cost of skis snowboard. The fibre composite fabricated skis snowboard are suitable for any types of snow and are easy to maintain. The carbon fibres strengthens the flexion degree of the skis. The glass fibre increases the toughness of skateboarding.

Industrial Construction

FRP composites are used for a wide fraction of civil applications. The confidence of usinf composites for construction purpose grew dramatically with significant investments from private and public companies. The FRP composites based on carbon, glass and aramid fibres with epoxy, polyester and vynilester resins have been used widely by the construction industry (Bunsell and Renard, 2005). High temperature applications have recently attracted the use of basalt fibres (Şerbescu et al., 2006).

The range of properties as well as the ease of formability offered by the FRP composites have led to fabrication of elements having large load bearing capacity. Plate elements, shell elements and folded structures can be easily fabricated for use in different building structures (Taranu, 2009).

Number of techniques have been used for the construction of industrial buildings using FRPs. For instance, pultrusion technique has been used to produce different shapes of bars and structures such as domes (Taranu, 2009). Filament winding technique has been used for the production of transport and storage elements such as pipes and tanks. Hand lay-up technique has been used for construction of various parts of roof covering (Decher, 1998).

The advanced polymeric composites are being used for the offshore platforms. Some of the examples of products for offshore applications include: firewater piping, grating, sea water piping, storage vessels, blast and fire walls. The use of fibre glass pipes have overcome the major disadvantages of leakage (Chakrabarti, 2005). The ropes and cables are made from using aramid or polyethylene fibres (Bunsell and Renard, 2005).

Another application in the civil engineering area is that of chimneys operating under severe wind, temperature and seismic conditions. Chimneys are constructed using brick reinforced with concrete. Construction of chimney for beer factory have been proposed by the designers using the brick reinforced with concrete (Oprisan et al., 2005). The chimneys were strengthened using carbon fibre reinforced textile composites.

CONCLUSION

The present chapter presents an overview of the manufacturing techniques and applications of polymeric composites for industrial usage. The examples presented in different industrial context highlights the important role played by polymeric composites. The applications of polymeric composites is suggestive of the fact that the polymeric composites play a key role in the development of key industrial sectors. The use of polymeric composites offers convenience, competitiveness and savings to the manufacturer as well as the end users. The most demanding environment can be sustained with ease by the polymeric composites. The polymeric composites are characterised by low weight with durability and is therefore has successfully replaced metal alloys in different industrial applications.

REFERENCES

Bednarcyk, B. A. (2003). An inelastic micro/macro theory for hybrid smart/metal composites. *Composites. Part B, Engineering*, *34*(2), 175–197. doi:10.1016/S1359-8368(02)00067-7

Bell, S., & Arcuri, M. (2012). A speedboat with a vacuum-infused and bonded foam sandwich design. *JEC composites*, (75), 30-31.

Bielawski, R. (2014). Konstrukcje lotnictwa wojskowego. *Materiały Kompozytowe*, 11-14.

Bosser, L., Poisson, E., & Davies, P. (2012). Bio-sourced sandwich materials for marine applications. *JEC Composites*, (75), 25-27.

Bunsell, A. R., & Renard, J. (2005). *Fundamentals of fibre reinforced composite materials*. CRC Press. doi:10.1201/9781420056969

Chakrabarti, S. K. (2005). *Handbook of Offshore Engineering*. Amsterdam: Elsevier.

Choqueuse, D., Davies, P., & Bigourdan, B. (2012). Composite pressure vessels for deep-sea applications. *JEC Composites Magazine*, 75.

Decher, E. (1998). *Elemente şi structuri din materiale compozite pentru acoperişurile construcţiilor inginereşti*. Academic Press.

Friedrich, K., Zhang, Z., & Schlarb, A. K. (2005). Effects of various fillers on the sliding wear of polymer composites. *Composites Science and Technology*, *65*(15), 2329–2343. doi:10.1016/j.compscitech.2005.05.028

Ghilai, G., Green, A. K., Leibovich, H., & Miller, Z. (2012). Low-cost composite structures: how to avoid pitfalls. *JEC Composites*, (74), 41-45.

Gibas, E. (2013). Plastics in the automotive industry can help reduce the weight of the car. *PlastNews*, (2), 29-31.

Grealish, B., & Gray, N. (2013). World Rally Champion's super-car benefits from lightweight composite body panels. *JEC Composites Magazine, 82*, 36-37. Retrieved from http://www.jeccomposites.com/news/composites-news/hexmc-carbonfibreepoxy-moulded-parts-boeing-787-0

Huang, H., & Talreja, R. (2006). Numerical simulation of matrix micro-cracking in short fiber reinforced polymer composites: Initiation and propagation. *Composites Science and Technology*, *66*(15), 2743–2757. doi:10.1016/j.compscitech.2006.03.013

Iyer, R. (2013). New long-glass-fibre-reinforces polypropylene for automakers'. *JEC Composites Magazine*, *82*, 33.

Kollmannsberger, A., Brand, M., & Drechsler, K. (2012). Braiding a convertible windscreen frame with concave sections. *JEC Composites*, (74), 24-25.

Mayer, C., Wang, X., & Neitzel, M. (1998). Macro-and micro-impregnation phenomena in continuous manufacturing of fabric reinforced thermoplastic composites. *Composites. Part A, Applied Science and Manufacturing*, *29*(7), 783–793. doi:10.1016/S1359-835X(98)00056-6

Merkisz, J., & Bajerlein, M. (2011). Composite materials used in the aerospace industry. *Logistyka*, 2829-2837

Meyer, B. (2012). Lighter than water. *JEC Composites Magazine*, *74*, 40.

Mkaddem, A., Demirci, I., & El Mansori, M. (2008). A micro–macro combined approach using FEM for modelling of machining of FRP composites: Cutting forces analysis. *Composites Science and Technology*, *68*(15), 3123–3127. doi:10.1016/j.compscitech.2008.07.009

Moniruzzaman, M. (2012). PEI/carbon as metal substitute in aircraft food tray arms. *JEC Composites*, (70), 48-53.

Munshi, K. (2012). A modular composite houseboat for tourism in Kerala. *JEC Composites*, (75), 36-37.

Mustakangas, M. (2012). Milestone celebration of Finnish aerospace composites expertise. *JEC Composites*, (72), 49-51.

Oprisan, G., Taranu, N., Munteanu, V., & Entuc, I. (2010). Application of modern polymeric composite materials in industrial construction. Buletinul Institutului Politehnic din lasi. Sectia Constructii. *Arhitectura*, *56*(3), 121.

Padey, D. (2012). PEEK thermoplastic powers state-of-the-art aerospace engineering for metal replacement. *JEC Composites*, (72), 34-37.

Perraudin, M., & Hummer, E. (2012). BioMobile. ch: incorporating green reinforcements into an automotive body frame. *JEC Composites*, (74).

Şerbescu, A., Kypros, P., & Ţăranu, N. (2006). The Efficiency of Basalt Fibres in Strengthening the Reinforced Concrete Beams. The Bulletin of the Polytechnic Institute of Jassy, Construction. *Architecture Section*, *52*(3-4), 47–58.

Shaw, A., Sriramula, S., Gosling, P. D., & Chryssanthopoulos, M. K. (2010). A critical reliability evaluation of fibre reinforced composite materials based on probabilistic micro and macro-mechanical analysis. *Composites. Part B, Engineering*, *41*(6), 446–453. doi:10.1016/j.compositesb.2010.05.005

Smith, G. (2012). Flax fabric used to create an eco-friendly surfboard. *JEC Composites*, (75).

Solano, J. (2012). OAK Aerocomposit chooses innovative solutions to build the MS-21 composite wing. *JEC Composites*, (72), 44-46.

Taranu, N. (2009). *Composite materials-course notes*. The University of Sheffield Printing Office.

Tay, T. E., Tan, V. B., & Liu, G. (2006). A new integrated micro–macro approach to damage and fracture of composites. *Materials Science and Engineering B*, *132*(1), 138–142. doi:10.1016/j.mseb.2006.02.023

Weller, S. (2012). Out-of-autoclave CF monocoque sports car cockpit chassis. *JEC Composites*, (76), 81-82.

Chapter 2
A Review of Chemical Treatments on Natural Fibers-Based Hybrid Composites for Engineering Applications

Sudeepan Jayapalan
Birla Institute of Technology, India

ABSTRACT

The development of hybrid-fiber-reinforced composites has increased in recent decades because of its abundance, low cost, low weight, high strength, stiffness, and bio-degradability, thereby increasing its engineering applications. However, the major drawbacks of natural fibers in composites are their high moisture absorption and poor compatibility between fiber and matrix. Hence, chemical treatments are primarily considered to modify the fiber surfaces with the objective to improve interfacial bonding between fiber and matrix. This chapter addresses an overview of chemical treatments and their effects on natural fibers-based hybrid composites are reviewed. The chemical treatments include alkali, silane, maleated, and others, focused mainly on hybrid natural fiber composites. The significance of chemical treatment of natural fibers aimed to improve adhesion between fiber surface and matrix along with reduction in water absorption property to improve physical and mechanical properties as compared with untreated fibers for use in components of engineering applications is explored.

DOI: 10.4018/978-1-5225-5216-1.ch002

INTRODUCTION

For the past few decades, the use of natural fiber reinforced composite materials has increased considerably due to its low price and biodegradable compared to synthetic fibers and thereby gained attention from both scientific and industrial communities. Natural fiber reinforced composites are primarily used in applications such as civil engineering and automobile sectors in seat backs, dashboards, door panels, package trays, headliners, and trunk liners etc., (Puglia et al., 2005). The classification of composites is given in Figure 1. Traditional synthetic fiber reinforced composites use different types of glass, carbon, aluminum oxide, aramid, Kevlar and many others as reinforcing materials because of high stiffness and strength properties (Rout et al., 2001). However, synthetic fibers have disadvantages such as processing cost, recyclability, biodegradability, energy consumption, health hazards, environmental safety etc., (Bledzki et al., 1999). Hence, natural fibers like bamboo, flax, hemp, sisal, kenaf, and jute have been well recognized amongst researchers as a potential reinforcement material for engineering fiber components. The mechanical properties of some natural and synthetic fibers are listed in Table 1. Natural fibers have advantages over their synthetic counterparts such as low cost, acceptable specific strength, lightweight, non-toxic, low density, high specific modulus, bio-degradable and easy processing (Abdelmouleh et al., 2007; Hapuarachchi et al., 2007). These advantages make natural fibers to widen its applications as well as it becomes potential replacement of synthetic fibers in composite sector materials. However, natural fibers possess certain drawbacks in their structural compositions such as cellulose, hemicelluloses, lignin, pectin and waxy substances which allow excess moisture absorption from the surrounding which tends to weak interface bonding between the fibers and matrix (Doan et al., 2006). The composition of some of the natural fibers are shown in Table 2. The hydrophilic nature of fibers and hydrophobic nature of matrix makes the couplings between the two phases are difficult and this tends to ineffective stress transfer at the interface of the composites which results in reduced physical and mechanical properties. Therefore, chemical treatments are certainly needed on the surface of natural fibers to improve interfacial adhesion. Chemical treatments may activate hydroxyl functional groups in their structure that can react with the fiber surface which changes their composition and can effectively interlock with the matrix as well as reduction of moisture absorption properties (Abdelmouleh et al., 2007). The chemically treated natural fiber reinforced composites are investigated by several researchers aiming to enhance physical, mechanical, and thermal properties through proper adhesion of fibers with the polymer matrix (Ray et al., 2001; Mishra et al., 2001; Weyenberg et al., 2003). The commonly available natural fibers reinforced composites are not the primary choice for advanced engineering applications due to its lack in specific mechanical and thermal properties compared to synthetic fibers reinforced composites. Therefore, hybridization of natural-natural or natural-synthetic fibers are well suited for advanced engineering components because of better load carrying capacity, impact energy absorption, high strength and stiffness compared to single fiber reinforced composites. Hybrid reinforced composites comprised only natural-natural fibers are potentially useful materials with respect to environmental concerns (Saw and Datta, 2009; Idicula et al., 2010). However, researchers are exploring the use of hybrid natural and synthetic fibers in composites to widen its applications and to improve good mechanical performance and reduce the processing cost (Hariharan and Khalil, 2005; Rosa et al., 2009).

The objective of this review paper is to combine all research works related to different chemical treatments of natural fibers based hybrid composites to provide researchers with an overview of engineering applications of hybrid fibers reinforced polymer composites. This review paper also provides the fiber matrix interfacial adhesion and its influence on mechanical and thermal properties of different natural fiber reinforced composites.

HYBRID FIBERS REINFORCED COMPOSITES

Hybrid fibers reinforced polymer composites are gaining more importance in engineering applications because they offer wide variety of properties compared to single fiber reinforced polymer composites. The term hybrid fibers reinforced polymer composites mean the combination of two or more reinforcing fibers into

Table 1. Mechanical properties of various natural and synthetic fibers (Mohanty et al., 2000; Dhakal et al., 2007; Li et al., 2007; Kabir et al., 2012)

Fibers	Density (g/cm^3)	Tensile strength (MPa)	Young's modulus (GPa)	Specific strength (GPa/g/cm^3)	Specific modulus (GPa/g/cm^3)	Elongation at break (%)
Jute	1.3–1.4	393–773	13–26.5	0.3–0.5	10–18.3	1.16–1.5
Flax	1.50	345–1100	27.6	0.2–0.7	18.4	2.7–3.2
Hemp	1.14	690	30–60	0.6	26.3–52.6	1.6
Ramie	1.50	400–938	61.4–128	0.3–0.6	40.9–85.3	1.2–3.8
Sisal	1.45	468–640	9.4–22.0	0.3–0.4	6.4–15.2	3–7
Cotton	1.5–1.6	287–800	5.5–12.6	0.2–0.5	3.7–7.8	7.0–8.0
E-glass	2.5	2000–3500	70	0.8–1.4	28	2.5
S-glass	2.5	4570	86	1.8	34.4	2.8
Aramid	1.4	3000–3150	63–67	2.1–2.2	45–47.8	3.3–3.7

Figure 1. Classification of Composites (Jawaid and Khalil, 2011)

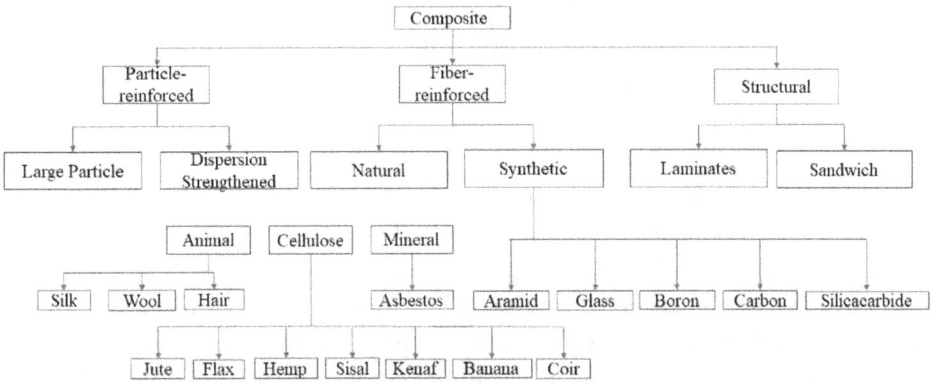

Table 2. Chemical compositions of various natural fibers (Mohanty et al., 2000; Dhakal et al., 2007; Li et al., 2007; Kabir et al., 2012)

Fibers	Cellulose (wt.%)	Lignin (wt.%)	Hemicellulose (wt.%)	Pectin (wt.%)	Wax (wt.%)	Moisture content (wt.%)
Jute	61–71.5	12–13	13.6–20.4	0.2	0.5	12.6
Flax	71	2.2	18.6–20.6	2.3	1.7	10.0
Hemp	70.2–74.4	3.7–5.7	17.9–22.4	0.9	0.8	10.8
Ramie	68.6–76.2	0.6–0.7	13.1–16.7	1.9	0.3	8.0
Sisal	67–78	8.0–11.0	10.0–14.2	10.0	2.0	11.0
PALF	70–82	5–12	–	–	–	11.8
Cotton	82.7	0.7–1.6	5.7	–	0.6	33–34

a single polymeric matrix. It can be the combination of different natural-natural fibers, or combination of different synthetic-synthetic fibers, or by the combination of natural-synthetic fibers. The advantages of using hybrid fibers reinforced composites in engineering based applications is that it can be a good balance between performance and cost (Dong et al., 2016). Generally, the high performance of one type of expensive fibers in the hybrid composites could complement with the low performance of the other inexpensive fibers. Hence, hybrid fibers composites may offer high performance and reduce overall costs of the composite materials (John & Anandjiwala, 2008). Bunsell and Harris (1974) studied hybrid composites materials consists of carbon and glass fibers into an epoxy resin. They found that the hybrid composites can create a material having combined advantages and increase in performance as well as increasing the use of expensive materials in engineering applications. Somerscale and Short (1978) published a literature review based on carbon and glass fiber hybrid reinforced plastics. From the review, they found that the incorporation of hybrid fibers into a single polymer matrix may offer better overall mechanical performance. Kretsis (1987) surveyed a research work related to tensile, compressive, flexural and shear properties of hybrid fiber reinforced polymeric composites considering mostly unidirectional fibers of specially carbon and glass fibers. Jawaid and Khalil (2011) published a review paper on cellulosic-synthetic hybrid fibers focusing on processing, mechanical, physical, electrical, thermal, and dynamic mechanical properties of various fibers. Swolfs et al., (2014) reviewed about fiber hybridization and their hybrid effects and properties in hybrid polymer composites such as tensile, flexural, impact and fatigue. Recently, Safri et al., (2018) presented a review paper on the impact behaviour of hybrid composites focusing mainly on impact resistance and penetration behaviour and its suitability for structural applications. From the point of literature review, most of the published review papers are discussing the mechanical performance of hybrid fiber composites. However, a limited published review paper on the different chemical modifications of hybrid natural fibers reinforced polymer composites for engineering applications. The present review paper is primarily focusing on the different chemical treatments of hybrid natural fibers, especially natural-natural and natural-synthetic hybrid fibers.

NATURAL-NATURAL HYBRID FIBERS REINFORCED COMPOSITES

In recent years, development of hybrid natural-natural fibers reinforced composites has attracted and received considerable attention of researchers and scientists in engineering applications because of the advantages such as low cost, bio-degradability, easy processing, and low density (Mishra et al., 2003). Hybrid natural fibers composites comprises of hybridization of two or more different cellulosic fibers reinforced in polymer matrix. Hybrid fiber composites may provide considerable mechanical properties, recyclability, environment safety, longer fatigue life, excellent impact resistance along with weight savings (Premalal et al., 2002). Ranjan et al., (2013) investigated the mechanical properties of banana/sisal fiber reinforced polylactic acid hybrid composites for structural applications. They found that the chemically treated fibers exhibited better mechanical properties than untreated composites. They also stated that chemical treatment improves the fiber matrix interaction. tensile, flexural, impact properties and water absorption tests were carried out using sisal/unsaturated polyester composite material. Initially the optimum fibre length and weight percentage are estimated. To improve the tensile, flexural and impact properties, sisal fibre was hybridised with bamboo fibre. This work shows that the addition of bamboo fibre in sisal/ unsaturated polyester composites of up to 50% by weight results in increasing the mechanical properties and decreasing the moisture absorption property. In this research work, the effects of fibre treatment and concentration on the mechanical properties of a short natural fibre reinforced polyester hybrid composite are investigated. The fibres were subjected to 10% sodium hydroxide solution treatment for 24 h. The mechanical properties of composites with treated fibres are compared with untreated fibre composites. The fractured surface of the treated fibre composite specimen was studied using Scanning Electron Microscopy (SEM).The treated hybrid composite was compared with an untreated hybrid composite, with the former showing a 30% increase in tensile strength, 27.4% - in flexural strength, and 36.9% - in impact strength, along with an extreme decrease in moisture absorption behaviour. Tensile, Flexual, Impact and Water Absorption Properties of Natural Fibre Reinforced Polyester Hybrid Composites

Właściwości wytrzymałościowe na rozciąganie, gięcie i uderzenie oraz absorpcja wody poliestrowych hybrydowych kompozytów wzmacnianych włóknami naturalnymi

The factors such as selection of fiber and matrix, interfacial strength, moisture content, and fiber orientation may affect the hybrid natural fiber performance (Pickering et al., 2016). Different researchers have investigated the various performance of different hybrid natural fibers reinforced composites (Alavudeen et al., 2015; Jawaid et al., 2014; Fiore et al., 2016; Manjunath and Udupa, 2016; Yusoff et al., 2016; Jamshaid et al., 2016; Gupta, 2017; Chaudhary et al., 2017).

NATURAL-SYNTHETIC HYBRID FIBERS REINFORCED COMPOSITES

For the past few decades, development of natural-synthetic hybrid fibers reinforced composites plays an important role in engineering based applications because of the superior properties along with cost reduction advantages. The hybrid natural-synthetic fibers will provide a good balance between performance and cost compared to single fiber reinforced composites (Panthapulakkal and Sain, 2007). Sarasini et al., (2016) investigated the carbon/flax hybrid composites and the results showed that most of the mechanical properties are enhanced. Cicala et al., (2009) studied the hybridisation of glass fibers with natural fibers such as hemp, flax and kenaf for piping industry applications. The results showed that the mechanical properties of hybrid composites are improved along with 20% cost reduction and 23% weight savings. Jarukumjorn and Suppakarn (2009) investigated the effect of glass fiber hybridisation on the physical properties of sisal fiber-polypropylene composites.

The study revealed that the hybrid composites enhanced tensile, flexural, and impact strength without having any significant effect on tensile and flexural modulus.

To improve the interfacial strength between the natural fibers and polymer matrix and to make the fibers less hydrophilic, different chemical modifications may be required. Hence, this present review paper is focused mainly on different chemical treatments of natural fibers.

CHEMICAL TREATMENT OF HYBRID FIBERS

The major concern of hybrid natural fiber reinforced composites are the hydrophilic and hydrophobic nature of the natural fiber and polymer matrix. The inherent incompatible between fiber-matrix makes weak interface bonding for the hybrid fiber composites. The compatibility of fiber-matrix may be improved, if the fiber surface is chemically modified (John and Anandjiwala, 2008). Hence, chemical modifications on fiber surfaces are necessary to reduce the hydrophilic nature of fibers and to improve interaction with the matrix (Abdelmouleh, 2004). For the past few decades, several researchers have investigated different types of chemical treatments of natural fibers aiming to improve the interaction between fiber and matrix. In the present review paper, the author will be concentrated mainly on different chemical modifications of natural fibers related to hybrid composites.

ALKALI TREATMENT OF NATURAL FIBERS

Alkali treatment is one of the most widely used chemical surface modifications of natural fibers. This type of chemical treatment is used to modify fiber surface and to reinforce natural fibers into thermoplastics or thermosets materials. The important modification occurring due to alkali treatment is the removal of hydrogen bonding in the network structure as well as fiber constituents such as hemicellulose, lignin, pectin, fat, and wax as a result surface roughness of fibers increases which results in improved interfacial bonding (Mohanty et al., 2001).

The alkali treatment starts with the immersion of fibers in Sodium hydroxide (NaOH), Potassium hydroxide (KOH) or Lithium hydroxide (LiOH) solution for a specific period. For the past few decades, use of hybrid natural-natural fibers reinforced composites in engineering applications are increased due to good physical properties, environmental safety, and economical. Some of the investigations occurred in the past few years related to chemical modifications of hybrid natural fiber composites are discussed. Jacob et al., (2004) investigated the mechanical properties of sisal/oil palm hybrid fiber reinforced natural rubber composites. They treated the fibers with varying concentrations viz., 0.5, 1, 2, 4, and 10% of NaOH solutions for 1 h at room temperature. They found that alkali treated fibers exhibited better tensile properties than untreated composites. Further, Jacob et al., (2006) focused on the dielectric constant of hybrid fibers containing different concentrations of alkali treated sisal-oil palm reinforced natural rubber bio composites. The results showed that the dielectric constant of alkali treated hybrid composites decreases with increase in concentration of alkali. The composites containing 4% alkali treated exhibits lowest dielectric constant. Saw et al., (2012) investigated the mechanical properties of chemically modified jute-coir hybrid fibers reinforced epoxy composites. They treated the fibers with 5% NaOH solution for 2 h and the results showed that the mechanical performance of hybrid composites are increased compared to untreated fibers. Ranjan et al. (2013) investigated the mechanical characteristics of banana/sisal hybrid fiber reinforced polylactic acid (PLA) composites. They treated the fibers with 2 wt% NaOH solution for 2 hrs. The researchers showed that the alkali treatment of fibers improved the interfacial bonding between fiber and matrix, hence better tensile, flexural, and impact strength than untreated fiber composites. Jawaid et al., (2011) immersed oil palm and jute fibers in a solution of 3% HEA and 1% dicumyl peroxide in methanol for 5 min. They found that the flexural and impact properties of treated hybrid fiber composites are improved as compared to untreated hybrid composites due to better fiber/matrix interfacial bonding. Alavudeen et al., (2015) studied the mechanical properties of banana/kenaf fiber reinforced polyester hybrid composites. The investigators treated the fibers with 5, 10, and 15 wt% NaOH solution for 2 h, 4 h, 6 h, and 8 h each. Due to hybridization of kenaf with banana fibers, the tensile, flexural, and impact strength increased compared to individual fibers. Kumar et al., (2016) prepared coconut sheath/banana fiber hybrid composites. They surface treated the fibers using 1 N alkali solution and found that the treated fibers showed a positive effect on the mechanical properties. Further, Venkatesh et al., (2016) studied the tensile, flexural, impact and water absorption properties of 10 wt% NaOH treated bamboo fiber hybridised sisal/unsaturated polyester hybrid composites. They found that the treated hybrid fibers show 30% increase in tensile strength, 27.4% in flexural strength, and 36.9% in impact strength along with decrease in moisture absorption behavior. Similarly, Otto et al., (2017) reported that 1.5% NaOH treated natural fibers reinforced polyurethane hybrid composite have a good potential candidate for shock absorption and padding materials due to high modulus and resilience.

Natural hybrid fibers have advantages of high strength and stiffness, low density, bio-degradable, processed at low cost, however the fibers have its limitations such as high moisture absorption, low durability, and low impact strength (Pickering et al., 2016). Hence, the researchers have mostly focused on synthetic-natural hybrid fibers for the enhancement of physical properties of fibers in engineering applications.

Some of the research articles published in the past few decades related to chemical treatment of synthetic-natural hybrid fibers are discussed. Arbelaiz et al., (2005) treated flax fibers with 20 wt% aqueous NaOH solution for 1 h at room temperature. They prepared glass and treated flax fibers into polypropylene hybrid composites for the mechanical properties and found that the hybrid composites mechanical performance increases than untreated composites. Patel et al., (2008) investigated the effect of alkali treatment of hybrid jute/glass fibers reinforced with epoxy composites and the results showed that the tensile, flexural, and electric strength improves. Jarukumjorn et al., (2009) studied the effect of glass fiber hybridisation on the physical properties of sisal fiber reinforced polypropylene composites. They treated the fibers with 2 wt% NaOH solution for 2 h. They found that the incorporation of glass fibers into sisal-polypropylene composites increases tensile, flexural, and impact strength without any significant effect on modulus along with improvement in thermal and water resistance properties. Kumar et al., (2009) studied the mechanical properties of alkali treated coir/glass fibers reinforced phenolic resin and the results showed that significant improvement in mechanical properties on alkali treated coir based hybrid composites. Further, Reddy et al., (2010) studied the effect of alkali treatment on bamboo/glass fiber reinforced polyester composites on tensile properties. They found that the mechanical performance of hybrid composites is enhanced with the alkali treated bamboo fibers compared to untreated fibers. Rao et al., (2010) investigated the flexural, and compressive properties of alkali treated bamboo/glass hybrid fibers shows enhanced properties compared to untreated bamboo fibers. Dhakal et al., (2012) investigated the thermal degradation of hemp/ glass fiber reinforced unsaturated polyester hybrid composites. The investigators treated the hemp fibers in a 10% NaOH and 90% distilled water solution bath for 48 h with a temperature of 18 ± 2 °C. They found that the surface treatment and glass fiber hybridization improves thermal stability of hybrid composites. Shanmugam and Thiruchitrambalam (2013) investigated the mechanical properties of alkali treated palmyra palm leaf stalk/jute fiber reinforced hybrid polyester composites. They treated the fibers in 5% NaOH solution for 30 min and they reported that the hybridization of alkali treated fibers enhanced mechanical performance which are comparable to natural/glass fiber composites and thereby increasing the hybrid fibers applications in production of light weight automotive components. Yahaya et al., (2015) investigated the mechanical properties of 6% NaOH for 3 h chemically treated kenaf-aramid fibers reinforced epoxy resin and the results showed that the tensile and flexural properties of treated fibers are better than untreated fibers. Another researcher, Davoodi et al., (2010) studied the mechanical properties of hybrid kenaf/ glass fiber reinforced epoxy composites. They treated the kenaf fibers with 6% NaOH solution and found that the treated kenaf fibers with glass fiber hybrid shows an increased impact and flexural properties. Asim et al., (2017) investigated the effect of alkali treatment on mechanical properties of kenaf/pine apple leaves reinforced phenolic hybrid composites. The results showed that alkali treated fibers improves mechanical performance of hybrid composites than untreated fibers. The effect of chemical treatments on natural-synthetic hybrid fibers reinforced composites are also studied by various researchers (John and Naidu, 2007; Athijayamani et al., 2010; Raghu et al., 2010).

From the review, it is concluded that alkali treatment of hybrid fibers improves the physical properties by improving the interfacial bonding between fibers and matrix. The alkali treatment of natural fibers composites significantly improves the mechanical and thermal behaviors. The excess delignification of fiber may take place and results in weakening of fibers, if concentration of alkali is higher than the optimum condition (Li et al., 2007). The alkali treated hybrid fibers may be utilized for engineering applications as it possesses similar properties required for engineering components.

SILANE TREATMENT OF NATURAL FIBERS

Silane coupling agents are used to modify surface of fibers and it is simple and cost-effective method (Kango et al., 2013). The silane coupling agents act as a bridge between natural fiber and polymer matrix interface and increasing the interfacial strength and may reduce the number of hydroxyl groups in the fiber-matrix interface. The investigations related to silane treatment of hybrid natural fibers are discussed. Arbelaiz et al., (2005) investigated the mechanical properties of silane modified flax/glass fiber reinforced polypropylene. The results showed that the silane modified hybrid composites enhanced the mechanical properties. Park et al., (2006) investigated the interfacial strength of jute/hemp fibers reinforced polypropylene hybrid composites. The results showed that the interfacial shear strength increases due to the modification of fiber by silane coupling agent. Foulk et al., (2006) evaluated the mechanical properties of silane treated flax/cotton fibers reinforced recycled polyethylene composites. They found that the silane treated hybrid fibers exhibited good mechanical properties compared to untreated fibers. Jacob et al., (2006) investigated the dielectric characteristics of silane treated sisal-oil palm hybrid fiber reinforced natural rubber bio composites. They found that silane modification of fibers resulted in decrease of dielectric constant and volume resistivity values. John and Naidu (2007) studied the effect of silane treatment of sisal/glass hybrid fibers reinforced polyester composites on the flexural and tensile properties and observed that tensile strength improved significantly while no significance on flexural properties. Varada Rajulu and Devi (2008) investigated the flexural properties of silane coupling agent modified glass/ridge gourd reinforced phenolic hybrid composites. It is observed that the coupling agent improved the mechanical properties of hybrid composites compared to untreated fibers. Uthayakumar et al., (2014) studied the mechanical properties of banana fiber/redmud reinforced polyester composites. They found that impact and flexural properties are increased for silane treated fibers but tensile strength showed a decreasing trend. Muhammad et al., (2015) fabricated the hybrid kenaf/glass fiber reinforced epoxy composites. The fibers are treated with silane coupling agent and the physical properties are determined for the treated fibers. It was found that the silane treatment increased impact and flexural properties respectively. Asim et al., (2017) investigated the effect of silane coupling agent on mechanical properties of pine apple leaf-kenaf fiber reinforced phenolic composites. They treated the fibers with silane having concentration of 2% for 3 h. The fibers are washed thoroughly with running water until pH are neutralized. The results showed that silane treated fibers exhibits better mechanical performance of hybrid composites than untreated fibers. They also suggested that the hybrid composites can be utilized to produce components for building structure and automobile applications.

MALEATED TREATMENT OF NATURAL FIBERS

Maleated coupling agents are predominantly used to achieve better interfacial bonding between matrix and fiber reinforcements (Keener, 2004). Among the various chemical modifications, maleated coupling agents treated fibers exhibits better mechanical performance (Rana, 1998). Rozman et al., (2001) investigated tensile and flexural properties of oil palm-glass fiber reinforced polypropylene hybrid composites. They found that the maleated coupling agent added with hybrid fibers resulted in an improvement in tensile and flexural properties of composites. Arbelaiz et al., (2005) developed hybrid composite by combining flax and glass fibers in maleic anhydride grafted polypropylene matrix. They found that the MAPP modified hybrid composites show higher mechanical properties than unmodified hybrid composites. Panthapulakkal and Sain (2007) investigated the mechanical and thermal properties of injection molded short hemp/glass fiber reinforced polypropylene hybrid composites. They used 5 wt% maleated polypropylene based compatibilizer in the preparation of composites. The hybridization of glass fiber with the composites enhanced the mechanical, thermal and water resistance performance. They also suggested that the prepared composites are promising candidates for structural applications. Haneefa et al., (2008) studied the tensile and flexural properties of short banana/glass hybrid fiber reinforced polystyrene composites. They modified the fiber using maleic anhydride and found that the hybridization increases the mechanical performance than untreated fibers. Nayak and Mohanty (2010) studied the effect of maleic anhydride grafted polypropylene (MAPP) reinforced with short sisal/glass fibers on mechanical properties. They reported that the mechanical properties such as tensile strength, flexural strength, and impact strength increased in the presence of 2% MAPP and the rate of water absorption decreased due to presence of coupling agent. Busu et al., (2010) investigated the mechanical and physical properties of thermoplastic natural rubber hybrid composites reinforced with kenaf and short glass fibers. They treated the glass fiber with silane coupling agent while used MAPP as compatibilizer for the composites. They found that the treated fiber has resulted in an increment of 100% of flexural modulus and impact strength. From the present review, it is concluded that maleated treatment may be used for hybrid fibers to enhance the mechanical properties such as tensile modulus, tensile strength, flexural modulus, flexural strength, hardness, and impact strength.

OTHER CHEMICAL TREATMENT OF NATURAL FIBERS

Different researchers are investigated different methods of other chemical treatments of single natural fibers reinforced composites. To the best of author's knowledge, no research work has been reported in the literature on hybrid natural fiber reinforced composites. Hence, research work carried out on other chemical treatment of single natural fibers are discussed below.

Acetylation treatment known as esterification method are used to treat natural fibers by introducing acetyl functional group into the fibers and thus reduce the hygroscopic nature of fibers which increases the dimensional stability of composites (Sreekala et al., 2000). Bledzki et al., (2008) studied the mechanical properties of acetylated flax fiber reinforced polypropylene. They found that the treated composites enhanced the tensile and flexural properties compared to untreated fiber composites.

Benzoylation treatment is another chemical treatment which uses benzoyl chloride to decrease hydrophilic nature of fibers and improves interfacial adhesion between fiber and matrix (Li et al., 2007). Wang et al., (2007) studied the effect of benzoylation treatment of flax fiber reinforced low density polyethylene composites and reported that increase in tensile strength and moisture resistance properties.

Other chemical treatments used to modify the structure of natural fibers and to improve interfacial bonding of fiber and matrix are peroxide treatment (Joseph et al., 1996; Li et al., 2007; Wang et al., 2007), sodium chlorite treatment (Misra et al., 2002; Li et al., 2007), isocyanate treatment (George et al., 2001; Kaliaa et al., 2009), stearic acid treatment (Torres and Ubillas, 2005; Kalaprasad et al., 2004; Paul et al., 2010), Permanganate treatment (Rahman et al., 2007; Paul et al., 2010), Triazine treatment (Li et al., 2007; Kaliaa et al., 2009).

CONCLUSION

Natural hybrid composites have been extensively used in engineering applications due to their excellent mechanical properties, processing advantages compared to single fiber reinforced composites. The present review described the research works related to different chemical treatments of natural fibers based hybrid composites. Natural hybrid composites composed of natural-natural fibers have excellent mechanical and thermal properties considering environmental safety, low cost, and bio-degradable. Researchers are investigating the natural-synthetic hybrid composites to enhance the properties compared to metal. However, hydrophilic nature of fibers makes it liable to higher water absorption that creates incompatible with the polymer matrix which affects the material properties. This incompatibility between fiber and matrix makes the composites interfacial strength weak and poor in its properties. Hence, chemical treatments of fibers are required to reduce hydrophilic nature and to improve surface adhesion with the matrix. The chemical treatment methods such as alkaline, silane, maleated, and other chemical treatments are discussed in this review paper. It is concluded from the review that chemical treatments for natural fibers may provide significant improvements in the fiber strength, and fiber-matrix adhesion for natural fiber reinforced composites. Hence, it can be concluded that the natural fibers reinforced hybrid composites have better mechanical properties compared to individual fibers reinforced composites that make it as a suitable reinforcement for engineering applications. Further, more research work related to electrical, thermal, and dynamic mechanical properties may be needed to increase its engineering applications for natural hybrid composites.

REFERENCES

Abdelmouleh, M., Boufi, S., Belgacem, M. N., Duarte, A. P., Salah, A. B., & Gandini, A. (2004). Modification of cellulosic fibres with functionalised silanes: Development of surface properties. *International Journal of Adhesion and Adhesives*, *24*(1), 43–54. doi:10.1016/S0143-7496(03)00099-X

Abdelmouleh, M., Boufi, S., Belgacem, M. N., & Dufresne, A. (2007). Short natural-fibre reinforced polyethylene and natural rubber composites: Effect of silane coupling agents and fibres loading. *Composites Science and Technology*, *67*(7), 1627–1639. doi:10.1016/j.compscitech.2006.07.003

Alavudeen, A., Rajini, N., Karthikeyan, S., Thiruchitrambalam, M., & Venkateshwaren, N. (2015). Mechanical properties of banana/kenaf fiber-reinforced hybrid polyester composites: Effect of woven fabric and random orientation. *Materials & Design (1980-2015), 66*, 246-257.

Arbelaiz, A., Fernández, B., Cantero, G., Llano-Ponte, R., Valea, A., & Mondragon, I. (2005). Mechanical properties of flax fibre/polypropylene composites. Influence of fibre/matrix modification and glass fibre hybridization. *Composites. Part A, Applied Science and Manufacturing*, *36*(12), 1637–1644. doi:10.1016/j.compositesa.2005.03.021

Asim, M., Jawaid, M., Abdan, K., & Ishak, M. R. (2017). The Effect of Silane Treated Fibre Loading on Mechanical Properties of Pineapple Leaf/Kenaf Fibre Filler Phenolic Composites. *Journal of Polymers and the Environment*, 1–8.

Athijayamani, A., Thiruchitrambalam, M., Natarajan, U., & Pazhanivel, B. (2010). Influence of alkali-treated fibers on the mechanical properties and machinability of roselle and sisal fiber hybrid polyester composite. *Polymer Composites*, *31*(4), 723–731.

Bledzki, A. K., & Gassan, J. (1999). Composites reinforced with cellulose based fibres. *Progress in Polymer Science*, *24*(2), 221–274. doi:10.1016/S0079-6700(98)00018-5

Bledzki, A. K., Mamun, A. A., Lucka-Gabor, M., & Gutowski, V. S. (2008). The effects of acetylation on properties of flax fibre and its polypropylene composites. *Express Polymer Letters*, *2*(6), 413–422. doi:10.3144/expresspolymlett.2008.50

Bunsell, A. R., & Harris, B. (1974). Hybrid carbon and glass fibre composites. *Composites*, *5*(4), 157–164. doi:10.1016/0010-4361(74)90107-4

Chaudhary, V., Bajpai, P. K., & Maheshwari, S. (2017). Studies on Mechanical and Morphological Characterization of Developed Jute/Hemp/Flax Reinforced Hybrid Composites for Structural Applications. *Journal of Natural Fibers*, 1–18.

Cicala, G., Cristaldi, G., Recca, G., Ziegmann, G., El-Sabbagh, A., & Dickert, M. (2009). Properties and performances of various hybrid glass/natural fibre composites for curved pipes. *Materials & Design*, *30*(7), 2538–2542. doi:10.1016/j.matdes.2008.09.044

Davoodi, M. M., Sapuan, S. M., Ahmad, D., Ali, A., Khalina, A., & Jonoobi, M. (2010). Mechanical properties of hybrid kenaf/glass reinforced epoxy composite for passenger car bumper beam. *Materials & Design*, *31*(10), 4927–4932. doi:10.1016/j.matdes.2010.05.021

De Rosa, I. M., Santulli, C., Sarasini, F., & Valente, M. (2009). Effect of loading-unloading cycles on impact-damaged jute/glass hybrid laminates. *Polymer Composites*, *30*(12), 1879–1887. doi:10.1002/pc.20789

Dhakal, H. N., Zhang, Z. Y., Richardson, M. O. W., & Errajhi, O. A. Z. (2007). The low velocity impact response of non-woven hemp fibre reinforced unsaturated polyester composites. *Composite Structures*, *81*(4), 559–567. doi:10.1016/j.compstruct.2006.10.003

Dong, J., Locquet, A., Declercq, N. F., & Citrin, D. S. (2016). Polarization-resolved terahertz imaging of intra-and inter-laminar damages in hybrid fiber-reinforced composite laminate subject to low-velocity impact. *Composites. Part B, Engineering*, *92*, 167–174. doi:10.1016/j.compositesb.2016.02.016

Fiore, V., Scalici, T., Calabrese, L., Valenza, A., & Proverbio, E. (2016). Effect of external basalt layers on durability behaviour of flax reinforced composites. *Composites. Part B, Engineering*, *84*, 258–265. doi:10.1016/j.compositesb.2015.08.087

Foulk, J. A., Chao, W. Y., Akin, D. E., Dodd, R. B., & Layton, P. A. (2006). Analysis of flax and cotton fiber fabric blends and recycled polyethylene composites. *Journal of Polymers and the Environment*, *14*(1), 15–25. doi:10.1007/s10924-005-8703-1

Gao, S. L., & Mäder, E. (2006). Jute/polypropylene composites I. Effect of matrix modification. *Composites Science and Technology*, *66*(7), 952–963.

George, J., Sreekala, M. S., & Thomas, S. (2001). A review on interface modification and characterization of natural fiber reinforced plastic composites. *Polymer Engineering and Science*, *41*(9), 1471–1485. doi:10.1002/pen.10846

Gupta, M. K. (2017). Effect of frequencies on dynamic mechanical properties of hybrid jute/sisal fibre reinforced epoxy composite. *Advances in Materials and Processing Technologies*, 1-14.

Haneefa, A., Bindu, P., Aravind, I., & Thomas, S. (2008). Studies on tensile and flexural properties of short banana/glass hybrid fiber reinforced polystyrene composites. *Journal of Composite Materials*, *42*(15), 1471–1489. doi:10.1177/0021998308092194

Hapuarachchi, T. D., Ren, G., Fan, M., Hogg, P. J., & Peijs, T. (2007). Fire retardancy of natural fibre reinforced sheet moulding compound. *Applied Composite Materials*, *14*(4), 251–264. doi:10.1007/s10443-007-9044-0

Hariharan, A. B. A., & Khalil, H. A. (2005). Lignocellulose-based hybrid bilayer laminate composite: Part I-studies on tensile and impact behavior of oil palm fiber-glass fiber-reinforced epoxy resin. *Journal of Composite Materials*, *39*(8), 663–684. doi:10.1177/0021998305047267

Idicula, M., Joseph, K., & Thomas, S. (2010). Mechanical performance of short banana/sisal hybrid fiber reinforced polyester composites. *Journal of Reinforced Plastics and Composites*, *29*(1), 12–29. doi:10.1177/0731684408095033

Jacob, M., Thomas, S., & Varughese, K. T. (2004). Mechanical properties of sisal/oil palm hybrid fiber reinforced natural rubber composites. *Composites Science and Technology*, *64*(7), 955–965. doi:10.1016/S0266-3538(03)00261-6

Jacob, M., Varughese, K. T., & Thomas, S. (2006). Dielectric characteristics of sisal–oil palm hybrid biofibre reinforced natural rubber biocomposites. *Journal of Materials Science*, *41*(17), 5538–5547. doi:10.1007/s10853-006-0298-y

Jamshaid, H., Mishra, R., Militky, J., Pechociakova, M., & Noman, M. T. (2016). Mechanical, thermal and interfacial properties of green composites from basalt and hybrid woven fabrics. *Fibers and Polymers*, *17*(10), 1675–1686. doi:10.1007/s12221-016-6563-z

Jarukumjorn, K., & Suppakarn, N. (2009). Effect of glass fiber hybridization on properties of sisal fiber–polypropylene composites. *Composites. Part B, Engineering*, *40*(7), 623–627. doi:10.1016/j.compositesb.2009.04.007

Jawaid, M., Alothman, O. Y., Paridah, M. T., & Khalil, H. S. A. (2014). Effect of oil palm and jute fiber treatment on mechanical performance of epoxy hybrid composites. *International Journal of Polymer Analysis and Characterization*, *19*(1), 62–69. doi:10.1080/1023666X.2014.858429

Jawaid, M., Khalil, H. A., Bakar, A. A., & Khanam, P. N. (2011). Chemical resistance, void content and tensile properties of oil palm/jute fibre reinforced polymer hybrid composites. *Materials & Design*, *32*(2), 1014–1019. doi:10.1016/j.matdes.2010.07.033

Jawaid, M. H. P. S., & Khalil, H. A. (2011). Cellulosic/synthetic fibre reinforced polymer hybrid composites: A review. *Carbohydrate Polymers*, *86*(1), 1–18. doi:10.1016/j.carbpol.2011.04.043

John, K., & Naidu, S. V. (2007). Chemical resistance of sisal/glass reinforced unsaturated polyester hybrid composites. *Journal of Reinforced Plastics and Composites*, *26*(4), 373–376. doi:10.1177/0731684406072524

John, M. J., & Anandjiwala, R. D. (2008). Recent developments in chemical modification and characterization of natural fiber-reinforced composites. *Polymer Composites*, *29*(2), 187–207. doi:10.1002/pc.20461

Joseph, K., Thomas, S., & Pavithran, C. (1996). Effect of chemical treatment on the tensile properties of short sisal fibre-reinforced polyethylene composites. *Polymer*, *37*(23), 5139–5149. doi:10.1016/0032-3861(96)00144-9

Kabir, M. M., Wang, H., Lau, K. T., & Cardona, F. (2012). Chemical treatments on plant-based natural fibre reinforced polymer composites: An overview. *Composites. Part B, Engineering*, *43*(7), 2883–2892. doi:10.1016/j.compositesb.2012.04.053

Kalaprasad, G., Francis, B., Thomas, S., Kumar, C. R., Pavithran, C., Groeninckx, G., & Thomas, S. (2004). Effect of fibre length and chemical modifications on the tensile properties of intimately mixed short sisal/glass hybrid fibre reinforced low density polyethylene composites. *Polymer International*, *53*(11), 1624–1638. doi:10.1002/pi.1453

Kalia, S., Kaith, B. S., & Kaur, I. (2009). Pretreatments of natural fibers and their application as reinforcing material in polymer composites—a review. *Polymer Engineering and Science*, *49*(7), 1253–1272. doi:10.1002/pen.21328

Kango, S., Kalia, S., Celli, A., Njuguna, J., Habibi, Y., & Kumar, R. (2013). Surface modification of inorganic nanoparticles for development of organic–inorganic nanocomposites—a review. *Progress in Polymer Science*, *38*(8), 1232–1261. doi:10.1016/j.progpolymsci.2013.02.003

Keener, T. J., Stuart, R. K., & Brown, T. K. (2004). Maleated coupling agents for natural fibre composites. *Composites. Part A, Applied Science and Manufacturing*, *35*(3), 357–362. doi:10.1016/j.compositesa.2003.09.014

Kretsis, G. (1987). A review of the tensile, compressive, flexural and shear properties of hybrid fibre-reinforced plastics. *Composites*, *18*(1), 13–23. doi:10.1016/0010-4361(87)90003-6

Kumar, N. M., Reddy, G. V., Naidu, S. V., Rani, T. S., & Subha, M. C. S. (2009). Mechanical properties of coir/glass fiber phenolic resin based composites. *Journal of Reinforced Plastics and Composites*, *28*(21), 2605–2613. doi:10.1177/0731684408093092

Li, X., Tabil, L. G., & Panigrahi, S. (2007). Chemical treatments of natural fiber for use in natural fiber-reinforced composites: A review. *Journal of Polymers and the Environment*, *15*(1), 25–33. doi:10.1007/s10924-006-0042-3

Manjunath, V., & Udupa, N. G. S. (2016). A Study on Hybrid Composite using Areca and Eucalyptus Fiber by using Epoxy Resin. *Journal of Industrial Mechanics*, *1*(1-2).

Mishra, S., Misra, M., Tripathy, S. S., Nayak, S. K., & Mohanty, A. K. (2001). Graft copolymerization of acrylonitrile on chemically modified sisal fibers. *Macromolecular Materials and Engineering*, *286*(2), 107–113. doi:10.1002/1439-2054(20010201)286:2<107::AID-MAME107>3.0.CO;2-0

Mishra, S., Mohanty, A. K., Drzal, L. T., Misra, M., Parija, S., Nayak, S. K., & Tripathy, S. S. (2003). Studies on mechanical performance of biofibre/glass reinforced polyester hybrid composites. *Composites Science and Technology*, *63*(10), 1377–1385. doi:10.1016/S0266-3538(03)00084-8

Misra, S., Misra, M., Tripathy, S. S., Nayak, S. K., & Mohanty, A. K. (2002). The influence of chemical surface modification on the performance of sisal-polyester biocomposites. *Polymer Composites*, *23*(2), 164–170. doi:10.1002/pc.10422

Mohanty, A. K., Misra, M., & Drzal, L. T. (2001). Surface modifications of natural fibers and performance of the resulting biocomposites: An overview. *Composite Interfaces*, *8*(5), 313–343. doi:10.1163/156855401753255422

Mohanty, A. K., Misra, M., & Hinrichsen, G. (2000). Biofibers, biodegradable polymers and biocomposites: An overview. *Macromolecular Materials and Engineering*, *276*(1), 1–24. doi:10.1002/(SICI)1439-2054(20000301)276:1<1::AID-MAME1>3.0.CO;2-W

Nayak, S. K., & Mohanty, S. (2010). Sisal glass fiber reinforced PP hybrid composites: Effect of MAPP on the dynamic mechanical and thermal properties. *Journal of Reinforced Plastics and Composites*, *29*(10), 1551–1568. doi:10.1177/0731684409337632

Otto, G. P., Moisés, M. P., Carvalho, G., Rinaldi, A. W., Garcia, J. C., Radovanovic, E., & Fávaro, S. L. (2017). Mechanical properties of a polyurethane hybrid composite with natural lignocellulosic fibers. *Composites. Part B, Engineering*, *110*, 459–465. doi:10.1016/j.compositesb.2016.11.035

Panthapulakkal, S., & Sain, M. (2007). Injection-molded short hemp fiber/glass fiber-reinforced polypropylene hybrid composites—Mechanical, water absorption and thermal properties. *Journal of Applied Polymer Science*, *103*(4), 2432–2441. doi:10.1002/app.25486

Park, J. M., Quang, S. T., Hwang, B. S., & DeVries, K. L. (2006). Interfacial evaluation of modified Jute and Hemp fibers/polypropylene (PP)-maleic anhydride polypropylene copolymers (PP-MAPP) composites using micromechanical technique and nondestructive acoustic emission. *Composites Science and Technology*, *66*(15), 2686–2699. doi:10.1016/j.compscitech.2006.03.014

Patel, V. A., Vasoya, P. J., Bhuva, B. D., & Parsania, P. H. (2008). Preparation and physicochemical study of hybrid glass-jute (treated and untreated) bisphenol-C-based mixed epoxy phenolic resin composites. *Polymer-Plastics Technology and Engineering*, *47*(8), 842–846. doi:10.1080/03602550802188839

Paul, S. A., Joseph, K., Mathew, G. G., Pothen, L. A., & Thomas, S. (2010). Influence of polarity parameters on the mechanical properties of composites from polypropylene fiber and short banana fiber. *Composites. Part A, Applied Science and Manufacturing*, *41*(10), 1380–1387. doi:10.1016/j.compositesa.2010.04.015

Pickering, K. L., Efendy, M. A., & Le, T. M. (2016). A review of recent developments in natural fibre composites and their mechanical performance. *Composites. Part A, Applied Science and Manufacturing*, *83*, 98–112. doi:10.1016/j.compositesa.2015.08.038

Premalal, H. G., Ismail, H., & Baharin, A. (2002). Comparison of the mechanical properties of rice husk powder filled polypropylene composites with talc filled polypropylene composites. *Polymer Testing*, *21*(7), 833–839. doi:10.1016/S0142-9418(02)00018-1

Puglia, D., Biagiotti, J., & Kenny, J. M. (2005). A review on natural fibre-based composites—Part II: Application of natural reinforcements in composite materials for automotive industry. *Journal of Natural Fibers*, *1*(3), 23–65. doi:10.1300/J395v01n03_03

Raghavendra Rao, H., Varada Rajulu, A., Ramachandra Reddy, G., & Hemachandra Reddy, K. (2010). Flexural and compressive properties of bamboo and glass fiber-reinforced epoxy hybrid composites. *Journal of Reinforced Plastics and Composites*, *29*(10), 1446–1450. doi:10.1177/0731684409105077

Raghu, K., Noorunnisa Khanam, P., & Venkata Naidu, S. (2010). Chemical resistance studies of silk/sisal fiber-reinforced unsaturated polyester-based hybrid composites. *Journal of Reinforced Plastics and Composites*, *29*(3), 343–345. doi:10.1177/0731684408097770

Rahman, M. M., Mallik, A. K., & Khan, M. A. (2007). Influences of various surface pretreatments on the mechanical and degradable properties of photografted oil palm fibers. *Journal of Applied Polymer Science, 105*(5), 3077–3086. doi:10.1002/app.26481

Ranjan, R., Bajpai, P. K., & Tyagi, R. K. (2013). Mechanical Characterization of Banana/Sisal Fibre Reinforced PLA Hybrid Composites for Structural Application. *Engineering International, 1*(1), 39–48. doi:10.18034/ei.v1i1.118

Ray, D., Sarkar, B. K., Rana, A. K., & Bose, N. R. (2001). Effect of alkali treated jute fibres on composite properties. *Bulletin of Materials Science, 24*(2), 129–135. doi:10.1007/BF02710089

Rout, J., Misra, M., Tripathy, S. S., Nayak, S. K., & Mohanty, A. K. (2001). The influence of fibre treatment on the performance of coir-polyester composites. *Composites Science and Technology, 61*(9), 1303–1310. doi:10.1016/S0266-3538(01)00021-5

Rozman, H. D., Tay, G. S., Kumar, R. N., Abusamah, A., Ismail, H., & Ishak, Z. M. (2001). Polypropylene–oil palm empty fruit bunch–glass fibre hybrid composites: A preliminary study on the flexural and tensile properties. *European Polymer Journal, 37*(6), 1283–1291. doi:10.1016/S0014-3057(00)00243-3

Safri, S. N. A., Sultan, M. T. H., Jawaid, M., & Jayakrishna, K. (2017). Impact behaviour of hybrid composites for structural applications: A review. *Composites. Part B, Engineering*.

Sarasini, F., Tirillò, J., D'Altilia, S., Valente, T., Santulli, C., Touchard, F., ... Gaudenzi, P. (2016). Damage tolerance of carbon/flax hybrid composites subjected to low velocity impact. *Composites. Part B, Engineering, 91*, 144–153. doi:10.1016/j.compositesb.2016.01.050

Saw, S. K., & Datta, C. (2009). Thermo mechanical properties of jute/bagasse hybrid fibre reinforced epoxy thermoset composites. *BioResources, 4*(4), 1455–1475.

Saw, S. K., Sarkhel, G., & Choudhury, A. (2012). Preparation and characterization of chemically modified Jute–Coir hybrid fiber reinforced epoxy novolac composites. *Journal of Applied Polymer Science, 125*(4), 3038–3049. doi:10.1002/app.36610

Shanmugam, D., & Thiruchitrambalam, M. (2013). Static and dynamic mechanical properties of alkali treated unidirectional continuous palmyra palm leaf stalk fiber/jute fiber reinforced hybrid polyester composites. *Materials & Design, 50*, 533–542. doi:10.1016/j.matdes.2013.03.048

Sreekala, M. S., Kumaran, M. G., Joseph, S., Jacob, M., & Thomas, S. (2000). Oil palm fibre reinforced phenol formaldehyde composites: Influence of fibre surface modifications on the mechanical performance. *Applied Composite Materials*, *7*(5), 295–329. doi:10.1023/A:1026534006291

Summerscales, J., & Short, D. (1978). Carbon fibre and glass fibre hybrid reinforced plastics. *Composites*, *9*(3), 157–166. doi:10.1016/0010-4361(78)90341-5

Swolfs, Y., Gorbatikh, L., & Verpoest, I. (2014). Fibre hybridisation in polymer composites: A review. *Composites. Part A, Applied Science and Manufacturing*, *67*, 181–200. doi:10.1016/j.compositesa.2014.08.027

Torres, F. G., & Cubillas, M. L. (2005). Study of the interfacial properties of natural fibre reinforced polyethylene. *Polymer Testing*, *24*(6), 694–698. doi:10.1016/j.polymertesting.2005.05.004

Uthayakumar, M., Manikandan, V., Rajini, N., & Jeyaraj, P. (2014). Influence of redmud on the mechanical, damping and chemical resistance properties of banana/polyester hybrid composites. *Materials & Design*, *64*, 270–279. doi:10.1016/j.matdes.2014.07.020

Van de Weyenberg, I., Ivens, J., De Coster, A., Kino, B., Baetens, E., & Verpoest, I. (2003). Influence of processing and chemical treatment of flax fibres on their composites. *Composites Science and Technology*, *63*(9), 1241–1246. doi:10.1016/S0266-3538(03)00093-9

Varada Rajulu, A., & Devi, R. R. (2008). Flexural properties of ridge gourd/phenolic composites and glass/ridge gourd/phenolic hybrid composites. *Journal of Composite Materials*, *42*(6), 593–601. doi:10.1177/0021998307086197

Venkata Subba Reddy, E., Varada Rajulu, A., Hemachandra Reddy, K., & Ramachandra Reddy, G. (2010). Chemical resistance and tensile properties of glass and bamboo fibers reinforced polyester hybrid composites. *Journal of Reinforced Plastics and Composites*, *29*(14), 2119–2123. doi:10.1177/0731684409349520

Venkatesh, R. P., Ramanathan, K., & Raman, V. S. (2016). Tensile, Flexual, Impact and Water Absorption Properties of Natural Fibre Reinforced Polyester Hybrid Composites. *Fibres & Textiles in Eastern Europe*, *24*(3), 90–94. doi:10.5604/12303666.1196617

Wan Busu, W. N., Anuar, H., Ahmad, S. H., Rasid, R., & Jamal, N. A. (2010). The mechanical and physical properties of thermoplastic natural rubber hybrid composites reinforced with Hibiscus cannabinus, L and short glass fiber. *Polymer-Plastics Technology and Engineering*, *49*(13), 1315–1322. doi:10.1080/03602559.2010.496408

Wang, B., Panigrahi, S., Tabil, L., & Crerar, W. (2007). Pre-treatment of flax fibers for use in rotationally molded biocomposites. *Journal of Reinforced Plastics and Composites*, *26*(5), 447–463. doi:10.1177/0731684406072526

Yahaya, R., Sapuan, S. M., Jawaid, M., Leman, Z., & Zainudin, E. S. (2015). Effect of layering sequence and chemical treatment on the mechanical properties of woven kenaf–aramid hybrid laminated composites. *Materials & Design*, *67*, 173–179. doi:10.1016/j.matdes.2014.11.024

Yusoff, R. B., Takagi, H., & Nakagaito, A. N. (2016). Tensile and flexural properties of polylactic acid-based hybrid green composites reinforced by kenaf, bamboo and coir fibers. *Industrial Crops and Products*, *94*, 562–573. doi:10.1016/j.indcrop.2016.09.017

Chapter 3
Fabrication of Porous NiTi Alloy Using Organic Binders

Neeraj Sharma
Maharishi Markandeshwar University, India

Kamal Kumar
Punjab Engineering College, India

ABSTRACT

Nitinol has growing applications in aerospace industries, MEMS, and bio-medical industries due to its unique properties of pseudo-elasticity, bio-compatibility, and shape-memory effect. Behaviour of NiTi alloy can be changed by altering the composition, modifying the porosity, and applying external thermal and mechanical treatment. In this chapter, porous NiTi alloy with powder metallurgy is fabricated by varying the composition of polypropylene as an organic binder from 0% to 15%, and Young's modulus and porosity of porous alloy has been evaluated. The effect of process parameters—compaction pressure, sintering temperature, and sintering time—are evaluated using Taguchi L16 orthogonal array. These particles initially act as a binder but with the increase of temperature, the organic particles evaporate and create pores. With the increase of organic particle percentage, the porosity increases while Young's modulus decreases. SEM was used to characterize the fabricated porous NiTi alloy.

DOI: 10.4018/978-1-5225-5216-1.ch003

INTRODUCTION AND LITERATURE REVIEW

Equi-atomic Nitinol (Ni_{50}-Ti_{50}) is categorized as a smart material due to its distinctive characteristics viz., shape memory effect and super elasticity, etc. Therefore, this alloy has many applications in engineering fields' (viz. micro-electromechanical-systems (MEMS) in aerospace, aircrafts, electrical switches, actuators and vibration absorbers, etc). Nitinol is also used in orthopaedic implants, cardio stents, orthodontic devices, etc. because of its good biocompatibility, self expanding behavior and high strain recovery (Castleman et al., 1976; Sharma et al., 2015a; 2015b).

Shape memory effect (SME) is the property by virtue of which a material can regain its original shape from permanent position after some thermal changes. The alloy representing this characteristic is known as shape memory alloys (SMA). Ni-Ti exhibit equi-atomic intermetallic compound of nickel and titanium, which can undergo a reversible solid state phase transformation from ordered cubic crystal structure (B2) called austenite to distorted monoclinic (B19') called martensite. Shape memory behaviour is of two types

1. One way (1WE)
2. Two Way (2WE)

When only the parent phase regain its original position it is 1WE while in 2WE, SMA remember the shape of both parent and product phase and it can be accomplished by heating and cooling the specimen. So, 1WE SMA remembers only one position, which it can attain after some appropriate thermal changes. In 2WE SMA, material remembers two positions, one for low temperature and another for high temperature. Figure 1 represents the transformation temperature range for NiTi alloy, which varies from 143K-351K.

Biocompatibility is a property by which a material can be used in a bio-fluid environment for a long time without any reaction with the bio-media (Catauro et al., 2004). In case of equi-atomic NiTi, a protective layer of TiO_2 is formed that prevents the release of Ni^+ ions and improves the corrosion, and hence shows good biocompatibility as verified by Es-Souni et al. (2005) through in-vitro cytotoxicity test. NiTi is mostly utilized in surgical devices and coronary stents due to its self-expanding nature and anti-thrombosis property. Thrombosis is a property of material in contact with bio-fluid to clot the white/red thrombus on its surface. NiTi stents exhibit negligible thrombosis as compared to stent made of stainless steel, wherein the later material, the probability of corrosion due to hostile electrolytic environment of the human body is high (Whitcher, 1997; Thierry et al., 2002; Furie and Furie, 2008).

Figure 1.

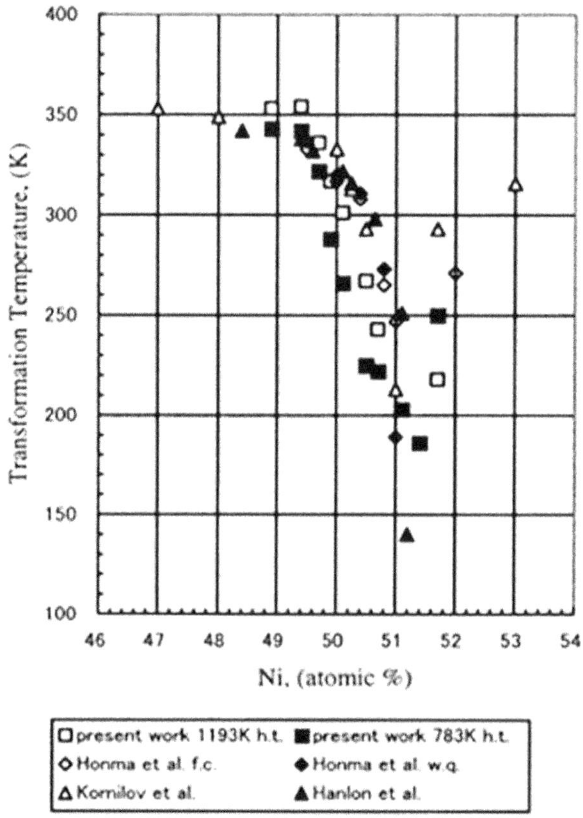

Due to high composition sensitivity, precise control of NiTi composition is a major issue. Using conventional methods, the properties of NiTi are severely affected because of the contamination of material due to the presence of oxygen and carbon in the surrounding environment (Frenzel et al., 2007). Additive manufacturing methods (Elahinia & Ahmadian, 2005; Lee et al., 2006; Williams, Shaw & Elahinia, 2005) are the newly developed advanced methods for the fabrication of net-shape NiTi geometries with highly controlled composition. But these methods are not commercialized till now due to their higher investment. NiTi can be fabricated with appreciating properties using inert atmosphere in conventional casting or powder metallurgy processes. But some post processing of NiTi is further required to remove the excess material and to eliminate any chance of release of Ni^{+} ions in human bodies. This short discussion indicates that the fabrication of net shape NiTi implants and other devices are highly challenging.

Figure 2.

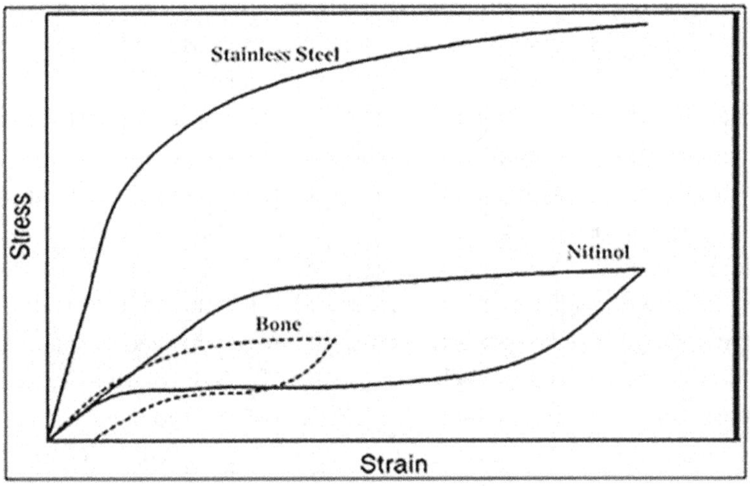

Figure 3.

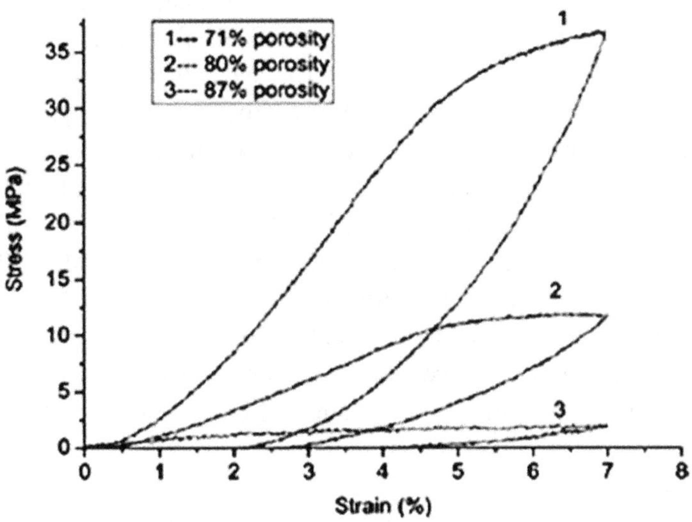

Stress shielding on human bones is a major problem for the medical implants made of steel alloy and Ti. Due to high value of young's modulus of these metallic materials, there is a mismatch between moduli of bone and steel alloy which causes implant loosening. Figure 2 shows the strain recovery behaviour of stainless steel, NiTi and bone. NiTi shows a high value of pseudo-elasticity and high value of strain recovery. The Young's modulus of NiTi is also comparable with that of bone

(Krishna et al., 2007). Therefore, NiTi has replaced the steel alloys for medical implants. Nowadays, porous implants of NiTi are gaining very high acceptance. Porosity influences the young's modulus and strain recovery of NiTi as shown in Figure 3, representing the stress-strain graph of loading and unloading compressive experiment, performed by Xiong et al. (2008). Also, porous structure favours high bone tissue ingrowth in pores of metallic implants.

Due to the distinctive quality of SME, NiTi has gained wide acceptance as actuators in mechanical and aerospace industries. Micro-electromechanical systems (MEMS) are developed for micro-level instruments since Ni-Ti alloys have high damping properties along with high abrasion resistance. NiTi in the form of thin films has technological importance in the fabrication of MEMS. Film thickness has a great role in controlling the transformation behaviour of NiTi from the martensite to austenite phase (Otsuka and Ren, 2005; Kumar et al., 2012). Micro and nano-finishing plays an important role for NiTi to be used for the applications in μ-TAS (Total Analysis System), fluidic devices and medical instruments. Controlling the chemical reaction of NiTi with working environment is a quite challenging research opportunities. In case, highly smooth NiTi film is used as MEMS in laser operated devices, then energy loss may occur and response of NiTi may get affected. Therefore, some surface modification may be allowed for improving the SME. Recently, NiTi alloys have been used in civil structures for vibration controls and for seismic isolations because of high energy dissipation through hysteresis, corrosion resistance, shape recovery after large strain and high fatigue resistance. Reinforcement and repair of damaged structure using SMAs introduced in civil engineering (Shahin et al., 2005; Burton et al., 2006; Savi et al., 2011; Wang and Mak, 2014; Guo and Kato, 2015), might be helpfull in crack healing for concrete structures (Hamada et al., 2003; Kirkby et al., 2008). In almost all previous surveys, NiTi SMAs were embedded inside a matrix and have excellent corrosion resistance. High performance structures can be fabricated by SMA due to active vibration control (Shahin et al., 2005). But still design guidelines are in a pending state to use SMA as seismic protective system (Ozbulut et al., 2011).

In this research work, the equi-atomic alloy has been fabricated using powder metallurgical technique. Polypropylene was used as a organic binder particles in different composition to study the effect on characteristics of NiTi alloy. More specifically the below mentioned work has been undertaken.

- Fabrication of homogeneous and porous structures of Ni-Ti: a titanium alloy, using conventional powder compaction technique.
- Experimentally determining the influence of process parameters on Young's modulus and porosity of Ni-Ti. Input process parameters are:
    - Compaction pressure

  - Sintering temperature
  - Sintering time
  - Concentration of binder particles
- Surface characterization of material using scanning electron microscopy (SEM).

MATERIALS AND METHODS

Materials

The work material selected in this work was Titanium powder (average size ~ 35μm, purity > 98.5%) and Nickel powder (average size ~20 μm, purity > 99.8%). These powders were mixed and blended with polypropylene, organic binder particles in different compositions for 4 to 8 hrs in ball mill. The mechanical and physical properties of NiTi material are given in Table 1. The chemical composition of Ni and Ti powders as envisaged from Spectro test is given in Table 2.

Specimen Preparation and Process Parameters

There are three major steps of specimen preparation viz.: blending, compaction and sintering. Mechanical alloying of elemental powder is carried by rotary ball mill at a speed of 300 R.P.M. Steatite ceramic balls (25 no. of 15mm diameter each) were used for mechanical alloying. A standard BPR (ball to powder ratio) 10:1 was upheld throughout the ball milling. This is a mandatory step for the homogeneous mixing of elemental powders. Powder metallurgy was used to fabricate the work-specimen. 15kN powder compaction press (Kimaya make) was employed to press the powder in a die. After the compaction, the green compact comes out from the die. It is then

Table 1. Mechanical and Physical Properties of NiTi alloys

Properties	Ni-Ti
Recovered Elongation	8%
Bio-compatibility	Excellent
Density (g/cm^3)	6.45
Magnetic	No
Ultimate tensile Strength	1240 MPa (Approx.)
Coefficient of Thermal Expansion	Martensite – 6.6×10^{-6} (cm/cm/^{O}C) Austenite – 11.0×10^{-6} (cm/cm/^{O}C)
Resistivity	80-100 (μΩ cm)

Table 2. Nickel and Chemical Powder chemical composition

Nickel Powder		Titanium Powder	
Element	**% Composition**	**Element**	**% Composition**
Fe	0.031	Ni	0.086
Cr	0.024	Fe	0.930
Co	0.090	V	0.341
Ni	99.84	Mo	0.050
		Ti	98.590

sintered at a temperature of ~60-70% melting temperature for the diffusion and other atomic transport mechanism of powder particles. Highly pure Ar (> 99.999% purity) is supplied throughout the sintering. Table 3 shows the values of selected input parameters and their levels.

Measurement of Responses

Young's Modulus

Solid cylindrical specimens were fabricated in cylindrical shape of length 12 mm and diameter 6mm (using ASTM-E9-09; L/D= 2). Young's modulus of specimen was measured using Universal testing machine at a constant strain rate of 0.05mm/min at room temperature. Young's modulus was measured in 'GPa'.

Porosity

Porosity can be easily measured by weighing method, in which the weight of specimen is measured. Equation 1 is used

Table 3. Level of Process Parameters

Sr. No	Input Parameters	Symbol	(Units)	Levels			
1.	Compaction Pressure	A	MPa	110	130	150	170
2.	Sintering Temperature	B	°C	850	950	1050	1150
3.	Sintering Time	C	min	20	30	40	50
4.	Organic Particles	D	%	0	5	10	15

$$P = 1 - \frac{d}{d_a} \tag{1}$$

where:

d: density of specimen (mass/volume of specimen)
d_a: density of the alloy (6.45g/cm^2)

The closed porosity can also be calculated by this method.

Surface Preparation and Characterization

Surface preparation is one of the most important parts to observe the microstructure of the alloy. After the fabrication of sintered alloy, the surface of material was prepared by using 800, 1200 and 1600 grit SiC abrasive paper. After the preparation of surface, it was cleaned by acetone to remove debris, dust and any other particle. At this condition, the microstructure observation reveals that the macrostructure has no grain boundaries. To monitor the grain boundary, the material must be etched with a proper reagent. In the present work, NiTi alloy was etched with a reagent formed by 40ml HNO_3, 10ml HF and 50ml H_2O. Jeol make scanning electron microscope was used for the examination of morphology and particle side at 20 kV.

RESULTS AND DISCUSSIONS

The experiments were carried according to the selected scheme of Taguchi's L_{16} orthogonal array (Khanna et al., 2015; Jangra et al., 2016; Gupta et al., 2016a; 2016b) to find out the effect of input parameters on performance characteristics i.e. Young's modulus and porosity. The experimental results are discussed subsequently in the following sections.

Taguchi Based Response Analysis for Young's Modulus

Taguchi proposes an "off-line" strategy for quality improvement as an alternative to an attempt to inspect quality into a product on the production line. He observes that poor quality cannot be improved by the process of inspection, screening and salvaging. No amount of inspection can put quality back into the product. Taguchi recommends a three-stage process for improving quality through offline quality control: system design, parameter design and tolerance design. For each trial in the

Table 4. Planning of Experiments According to L_{16} OA

Sr. No.	Compaction Pressure	Sintering Temperature	Sintering Time	Organic Particles
1	110	850	20	0
2	110	950	30	5
3	110	1050	40	10
4	110	1150	50	15
5	130	850	30	10
6	130	950	20	15
7	130	1050	50	0
8	130	1150	40	5
9	150	850	40	15
10	150	950	50	10
11	150	1050	20	5
12	150	1150	30	0
13	170	850	50	5
14	170	950	40	0
15	170	1050	30	15
16	170	1150	20	10

L_{16} array, the levels of the process parameters are indicated in Table 3. Design matrix of L_{16} OA given in Table 4, according to which all experiments will be performed. The results of Young's Modulus are given in Table 5.

The collected data after experimentation were analysed through the application of Taguchi methods to ascertain the significant factors and their contribution for the variability of response quality characteristics. In the present study all the designs, plots and analysis have been carried out using Minitab statistical software (Minitab 16).

Table 6 shows the ANOVA (means) for Young's modulus of NiTi alloy. From Table 6, it is clear that the p-value for sintering temperature is larger than 0.05 i.e. this process parameter is non-significant under 95% confidence level. Also p-value < 0.05 for compaction pressure, sintering time and organic particles, shows that these parameters are significant. Compaction pressure gave a major contribution into the process with 66.85%, while organic particles contribution into the process is 18.96% followed by sintering time 9.01%. Table 7 gives ANOVA (S/N values) of Young's modulus of NiTi. Mean square can be easily obtained by dividing sum of square to the degree of freedom value. Larger will be the value of F-ratio larger be the contribution of that parameter in the process. The percentage variability of this model due to significant and non-significant factors can be explained by R-square.

Table 5. Results according to Design Matrix

Sr. No.	Porosity				Young's Modulus			
	Porosity (%)			S/N	Young's Modulus (GPa)			S/N
1	13	13	12	22.035	5.1	5	5.2	-14.153
2	22	23	21	26.830	4.3	4.3	4.2	-12.602
3	43	44	42	32.665	3.2	3.2	3.2	-10.103
4	57	57	56	35.066	1.7	1.8	1.9	-5.114
5	20	19	19	25.719	5.3	5.4	5.2	-14.487
6	19	20	18	25.551	5.3	5.1	5.2	-14.321
7	16	15	16	23.887	5.5	5.5	5.4	-14.755
8	35	34	36	30.874	5.3	5.2	5.2	-14.376
9	24	24	23	27.477	5.8	5.7	5.6	-15.118
10	26	26	26	28.299	5.3	5.1	5.2	-14.321
11	22	23	23	27.102	6.2	6.1	6.4	-15.896
12	16	16	17	24.251	6.6	6.6	6.7	-16.435
13	20	19	20	25.867	6.2	6.0	6.1	-15.707
14	12	11	12	21.317	7.1	7.2	7.3	-17.147
15	23	24	24	27.477	6.0	6.2	6.1	-15.707
16	24	25	24	27.719	5.8	5.7	5.7	-15.168

Table 6. Analysis of Variance for Means ($Ni_{50}Ti_{50}$-Young's Modulus)

Source	DF	SS	MS	F	P	p (%)
Compaction Pressure	3	17.181	5.727	112.660	0.001	66.854
Sintering Temperature	3	1.174	0.391	7.700	0.064	4.568
Sintering Time	3	2.318	0.772	15.200	0.026	9.019
Organic Particles	3	4.873	1.624	31.960	0.009	18.963
Residual Error	3	0.152	0.050			0.596
Total	15	25.699	**R-Sq = 99.4%; R-Sq(adj) = 97.0%**			

It is a statistical measure and is also known as coefficient of determination. Larger is value of R^2 will be considered in favorable side for the investigation of response. Adjusted R-square indicates the percentage variability explained by model due to significant terms only in a process. 'S' is termed as standard error of regression and indicates the average distance of observed value from the regression line. Lower

Table 7. Analysis of Variance for S/N ratios (Young's Modulus)

Source	DF	SS	MS	F	P
Compaction Pressure	3	73.274	24.425	14.420	0.027
Sintering Temperature	3	10.378	3.459	2.040	0.286
Sintering Time	3	15.063	5.021	2.960	0.198
Organic Particles	3	21.225	7.075	4.180	0.135
Residual Error	3	5.081	1.694		
Total	15	125.022	R-Sq = 95.9%; R-Sq(adj) = 79.7%		

Table 8. Response Table for Means (Young's Modulus)

Level	Compaction Pressure	Sintering Temperature	Sintering Time	Organic Particles
1	3.592	5.550	5.567	6.100
2	5.300	5.467	5.575	5.458
3	5.942	5.250	5.333	4.858
4	6.283	4.850	4.642	4.700
Delta	2.692	0.700	0.933	1.400
Rank	**1**	**4**	**3**	**2**

value will be favorable, as the observed value will be near by the fitted line. Figure 4 represent that with the raise of compaction pressure the value of Young's modulus found to be raised. The lower value of Young's modulus is required to match it with the moduli of bones.

Higher the value of compaction pressure, more closely the particles are bind together and hence a larger force is required to distort them, hence a larger value of Young's modulus. Increase in sintering temperature time decreases the modulus value as shown in Figure 4. The main reason behind this is the temperature removes the organic particles between the powders and makes it porous (Christine et al., 1998; Sharma et al., 2016b; 2016c). More time the green compact remains in the furnace, more number of binder particles assumed to be evaporate. Hence larger pores to be formed which decreases the Young's modulus of NiTi. Figure 5 presents the variation of S/N values (Young's modulus) with the change of process parameters in case of NiTi. Before going to start discussion on this it must be understood that either the response "Higher the better" or "Lower the best" type quality characteristics, but S/N values must be "Higher the better" type. The main reason behind this is that S/N value represents the signal to noise ratio or the ratio of significant factor to non-

Table 9. Response Table for S/N Ratio (Young's Modulus)

Level	Compaction Pressure	Sintering Temperature	Sintering Time	Organic Particles
1	-10.490	-14.870	-14.880	-15.620
2	-14.480	-14.600	-14.810	-14.650
3	-15.440	-14.120	-14.190	-13.520
4	-15.930	-12.770	-12.470	-12.570
Delta	5.440	2.090	2.410	3.060
Rank	**1**	**4**	**3**	**2**

Figure 4.

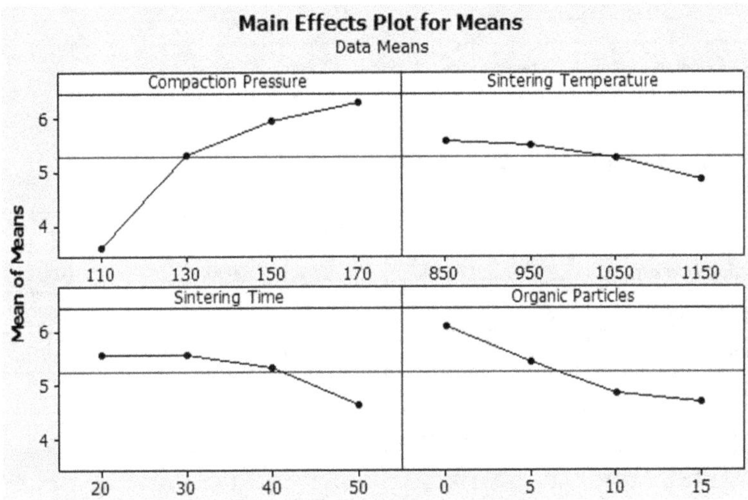

significant factor, which must be larger and shows that significant factor maximum contribution and non-significant factors minimum contribution.

Here in Figure 5 represents the results, which are in line with the results obtained from Figure 4. As low value of compaction pressure and high value of sintering temperature and sintering time suggested by Figure 4 and Figure 5 both.

Table 8 and 9 reported the response table for mean value (raw data) and S/N values of Young's modulus in case of NiTi alloy. In Table 8, "lower is better" type characteristic consider while in Table 9 "higher the better" type quality characteristic is considered. After considering these conditions, Table 8 and Table 9 represent the same settings of process parameters for obtaining the lowest Young's modulus. Also the order of process parameters (Rank) affecting response is same in both tables.

Figure 5.

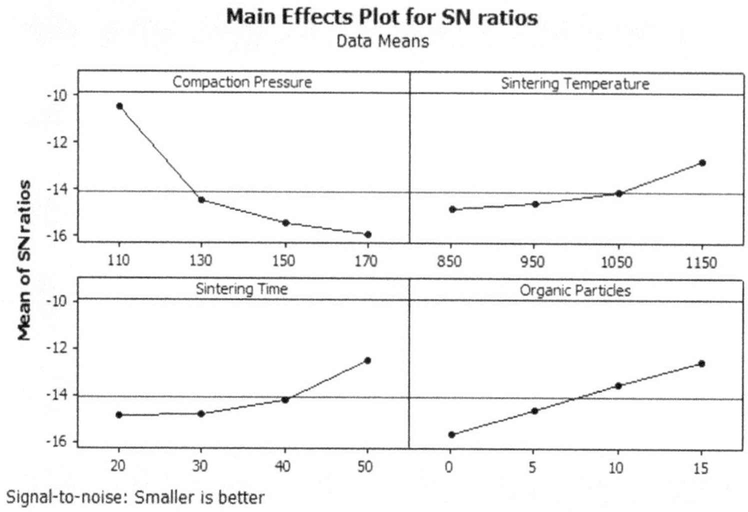

Table 10. Pooled ANOVA for Means

Source	DF	SS	MS	F	P
Compaction Pressure	3	17.181	5.727	25.900	0.001
Sintering Time	3	2.318	0.772	3.490	0.090
Organic Particles	3	4.874	1.624	7.350	0.020
Residual Error	6	1.327	0.221		
Total	15	25.700			

Figure 6 represent the variation of Young's modulus after pooling the sintering temperature. Figure 7 gives the residual plots for young's modulus. It consists of some test for the verification of ANOVA. First test is normality test or normal distribution test, in which residuals fall on a straight line. This indicates the normal distribution of errors. In another test the errors are randomly distributed, which are required for a good ANOVA. In Figure 7 all the errors are randomly distributed and ANOVA is good. Third Test is histogram and the last one is observation order or variance test. Both test indicates a good model and verifies ANOVA.

Table 11. .Response Table of Mean for Young's Modulus

Level	Compaction Pressure	Sintering Time	Organic Particles
1	**3.592**	5.567	6.100
2	5.300	5.575	5.458
3	5.942	5.333	4.858
4	6.283	**4.642**	**4.700**
Delta	**2.692**	**0.933**	**1.400**
Rank	**1**	**3**	**2**

Figure 6.

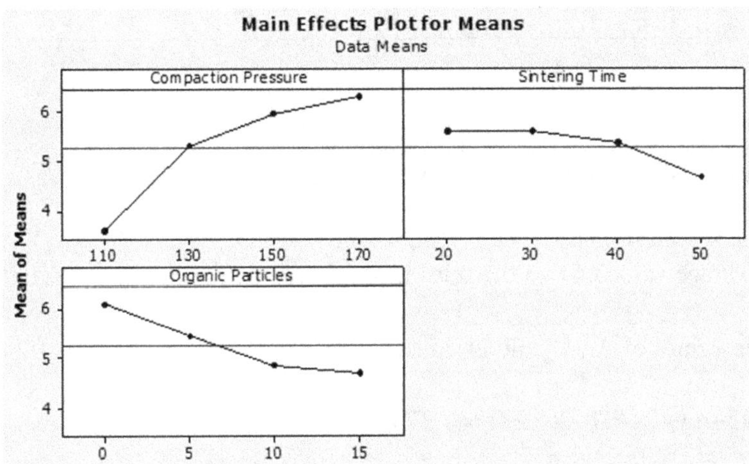

Optimal Setting for Young's Modulus

The optimum value of 'E' is envisaged at the chosen levels of parameters viz.: compaction pressure (A_1), sintering time (C_3) and organic particles (D_3) (Table 11). The approximated mean of the Young's modulus can be computed as:

$$\mu_E = A_1 + C_4 + D_4 - 2T \quad (2)$$

where, $\overline{T}$ = overall mean of Young's Modulus = $(\sum R)/48 = 5.28$ GPa

R values are chosen from the Table 5 and the values of A_1, C_3 and D_3 are estimated from the experimental data reported in the Table 11.

Figure 7.

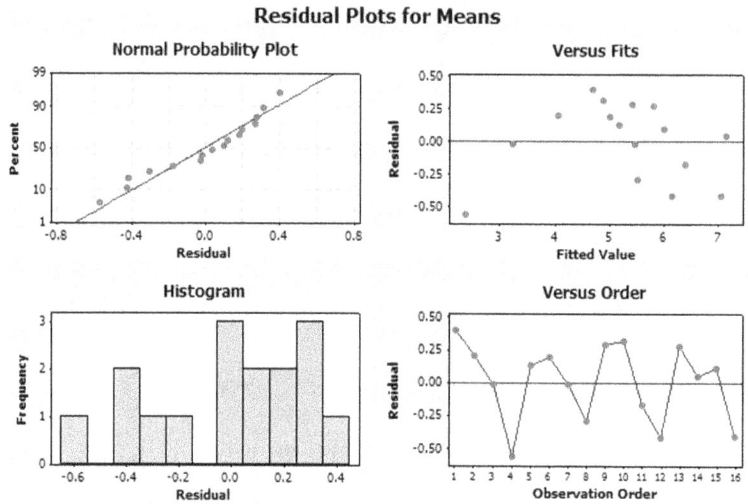

$\overline{A}_1$ = average value of E at the first level of compaction =3.592 GPa

$\overline{C}_4$ = average value of E at the third level of sintering time =4.642 GPa

$\left(\overline{D}_4\right)$ = average value of E at the third level of organic particles =4.7 GPa

Put the values of A_1, C_4 and D_4 in equation 5.6,

μ_E = 3.592+4.642+4.7- 2(5.28) = 2.374MPa

The 95% confidence intervals of confirmation experiments (CI_{CE}) and population (CI_{POP}) are calculated as:

$$CI_{CE} = \sqrt{F_\alpha(1,f_e)V_e\left[\frac{1}{n_{eff}}+\frac{1}{R}\right]} \quad \text{and} \quad CI_{POP} = \sqrt{\frac{F_\alpha(1,f_e)V_e}{n_{eff}}}$$

where, F_α (1, f_e) = The F-ratio at the confidence level of (1-α) against DOF 1 and error degree of freedom f_e.

$$n_{eff} = \frac{N}{1+\left[\text{DOF associated in the estimate of mean response}\right]} = 48/(1+9) =$$

4.8

Error variance = V_e = 0.2211; Total number of results = N = 48; Error DOF = f_e = 6;Sample size of experiments = R = $3F_{0.05}$ (1, 6) = 5.99So, CI_{CE} = ± 0.846 and CI_{POP} = ± 0.525

Therefore, the predicted confidence interval for confirmation experiments is:

$\mu_E - CI_{CE} < \mu_E < \mu_E + CI_{CE}$ i.e. $1.528 < \mu_E < 3.22$

The 95% confidence interval of the population is:

$\mu_E - CI_{POP} < \mu_E < \mu_E + CI_{POP}$ i.e. $1.849 < \mu_E < 2.899$

The optimal values of input control factors are as follows:

A_1 = 110MPa; (C_4): 50min; D_4: 15%.

Taguchi Based Response Analysis for Porosity

ANOVA for Porosity

Tables 12 and 13 give the ANOVA of mean values and S/N ratio for porosity. It is evident from the table that compaction pressure contributes 22.62% in porosity. P-value in Table 12 reveals that contribution of each factor is greater than 15%, while P-value is greater than 0.05. For the investigation of optimal setting, further analysis is carried for predicted values. Organic particle plays a major contribution of 31.74% trailed by sintering temperature of 26.05%. Table 13 (S/N ratio) pursue the same pattern of as followed by raw data. The P-values in case of S/N ratio are less than 0.05 and hence represent significant role of each imput parameter in the response. In Table 12 and 13, the R-square and adjusted R-square values are greater than 80% show a good variability. Figure 8 and 9 show the variation of porosity mean values and S/N values with process parameters.

From Figure 8, it is clear that with the increase of compaction pressure the porosity decreases. The main reason behind this is that with the increase of compaction pressures the space in the powder particles decrease.

Sintering temperature favours the porosity i.e. porosity increases with the sintering temperature. Increment in sintering time and temperature, the binder present in the core of material tries to evaporate, which favors porosity. Figure 9 follows the same pattern as explained in Figure 8. With the increase of organic particles in the alloy, the porosity found to be increased. The probable reason is the evaporation of organic particles with sintering temperature.

Figure 8.

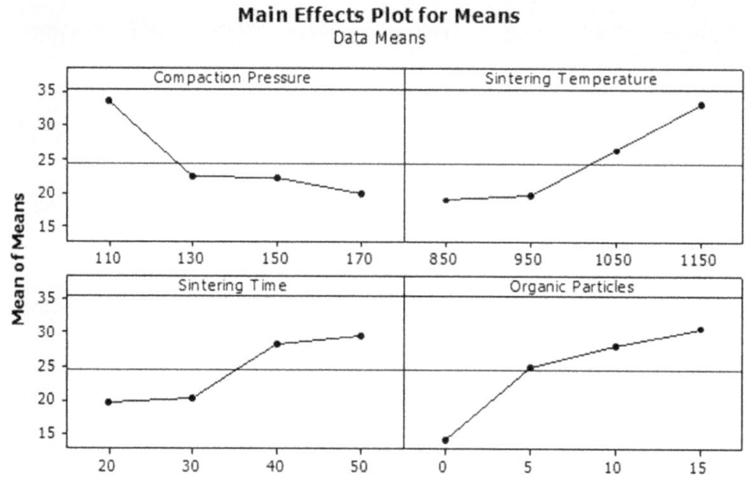

Figure 9.

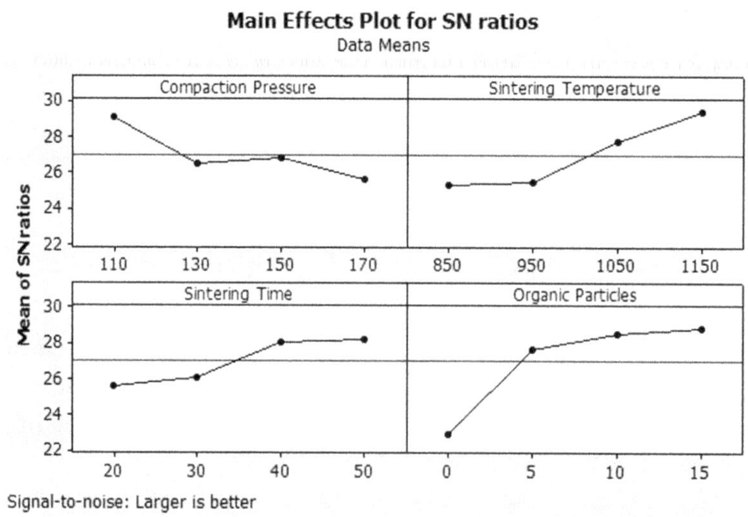

Table 14 and 15 give the response table for mean and S/N value of porosity in case of NiTi. Larger value of porosity will be beneficial for preventing stress shielding. Both Tables (Table 14 and 15) suggest the same optimal values of process parameters for porosity. Sintering temperature is affecting the porosity most followed by sintering time and compaction pressure.

Table 12. ANOVA for Means

Source	DF	SS	MS	F	P	p (%)
Compaction Pressure	3	459.140	153.050	6.030	0.087	22.620
Sintering Temperature	3	528.810	176.270	6.950	0.073	26.050
Sintering Time	3	321.640	107.210	4.220	0.134	15.840
Organic Particles	3	644.470	214.820	8.460	0.056	31.740
Residual Error	3	76.140	25.380			3.750
Total	15	2030.190	**R-Sq = 96.2%; R-Sq(adj) = 81.2%**			

Table 13. ANOVA for S/N ratios

Source	DF	SS	MS	F	P
Compaction Pressure	3	27.522	9.174	17.270	0.021
Sintering Temperature	3	47.921	15.973	30.080	0.010
Sintering Time	3	22.531	7.510	14.140	0.028
Organic Particles	3	94.512	31.504	59.320	0.004
Residual Error	3	1.593	0.531		
Total	15	194.079	**R-Sq = 99.2%; R-Sq(adj) = 95.9%**		

Table 14. Response Table for Means (Porosity)

Level	Compaction Pressure	Sintering Temperature	Sintering Time	Organic Particles
1	**33.580**	18.830	19.670	14.080
2	22.250	19.670	20.330	24.830
3	22.170	26.250	28.330	28.170
4	19.830	**33.080**	**29.500**	**30.750**
Delta	13.750	14.250	9.830	16.670
Rank	3	2	4	1

Table 15. Response Table for S/N Ratio (Porosity)

Level	Compaction Pressure	Sintering Temperature	Sintering Time	Organic Particles
1	**29.150**	25.270	25.600	22.870
2	26.510	25.500	26.070	27.670
3	26.780	27.780	28.080	28.600
4	25.600	**29.480**	**28.280**	**28.890**
Delta	3.550	4.200	2.680	6.020
Rank	3	2	4	1

Optimal Setting for Porosity (NiTi)

The optimum value of porosity is found by using equation 3 as:

$$\mu_{\sigma} = A_1 + B_4 + C_4 + D_4 - 3T \tag{5.6}$$

where, $\overline{T}$ = mean value of Porosity = $(\sum R)/48 = 24.45$

R values chosen from the Table 3 and the values of input parameters are taken from Table 11.

$\overline{A_1}$ = Porosity value at the first level of compaction pressure =33.58

$\left(\overline{B_4}\right)$ = Porosity value at the forth level of sintering temperature =33.08

$\left(\overline{C_4}\right)$ = Porosity value at the forth level of sintering time =29.50

$\left(\overline{D_4}\right)$ = Porosity value at the forth level of organic particles =30.75

Put value of input parameters in equation 5.6,

$\mu_{porosity}$ = 33.58+33.08+29.50+30.75-3(24.45) = 53.56

The 95% confidence intervals of confirmation experiments (CI_{CE}) and population (CI_{POP}) are given as:

So, $CI_{CE} = \pm$ 12.46 and $CI_{POP} = \pm$ 8.34

Therefore, the predicted confidence interval for confirmation experiments is:

$\mu_{porosity} - CI_{CE} < \mu_{porosity} < \mu_{porosity} + CI_{CE}$ i.e. $41.1 < \mu_{porosity} < 66.02$

The 95% confidence interval of the population is:

$\mu_{porosity} - CI_{POP} < \mu_{porosity} < \mu_{porosity} + CI_{POP}$ i.e. $45.22 < \mu_{porosity} < 61.9$

The optimal values of input parameters is given as:

(A_1): 110MPa; (B_4): 1150°C; C_4: 50 min; D_4 =15%.

Validation Experiments

Table 16 gives the predicted and experimental values at the optimal condition provided by response table (Table 11 and 14). At the setting suggested by above method the validation experiments are executed to get the Young's Modulus and porosity. The error between predicted and experimental values varies with in the range of ±10%, which signifies good results reproducibility.

Surface Characterization

As evident from Table 3, the presence of organic binder in the alloy increases the porosity. Figure 3 also reveals the same thing and the reason of this is the low density value of binder as compared to nickel and titanium. The melting and evaporation temperature of binder is also low as compared to metal powder, so this makes shape memory alloy porous. The confirmation of results was carried out by the SEM of sintered NiTi alloy.

Figure 10 (a) shows that the average pore size is 5μm, and the pore size in Figure 10 (b) and (c) is 7 μm and 9 μm respectively.

Table 16. Predicted and Confirmation Experiments Results

Response	Optimal Machining Conditions		$\frac{\text{Exp.} - \text{Predicted}}{\text{Exp.}} \times 100$
	Predicted	Experimental	
Setting Level	$A_1C_4D_4$/ $A_1B_4C_4D_4$	$A_1B_4C_4D_4$	
Young's Modulus (GPa)	2.374	2.170	-9.4%
Porosity (%)	53.560	49.470	-8.26%

Figure 10.

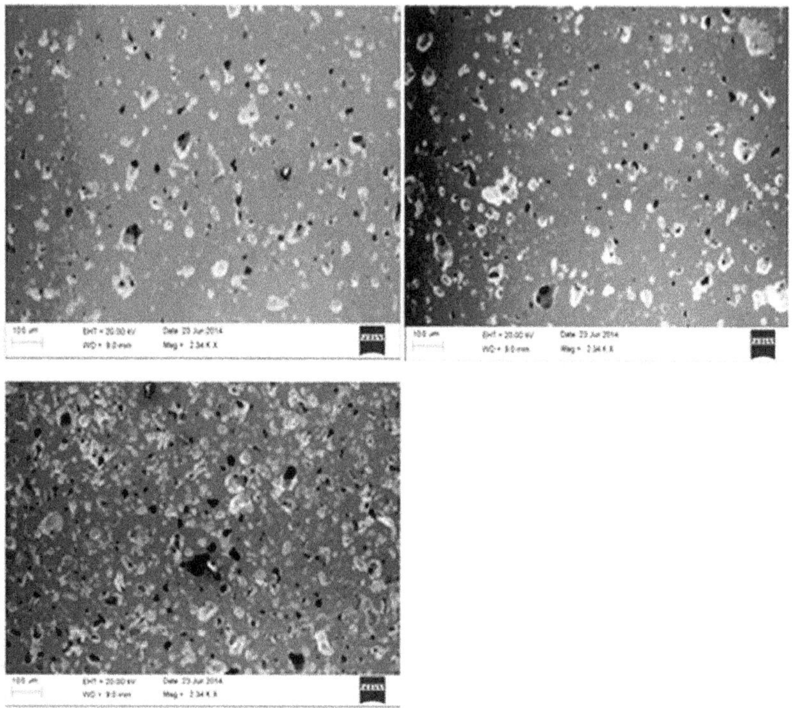

FUTURE WORK DIRECTIONS

In the present book chapter, performance characteristics of NiTi alloys viz. Young's modulus and porosity were considered. Other characteristics like tensile strength, machinability, transformation behavior etc. can also be studied. Effect of machining and heat treatment of NiTi alloy can be studied for shape memory effect and transition temperature range. NiTi composites (with soft metals and alloys) and alloys with composite coatings can also be considered for future investigation.

CONCLUSION

1. Porous NiTi structure is obtained using powder metallurgy technique with polypropylene as a binder.
2. The increase in binder from 0% to 15% decreases the alloy density from 6.5 g/cm^3 to 5.3g/cm^3.

3. Young's modulus value decreases with the increase in the value of organic binder percentage. Lower value of Young's modulus overcome the problem of Implant loosening.
4. Highest value of porosity (i.e. 51%) is observed at 15% binder percentage. The SEM micrographs revealed the porous structure.

REFERENCES

Burton, D. S., Gao, X., & Brinson, L. C. (2006). Finite element simulation of a self-healing shape memory alloy composite. *Mechanics of Materials*, *38*(5-6), 525–537. doi:10.1016/j.mechmat.2005.05.021

Castleman, L. S., Motzkin, S. M., Alicandri, F. P., Bonawit, V. L., & Johnson, A. A. (1976). Biocompatibility of nitinol alloy as an implant material. *Journal of Biomedical Materials Research*, *10*(5), 695–731. doi:10.1002/jbm.820100505 PMID:977602

Catauro, M., Raucci, M. G., Continenza, M. A., & Marotta, A. (2004). Biocompatibility Tests with Fibroblasts of Cao Rich Calcium Silicate Glasses. *Journal of Materials Science*, *39*(1), 373–375. doi:10.1023/B:JMSC.0000008093.18719.0b

Christine, E., Corbin, S. F. J., & Apte, S. P. (1998). *Porous metal structures and processes for their production.* US Patent, 5846664 A.

Elahinia, M. H., & Ahmadian, M. (2005). An enhanced SMA phenomenological model. Part II. The experimental study. *Journal of Smart Material and Structure*, *14*(1), 1309–1319. doi:10.1088/0964-1726/14/6/023

Es-Souni, M., Es-Souni, M., & Brandies, H. F. (2005). Assessing the Biocompatibility of NiTi shape memory alloys used for medical applications. *Analytical and Bioanalytical Chemistry*, *381*(3), 557–567. doi:10.1007/s00216-004-2888-3 PMID:15660223

Frenzel, J., Zhang, Z., Somsen, C., Neukin, K., & Eggeler, G. (2007). Influence of carbon on martensitic phase transformations in NiTi shape memory alloys. *Acta Materialia*, *55*(4), 1331–1341. doi:10.1016/j.actamat.2006.10.006

Furie, B., & Furie, B. C. (2008). Mechanisms of Thrombus Formation. *The New England Journal of Medicine*, *359*(9), 938–949. doi:10.1056/NEJMra0801082 PMID:18753650

Guo, W., & Kato, H. (2015). Submicron-porous NiTi and NiTiNb shape memory alloys with high damping capacity fabricated by a new top-down process. *Materials & Design*, *78*, 74–79. doi:10.1016/j.matdes.2015.04.028

Gupta, R.D., Khanna, R., Garg, M.P., & Sharma, N., & Preeti. (2016a). AWJ Circular Cutting and Advanced Parametric Optimization Using Taguchi, Grey Relational and Entropy Measurement Method. *Ciencia y Técnica (Buenos Aires)*, *31*, 57–86.

Gupta, R. D., Khanna, R., Kochar, G., & Sharma, N. (2016b). Experimental determination for optimal setting of parameters of EDM on Al/SiC Composite using Integrated Taguchi Grey Relational Analysis. *Journal of Machining and Forming Technologies*, *7*, 103–118.

Hamada, K., Kawano, F., & Asaoka, K. (2003). Shape recovery of shape memory alloy fiber embedded resin matrix smart composite after crack repair. *Dental Materials Journal*, *22*(2), 160–167. doi:10.4012/dmj.22.160 PMID:12873119

Jangra, K.K., Sharma, N., & Khanna, R., & Deepak. (2016). An Experimental Investigation and Optimization of Friction Stir Welding process for AA6082 T6 (cryogenic treated and untreated) using an integrated approach of Taguchi, GRA and Entropy Method. *Proc. of IMeche Part L: Part L: Journal of Materials, Design and Applications*, *230*, 454–469.

Kaieda, Y. (2003). Fabrication of composition-controlled TiNi shape memory wire using combustion synthesis process and the influence of Ni content on phase transformation behaviour. *Science and Technology of Advanced Materials*, *4*(3), 239–246. doi:10.1016/S1468-6996(03)00031-7

Khanna, R., Kumar, A., Garg, M. P., Singh, A., & Sharma, N. (2015). Multiple Performance Characteristics Optimization for Al 7075 on Electric Discharge Drilling by Taguchi Grey Relational Theory. *Journal of Industrial Engineering International*, *11*(4), 459–472. doi:10.1007/s40092-015-0112-z

Kirkby, E. L., Rule, J. D., Michaud, V. L., Sottos, N. R., White, S. R., & Manson, J. A. (2008). Embedded shape-memory alloy wires for improved performance of self-healing polymers. *Advanced Functional Materials*, *18*(15), 2253–2260. doi:10.1002/adfm.200701208

Krishna, B. V., Bose, S., & Bandyopadhyay, A. (2007). Laser processing of net-shape NiTi shape memory alloy. *Metallurgical and Materials Transactions. A, Physical Metallurgy and Materials Science*, *38*(5), 1096–1103. doi:10.1007/s11661-007-9127-4

Kumar, A., Singh, D., & Kaur, D. (2012). Variation in phase transformation paths of NiTi films as a function of film thickness. *Sensors and Actuators. A, Physical*, *178*, 57–63. doi:10.1016/j.sna.2012.02.014

Lee, K. S., Yang, D. Y., Park, S. H., & Kim, R. H. (2006). Recent Developments in the Use of Two-photon Polymerization in Precise 2D and 3D Microfabrication. *Polymers for Advanced Technologies*, *17*(2), 72–82. doi:10.1002/pat.664

Otsuka, K., & Ren, X. (2005). Physical metallurgy of Ti–Ni-based shape memory alloys. *Progress in Materials Science*, *50*(5), 511–678. doi:10.1016/j.pmatsci.2004.10.001

Ozbulut, O. E., Hurlebaus, S., & Desroches, R. (2011). Seismic Response Control Using Shape Memory Alloys: A Review. *Journal of Intelligent Material Systems and Structures*, *22*(14), 1531–1549. doi:10.1177/1045389X11411220

Savi, M. A., De Paula, A. S., & Lagoudas, D. C. (2011). Numerical investigation of an adaptive vibration absorber using shape memory alloys. *Journal of Intelligent Material Systems and Structures*, *22*(1), 67–80. doi:10.1177/1045389X10392612

Shahin, K., Zou, G. P., & Taheri, F. (2005). Shape memory alloy wire reinforced composites for structural damage repairs. *Mechanics of Advanced Materials and Structures*, *12*(6), 425–435. doi:10.1080/15376490500259251

Sharma, N., Jangra, K.K., & Raj, T. (2015). Fabrication of NiTi Alloy: A Review. *Proc. IMechE Part L: Journal of Materials: Design and Applications*. doi: 10.1177/1464420715622494

Sharma, N., Kumar, K., Raj, T., & Kumar, V. (2016). Porosity exploration of SMA by Taguchi, Regression Analysis and Genetic Programming. *Journal of Intelligent Manufacturing*. doi:10.1007/s10845-016-1236-8

Sharma, N., Raj, T., & Jangra, K. K. (2015). Applications of NiTi: A Review. *Journal of Engineering Technology*, *5*(1), 1–7. doi:10.4103/0976-8580.149472

Sharma, N., Raj, T., & Jangra, K. K. (2016). Micro-structural Evaluation of NiTi powder, Steatite and Steel balls after different Milling Conditions. *Materials and Manufacturing Processes*, *31*(5), 628–632. doi:10.1080/10426914.2015.1004710

Sharma, N., Raj, T., & Jangra, K. K. (2017). Physical and Tribological Characterization of NiTi alloy Fabricated by P/M. *Particulate Science and Technology: An International Journal.*, *35*, 541–546. doi:10.1080/02726351.2016.1171814

Thierry, B., Merhi, Y., Bilodeau, L., Trépanier, C., & Tabrizian, M. (2002). Nitinol versus stainless steel stents: Acute thrombogenicity study in an ex vivo porcine model. *Biomaterials*, *23*(14), 2997–3005. doi:10.1016/S0142-9612(02)00030-3 PMID:12069342

Wang, J., & Mak, C. (2014). Adaptive-passive vibration isolation between non rigid machines and non rigid foundations using a dual-beam periodic structure with shape memory alloy transverse connection. *Journal of Sound and Vibration*, *333*(23), 6005–6023. doi:10.1016/j.jsv.2014.07.010

Whitcher, F. D. (1997). Simulation of in vivo loading conditions of nitinol vascular stent structures. *Computers & Structures*, *64*(5-6), 1005–1011. doi:10.1016/S0045-7949(97)00014-X

Williams, E., Shaw, G., & Elahinia, M. (2010). Control of an automotive shape memory alloy mirror actuator. *Mechatronics*, *20*(5), 527–534. doi:10.1016/j.mechatronics.2010.04.002

Xiong, J. Y., Li, Y. C., Wang, X. J., Hodgson, P. D., & Wen, C. E. (2008). Titanium-nickel shape memory alloy foams for bone tissue engineering. *Journal of the Mechanical Behavior of Biomedical Materials*, *1*(3), 269–273. doi:10.1016/j.jmbbm.2007.09.003 PMID:19627791

Chapter 4
Application of Silica-Gel-Reinforced Aluminium Composite on the Piston of Internal Combustion Engine:
Comparative Study of Silica-Gel-Reinforced Aluminium Composite Piston With Aluminium Alloy Piston

Anuj Dixit
Independent Researcher, India

ABSTRACT

The piston of the internal combustion engine is one of the most complex parts among all engine components. During the operation, the pistons of IC engines are typically subjected to high loading and wearing. To withstand these, they require high mechanical properties and excellent tribological properties. This chapter aims to compare the mechanical as well as tribological properties of silica-gel-reinforced aluminium composite with aluminium alloy, which is used in manufacturing of piston of IC engine. Initially silica-gel-reinforced aluminium composite was fabricated with base material aluminium and six different percentages of silica gel reinforcement by stir casting method. After that, mechanical and tribological properties of silica-gel-reinforced aluminium composite were estimated and the tremendous mechanical and tribological properties among all percentages by different optimization techniques were found. The authors then compared the admirable properties of aluminium composite with aluminium alloy for manufacturing of piston of IC engine.

DOI: 10.4018/978-1-5225-5216-1.ch004

INTRODUCTION

Generally piston is non-repairable part and requires lasting the life span of the engine. Piston operation induce in the material a complex combination of mechanical stresses and high cyclic loading, high wear rate responsible of cracking and abrasion in the weakest zone. Mostly IC engine pistons have been manufactured by aluminium because aluminium alloy has excellent lightness (Density: 2.7 g/cm3), nonmagnetic and non-sparking as compared to other materials like cast iron. Aluminium is available in plentiful amount in the Earth after oxygen and silicon. Aluminium is available in earth about 8% by weight of the Earth's solid surface. Machining, casting, drawing and extrusion can be conveniently done on aluminium. It has excellent corrosion resistance because of thin surface layer of aluminium oxide which forms in free air to prevent further oxidation. But along with various advantages there are some major disadvantages of aluminium like its softness, high ductility and malleability, high wear rate, less strength (Modulus of elasticity: 70-80GPa). To overcome these demerits, many researchers fabricated various aluminium based composite with different reinforcing materials.

There are various types of ceramic viz. silicon carbide, silicon oxide, titanium oxide, titanium boride etc. which are used to reinforce aluminium metal composite. In this study, silica gel used as reinforced ceramic because of its high hardness and melting point, which helps to improve wear resistance characteristic (Sulima et al., 2010; Abdel-Azim et al., 1995). Purified form of quartz crystal, it has application in glass industries and construction field. Density of silica gel is approximate equal to the liquefied aluminium (about 2.64 g/cm3) by which uniform metal matrix composites can be developed (Feng et al., 2008; Tjong and Tam, 2006). Silica gel has high melting point as compared to the aluminium which make it useful in preparing the particle reinforced metal composite. According to Venkatprasad and Subramanian (2011), wear characteristic of aluminium metal matrix composite is directly related to percentage of reinforced ceramic, sliding speed and applied load in dry sliding condition. According to Mixed influence of grain size of reinforcement and sliding distance helped in the reduction of wear of composite (Modi et al., 2001). The filler grain size of SiC with matrix aluminium has less abrasive wear (Sahin, 2005).

Our study is based on the optimization of characteristic parameters on mechanical properties (Tensile strength, flexural strength) and tribological properties (wear rate and coefficient of friction) of silica gel reinforced aluminium composite and define consolidate parameter which give the optimum result. Taguchi is a design tool which gives easy, effective and organized approach for optimization. According to Taguchi (1990), DOE is relevant statistical tool to analyze the effect of various inconstant simultaneously and define order of steps which must follow a specific order to improve the performance of process. Taguchi techniques select few numbers of tests and get information accurately. All tests contain a specific number of combinations of factor and level. According to Fisher (1951), In DOE process, to get desired result, it is necessary to determine proper combination of factor and level. In Taguchi method, standard orthogonal array develops the schedule of test. To calibrate quality feature, Taguchi endorsed loss function which is also called Signal to Noise ratio(S/N ratio) in statistical form. Taguchi method uses a certain design orthogonal array to deduct number of test (Roy, 1990). The test results are then validated by ANOVA of dominating factors (Taguchi, 1993). While gray relational analysis was employed to optimize the multiple responses is more intricate than single response. Gray theory was proposed by Deng Ju-long (1982, 1989) at China and this procedure was introduced to study problems with multiple responses.

Research Background

Prasad et al. (2014) employed the reinforcement of aluminium with hard ceramic particles and graphite solid lubricants in automobile applications. Jha et al. (1986) explained that the talc elements within the composite enhance the wear resistance properties up to 22% to 30% compared with the pure metal. Sannino et al. (1995) disclosed that the wear resistance of the composites significantly depends on the grit size, hardness, weight fraction and distribution of reinforcement particles. Rohatgi (1991) investigated that the wear resistance generally depends on experimental conditions like hardness of the surface, sliding distance, applied load and sliding speed. Surappa (2003) suggested the wear rate of Al-Si/SiC composite disc against the automobile brake pad material, which increased with raise in applied load and sliding speed, whereas coefficient of friction reduced with raise in applied load and sliding speed. Rebba and Ramanaiah (2014) stated that the tribological and mechanical properties of composite material depend upon the reinforcement. According to Reddy and Zitoun (2010) and Kala et al. (2014), aluminium composite is getting hardened and shows high mechanical properties when ceramic is used as reinforcement. Ceramic particles increased wear resistance property of aluminium metal composite up-to a certain limit of normal load (Zhang et al., 1995). Volume fraction of reinforced ceramic has high influence on wear resistance (Surappa et al., 1982).Himanshu Kala et al. (2014) described that reinforcing aluminum with ceramics particles have excellent mechanical properties but in case of tribological case, it is limited.

Rahul Kumar et al. (2014) optimized the mechanical properties (tensile and flexural strength) of epoxy based wood dust reinforced green composite by Taguchi method. In L18 orthogonal array, percentage of reinforcement and speed were used as process parameters. The results were validated by ANOVA at 95% confidence level. K.M. Subbaya et al. (2012) revealed that wear rate of the SiC filled carbon epoxy composites highly dominated by reinforcement, load and distance by Grey based Taguchi multi-response method. J. Sudeepan et al. (2014), selected reinforcement, normal load and sliding speed as design factors with three levels for each design parameter for analysis of coefficient of friction and specific wear rate of ABS / kaolin composites by Taguchi method with Analysis of variance (ANOVA) to find out the most significant factor. Ravindra Singh Rana et al. (2014) has been carried out dry sliding wear test by using Taguchi technique on Al5083-10 with weight percentage of SiC composites by using Pin-on-disk wear test to estimate the significance of applied load, sliding speed and sliding distance on wear rate.

Sonawane et al. (2017) fabricated the aluminum alloy metal matrix composite (AAMMCs) by reinforcing aluminum alloy with SiC, Al2O3 and graphite particles for piston ring application. This study revealed that tensile strength is improved due to SiC and Al2O3 while graphite has adverse effect on tensile strength and hardness and all three reinforcements lead to the improvement in wear resistance but graphite is the only one which has reducing effect on coefficient of friction, This study also used Taguchi method to determine the number of compositions to be manufactured and ANOVA to understand the effect of various reinforcements on the different characteristics of the composite. Arsha et al. (2015) revealed that functionally graded automotive piston made of silicon reinforced A390 aluminium composite have high hardness towards the head region as well as a remarkable improvement of the tribological properties of the piston head. Park et al. (2001) evaluated the hardness and tensile strength of powder forged aluminium alloy piston and compared with normally formed aluminium piston. The outcome shows that forged aluminium piston has high mechanical properties as compare to formed aluminium piston. Haque and Sharif (2001) disclosed that heat treated silicon reinforced aluminium pistons have less wear rate in comparison to locally available aluminium piston. Nicoletto et al. (2014) suggested that mechanical properties and fatigue rate of piston of IC engine depends upon test temperature, chemical composition, process route and mechanisms of fracture. Huang et al. (2011) examined the study of Al–Si alloy-based composite pistons which shows lower wear rate and high hardness. Srivathsan (2014) studied a comparative analysis between aluminium alloy piston manufactured by casting and powder forging. The result of this investigation reveals that mechanical properties of piston manufactured by casting leads to powder forged piston of IC engine.

Summary and Gap in Research

It was observed from the review of available research following gaps are identified:

Now a day generally piston is made of aluminium alloy, but some researchers have done a lot of work on the piston of IC engine. Many researchers developed aluminium based metal matrix composite with different types of reinforcements, different sizes and manufactured techniques either by stir casting or by powder technique. The mechanical and wear properties of composites were also estimated. All the study reveals that percentage of reinforcement, type of reinforcement and its condition influence the mechanical properties and wear rate. With increase in weight percentage of reinforcement in the matrix both the mechanical properties and wear resistance properties of composite increases.

A major outcome of various research work has been showed that different reinforcement were used in aluminium like zircon sand, alumina, SiC, Tunguston etc. but no one used silica gel as reinforcement in aluminium and secondly limited data is available related to silica gel. Now from last few decade interests has been increased on those aluminium which could use in manufacturing of piston of IC engine. Due to high competition in automobile sector, manufacturer wants to reduce the cost with increase the efficiency. So low cost reinforcement should be use in aluminium composite for fabrication of piston. Silica gel is one of the best examples of this type of reinforcement. Such reinforced composites are likely to overcome the cost barrier for wide spread applications in piston of IC engine and also control the environmental condition. In this study, the combination of silica gel as reinforcement and matrix is Aluminium was used with the help of simple and cheaper technique called stir casting technique as a material for piston of IC engine.

Experimental Procedure

Specimen Preparation

A stir casting setup consisted of an electric Furnace was used to synthesize the composite. The three blades stirrer assembly was connected to a variable speed controller motor with range of 80 to 900 rpm with steel shaft. The stirrer was made by graphite. Graphite crucible of 1.5 Kg capacity was placed inside the furnace. Approximately 1Kg of aluminium was melted at 800°C in the furnace. Preheating of reinforcement (Silica gel at 200°C) was done for one hour to remove moisture and gases. The stirrer was then gone down vertically up to 3 cm from the bottom of the crucible (total depth of the melt was 9 cm). The speed of the stirrer was slightly raised to 800 rpm and the preheated reinforced particles were added at the rate of 10- 20g/min into the melt (Vishal and Sanjeev, 2007). The stirrer speed was reduced when particulates were poured into the melt. After that, stirring was continued for 10 to 12 minutes for appropriate mixing of reinforcement in the matrix. The melt was poured in the mould.

Hardness Testing

Estimation of Hardness was conducted on the pure aluminium (0% filler) specimen and five different percentages (2%, 4%, 8%, and 10%) aluminium composite specimens by using both Brinell Hardness Machine and Rockwell Hardness Tester. Both hardness tests were conducted to determine the affect of reinforced ceramic on the aluminium composite on Hardness. In Brinell Hardness Test, applied load was 500KG for time interval of 16 second in correspondence with ASTM E10-10. In Rockwell Hardness Test, applied load was 100 KG and 1/16 inch diameter steel sphere indenter was used on B-scale.

Tensile Strength Testing

The tensile test was conducted on Universal Testing Machine (INSTRON) equipped with computer on the specimens as per ASTM E08 standard. Specimen of six different percentage reinforcement weight (0%, 2%, 4%, 6%, 8%, 10%) were examined at three various speeds of 2, 3, 4 mm/min at room temperature. Dimension of specimen is 100 X 15 X 10 mm. These Parameters are given in Table 1.

In this experiment the test was conducted with the following parameters:

(1) Volume Fraction (2) Speed

Flexural Strength Testing

The flexural test was also conducted on Universal Testing Machine (INSTRON) equipped with computer on rectangular specimens as per ASTM E290 standard. Every specimen of six different percentage of reinforcement were estimated at three various speeds 2, 3, 4 mm/min. Dimension of specimen is 100 X 15 X 10 mm. In this testing, specimen puts on two supports and force applied by nose midway between the anchors. Both nose and anchors have cylindrical surface. These Parameters are given in Table 1.

In this experiment the test was conducted with the following parameters:

(1) Volume Fraction (2) Speed

Wear Testing

Wear have been tested on Pin on Disc wear tribotester as per ASTM G9-04 standards. The loaded pin was placed against the rotating disc (EN32 steel disc) with wear track diameter 60mm through a dead weight loading system. The pin of tribotester was full cleaned and its initial weight was measured. In this test, rectangular specimen (15mm X 15mm X 10mm) of aluminium composite is gripped against rotating disc

Table 1. Parameter taken during tensile and flexural test

Level							
Control Factor	1	2	3	4	5	6	Units
Volume Fraction	0	2	4	6	8	10	%age
Speed	1	1	1	1	1	1	mm/min

basically made of steel.The hardness of disc was 64 HRC and surface roughness was 0.1 microns. The surface of the pins and the disc was ground by the emery paper (80 grit size) and acetone before the test for effective contact of fresh and flat surface with the steel disc. There are requirement to clean specimen continuously with woolen cloth for proper functioning. The wear rate of the composite specimens was estimated by ratio of volume loss to sliding distance.

In this experiment the test was conducted with the following parameters:

(1) Volume Fraction (2) Applied Load (3) Sliding Speed

In the present experimental, the parameters such as sliding distance and time kept constant throughout for all the experiments. These Parameters are given in Table 2.

METHODOLOGIES

The word optimization is derived from Latin word "optime", which means the best. Optimization is the process to get the finest result subjected to given conditions/ restraints. Depending on the conditions, the word "optimum" may be taken as "minimum" or "maximum". The primarily aim of optimization is to either maximize the desired benefit or to minimize the effort required.

There are various method of optimization technique namely Taguchi method, grey-fuzzy, Genetic algorithm, Artificial Neural Network, Fuzzy, RSM etc. In this study Taguchi and Grey-Taguchi methods are used to optimize the results of tensile, flexural, and wear test.

Table 2. Parameter taken during sliding wear test

Design Factors	Unit	Levels		
Volume Fraction	Wt %	2	6	10
Applied Load	N	50	75	100
Sliding Speed	RPM	180	200	220

Taguchi's Method for Parameter Design

Orthogonal arrays are used by means of Taguchi method, to examine a large number of variables with a lesser number of tests. The results generated from less number of tests are authentic over the whole tests consists of the control factors and their relative level settings. The Taguchi method evaluate the results with the help of utilization of statistical measurement of performances in the form of Signal-to-Noise (S/N) ratio. The S/N ratio contains the mean data along with variability of the response. After evaluation of S/N ratio, it is very important to apply an Analysis of Variance (ANOVA) for finding the cumulative significance of different factors and for estimating error variance. The confirmation test is performed by using the predicted optimum levels for the process parameters being examined. The aim is to verify that to be sure the optimum conditions implied by the matrix tests enhance the project work. If enhanced project work and observed data are similar then recommended optimum conditions will be taken (Edwin, 1992 and Tsui, 2007).

Taguchi's S/N ratio Approach to Performance Evaluation

The loss function illustrates the deviation from the target value and further converted into S/N ratio which defines quality assessment index. S/N ratio provides the optimal design and the least variation. The high S/N ratio means obtainable quality will be much stable. S/N ratio also decreases the sensitivity of the system performance to source of variation (Tsui, 2007 and Mahapatra et al., 2007). Three types of S/N ratios are utilized to optimize the static problem, which are:-

1. **Nominal-the-Best (NB):** Nominal-the-best is taken when closer to the target value required and the deviation is to be found quadratic. The formula for this approach is:

$$S / N = -10 \log y / sy^2 \qquad \text{(a)}$$

2. **Smaller-is-Better (SB):** The Smaller -is-Better is required when a system wants smaller values of haracteristics. The formula for this approach is:

$$S / N = -10\log[1 / n \sum y_i^2] \qquad \text{(b)}$$

3. **Larger-is-Better (HB):** Higher-is-Better approach held when a system wants higher values of a characteristic. The formula for this approach is:

$$S / N = -10\log(1 / n\sum_{i=1}^{n} 1 / y_i^2) \qquad (c)$$

Grey Relational Analysis (GRA)

The principal theory of grey relational analysis was developed by Deng in 1982. The combination between factors of system and their performance can be obtained by the help of grey relational analysis (Haq et al., 2008 and Kuo et al., 2008). To execute GRA, initially perform the grey relational generation. In this step, performances of all alternatives are transformed into a comparability sequence. Then, the grey relational coefficient is calculated between the reference and all the comparability sequence. Lastly from grey relational coefficients, grey relational grade is estimated between comparability and the reference sequence. The alternative is taken as best option, if the comparability sequence derived from that alternative has the highest grey relational grade between itself and the reference sequence (Horng et al., 2008 and Chiang et al., 2006).

Grey Relational Generation

It is necessary for conversion of every performance values for each substitute into a comparability sequence like normalization. This process is known as grey relational generation in GRA.

For a multi- option decision making problem, if there are m options and n characteristics, the i_{th} option can be expressed as $Y_i = (y_{i1}, y_{i2}, \ldots\ldots\ldots y_{ij}, \ldots\ldots y_{in})$ where y_{ij} is the performance value of characteristic j of option i. The term Y_i can be converted into the comparability sequence $Z_{ij} = (z_{i1}, z_{i2}, \ldots\ldots\ldots z_{ij}, \ldots\ldots z_{in})$ by the use of one of Equations.

$$Z_{ij} = \frac{y_{ij} - \min y_{ij}}{\max y_{ij} - \min y_{ij}} \qquad (d)$$

for i = 1,2,....... m, j = 1,2,....... n

$$Z_{ij} = \frac{\max y_{ij} - y_{ij}}{\max y_{ij} - \min y_{ij}} \qquad (e)$$

for i = 1,2,....... m, j = 1,2,....... n

Eq. (d) is employed for the larger-the-better attributes; Eq. (e) is employed for the smaller-the-better attributes.

Grey Relational Coefficient Calculation

The grey relational coefficient describes closeness between zij and z0j. If grey relational coefficient has higher value, means z_{ij} is nearer to be z_{0j}. The grey relational coefficient can be found by Eq. (f).

$$\mathbf{y}_i\left(\mathbf{k}\right) = \frac{\Delta_{min} + \varepsilon\Delta_{max}}{\Delta_{oi}\left(\mathbf{k}\right) + \varepsilon\Delta_{max}} \quad \text{(f)}$$

for i = 1,2,....... m, j = 1,2,....... nHere, y (z0j,zij) is the grey relational coefficient between zij and x0j and

$$\Delta_{ij} = \left|z_{oj} - z_{ij}\right| \quad \text{(g)}$$

$$\Delta\mathrm{max} = \mathrm{Max}\left(\mathrm{for\ i} = 1,2,\ldots\ldots.\ \mathrm{m},\ \mathrm{j} = 1,2,\ldots\ldots.\ \mathrm{n}\right) \quad \text{(h)}$$

$$\Delta\mathrm{min} = \mathrm{Min}\left(\mathrm{for\ i} = 1,2,\ldots\ldots.\ \mathrm{m},\ \mathrm{j} = 1,2,\ldots\ldots.\ \mathrm{n}\right) \quad \text{(i)}$$

ε is the distinguishing coefficient, ε ϵ [0,1].

By differentiating the coefficient, compression or expansion of the variety of the grey relational coefficient is obtained. Various outcome of GRA are found by differentiating coefficients. In present study, the differentiating coefficient was set as 0.5.

Grey Relational Grade Calculation

The grey relational grade are estimated after evaluating the entire grey relational coefficient, y (zoj, zij) using Eq. (5.10).

$$\gamma_i = \frac{1}{\mathbf{n}}\sum_{k=1}^{n}\varepsilon_i \quad \text{(j)}$$

The grey relational grade decides the degree of resemblance between the reference and comparability sequence. The comparability sequence for an option containing largest grey relational grade is most similar to the reference sequence, and that option would be the best alternative.

SOLUTIONS AND RECOMMENDATIONS

Hardness

Both hardness test (Brinell and Rockwell) disclosed that Brinell as well as Rockwell hardness of silica gel reinforced aluminium composite were increase with enhance in %age of reinforcement silica gel. Improvement of hardness with percentage of silica gel reinforcement is shown in Figure 1. There are two main reasons to increase hardness, one is that Silica gel reduced grain size by which hardness was increased, and second is that Silica gel has high hardness than aluminium which reduces plastic deformation.

Tensile and Flexural Strength Test Result

The graph of tensile strength and flexural strength with percentage of reinforcement is shown in Figure 2 & 3 respectively. The test disclosed that both tensile strength and flexural strength of composite increased by increase in percentage of reinforcement upto 8%. When ceramic is added up beyond 8% both tensile strength and flexural strength decreased as increase in the reinforcement loosens the adhesion between the matrix and the reinforcement and hence lower strength is observed.

Taguchi Analysis for Tensile Strength and Flexural Strength

Experimental result with estimated S/N ratio for tensile and flexural strength is shown in Table 3. Response tables of mean S/N ratio and the result as in Table 4 and 5 respectively provides the affect of every control factor (A and B) on the responses in the study. The main effects plot for S/N ratio is shown in Figure 4 and 5 for tensile strength and flexural strength respectively. The significance of every parameter is measured form the slope of main effect plot. The line of high slope angle denotes that such parameter will be more significant, so from the Figure 4 & 5, it can be inferred that percentage of reinforcement is more significant than speed. High value of S/N ratio conform high quality so from Figure 4 & 5 it can be referred that A5B3 is an optimum combination of design parameter for both tensile strength as well as flexural strength. Thus this study defines that percentage of reinforcement is most

Figure 1. Variation of Hardness with %age of silica gel

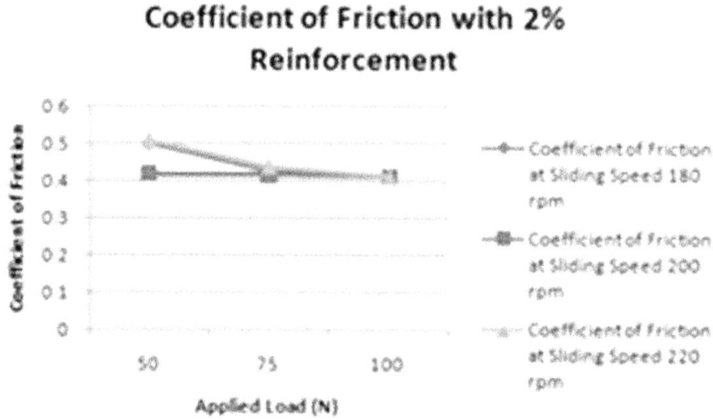

Figure 2. Tensile strength

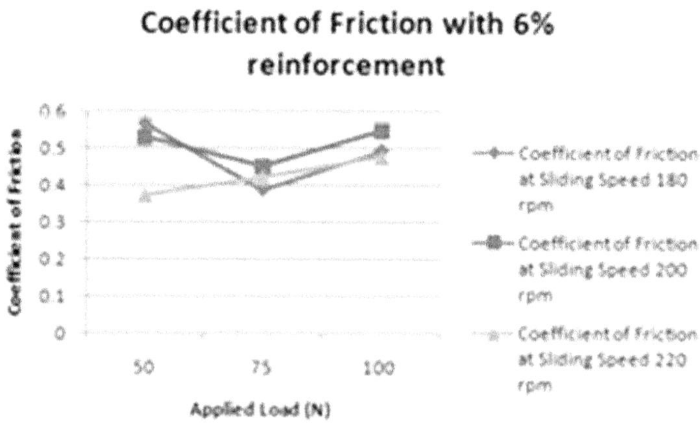

effective parameter for both tensile and flexural strength of aluminium with 8% of silica gel reinforcement as depicted in Table 4 & 5.

The control factors are very important in S/N ratio and it is noticed from Table 4 & 5 that percentage of reinforcement getting a rank 1, is obviously the dominating factor on tensile strength as well as flexural strength than speed.

Analysis of Variance for Tensile Strength and Flexural Strength

Analysis of Variance (ANOVA) used to determine significance of process parameters over the quality feature and proportion of participation of factors and also influence

Figure 3. Flexural strength

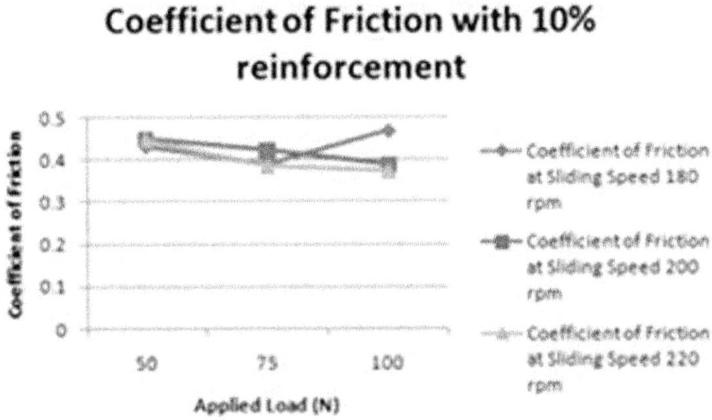

Figure 4. Main effects plot for S/N ratio for tensile strength

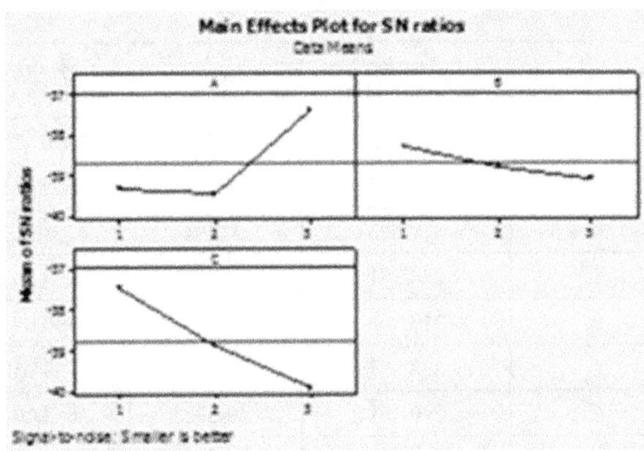

of interaction on responses (Montgomery, 2001). The summation of squared deviation from mean S/N ratio gives the percentage contribution of the factor. By executing ANOVA, it can be concluded which factor is most dominating. ANOVA table for tensile strength and flexural strength are presented in Table 6 and 7 respectively. From Table 6 & 7, it is noticed that percentage of filler silica gel is most contributing factor on tensile and flexural strength followed by speed.

Confirmation Test for Tensile Strength and Flexural Strength

Confirmation test validates the test results and to estimate the accuracy of test. Comparisons of actual value and estimated value are presented in table.8 and 9

Figure 5. Main effects plot for S/N ratio for flexural strength

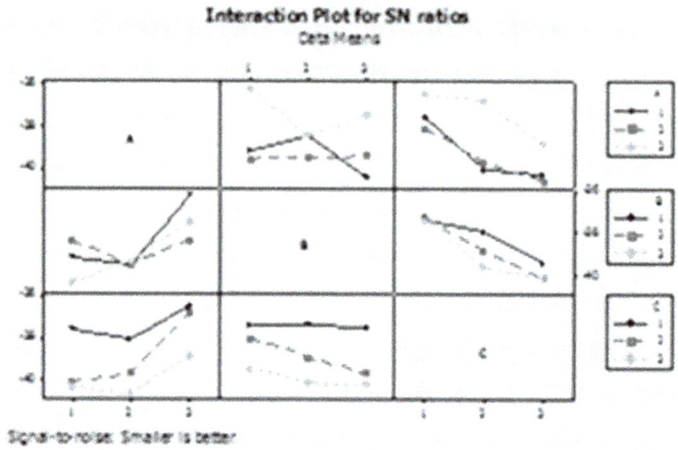

Table 3. Tensile strength test and Flexural Strength result with S/N ratio

Percentage of Reinforcement	Speed	Tensile Strength(MPa)	SNRA	Flexural Strength(GPa)	SNRA
0	2	265.145	48.470	446.616	52.999
0	3	302.668	49.619	509.821	54.148
0	4	327.123	50.294	551.012	54.823
2	2	274.251	48.763	639.562	56.118
2	3	313.063	49.913	730.072	57.267
2	4	338.357	50.588	789.059	57.942
4	2	299.301	49.522	682.609	56.683
4	3	341.658	50.672	779.212	57.833
4	4	369.262	51.347	842.168	58.508
6	2	306.976	49.742	707.345	56.993
6	3	350.419	50.892	807.448	58.142
6	4	378.732	51.567	872.686	58.817
8	2	347.684	50.824	994.278	59.950
8	3	396.888	51.973	1134.988	61.100
8	4	428.954	52.648	1226.690	61.775
10	2	302.931	49.627	732.146	57.292
10	3	345.802	50.777	835.759	58.442
10	4	373.741	51.451	903.285	59.117

Table 4. S/N ratio Response Table of Tensile Strength

Level	% Age of Reinforcement (A)	Speed (B)
1	**49.460**	**49.490**
2	49.750	50.640
3	50.510	51.320
4	50.730	
5	51.820	
6	50.620	
Delta	2.350	1.820
Rank	1	2

Table 5. S/N ratio Response Table of Flexural Strength

Level	% Age of Reinforcement (A)	Speed (B)
1	53.990	56.670
2	57.110	57.820
3	57.670	58.500
4	57.980	
5	60.940	
6	58.280	
Delta	6.950	1.820
Rank	1	2

Table 6. Analysis of variance table for Tensile Strength

Source	Degree of Freedom	Sequential Sum of Squares	Adjusted Sum of Squares	Adjusted Mean of Squares	F	P
A	5	591816	591816	118363	273.950	0.000
B	2	81608	81608	40804	94.440	0.000
Error	10	4321	4321	432		
Total	17	677746				

Table 7. Analysis of variance table for Flexural Strength

Source	Degree of Freedom	Sequential Sum of Squares	Adjusted Sum of Squares	Adjusted Mean of Squares	F	P
A	5	15976.200	15976.200	3195.200	273.950	0.000
B	2	14909.400	14909.400	7454.700	639.140	0.000
Error	10	116.600	116.600	11.700		
Total	17	31002.200				

respectively. The increment of S/N ratio for initial to optimal condition for tensile strength is 1.97637dB and for flexural strength is 3.94159 dB.

Grey-Taguchi Analysis for Tensile and Flexural Strength

The experimental results (tensile and flexural strength) are simplified in the span between zero and one in data preprocessing. Grey relational analysis depends on the quality performances of a data sequence. Different techniques of data preprocessing are available (Tosun, 2006).

The main effect plot for grey relation is depends on the values of grey relational grade with respect to controlling factors and shown in Figure 6. The main effect plot provides optimum level at 10 percent of reinforcement and 4 mm/min of testing speed for maximum tensile and flexural strength.

The controlling factors are very essential in S/N ratio and it can be observed from Table 11 that percentage of reinforcement receiving a rank 1, so obviously it is the most influencing factor on grey relational grade.

Analysis of Variance of Grey Relational Grade for Tensile and Flexural Strength

ANOVA is a statistical method to estimate significant factor. Analysis of Variance (ANOVA) used to evaluate significance of controlling factors over the quality feature

Table 8. Confirmation Table for Tensile Strength

Parameter	Mean Parameter	Optical Parameter	
		Experimental	Predicted
	A3B2		A5B3
Tensile Strength	341.657	396.887	428.954
S/N Ratio	50.671	51.973	52.648

Table 9. Confirmation Table for Flexural Strength

Parameter	Mean Parameter	Optical Parameter	
		Experimental	Predicted
	A3B2		A5B3
Flexural Strength	779.211	1134.988	1226.690
S/N Ratio	57.833	61.099	61.774

Table 10. S/N ratio values, normalized S/N ratio values, deviation sequences, Grey relational coefficient, and grey relational grade

Exp. No	S/N ratios		Normalised S/N Ratios		Deviation Sequences (Δoi)		Grey Relational Coefficient		Grey Relational Grade
	Tensile Strength	Flexural Strength	Tensile Strength	Flexural Strength	Tensile Strength	Flexural Strength	Tensile Strength	Flexural Strength	
1	51.451	59.117	0.308	0.338	0.692	0.662	0.420	0.431	0.425
2	50.777	58.442	0.482	0.424	0.518	0.576	0.491	0.465	0.478
3	49.627	57.292	0.778	0.571	0.222	0.429	0.692	0.538	0.615
4	52.648	61.775	0.000	0.000	1.000	1.000	0.333	0.333	0.333
5	51.973	61.100	0.174	0.086	0.826	0.914	0.377	0.354	0.365
6	50.824	59.950	0.470	0.232	0.530	0.768	0.485	0.394	0.440
7	51.567	58.817	0.278	0.377	0.722	0.623	0.409	0.445	0.427
8	50.892	58.142	0.452	0.463	0.548	0.538	0.477	0.482	0.480
9	49.742	56.993	0.748	0.609	0.252	0.391	0.665	0.561	0.613
10	51.347	58.508	0.335	0.416	0.665	0.584	0.429	0.461	0.445
11	50.672	57.833	0.509	0.502	0.491	0.498	0.504	0.501	0.503
12	49.522	56.683	0.805	0.648	0.195	0.352	0.719	0.587	0.653
13	50.588	57.942	0.530	0.488	0.470	0.512	0.516	0.494	0.505
14	49.913	57.267	0.704	0.574	0.296	0.426	0.628	0.540	0.584
15	48.763	56.118	1.000	0.720	0.000	0.280	1.000	0.641	0.821
16	50.294	54.823	0.606	0.885	0.394	0.115	0.559	0.813	0.686
17	49.619	54.148	0.780	0.971	0.220	0.029	0.694	0.945	0.820
18	49.401	53.920	0.836	1.000	0.164	0.000	0.753	1.000	0.876

Figure 6. Main effect plot for S/N ratio for grey relational grade

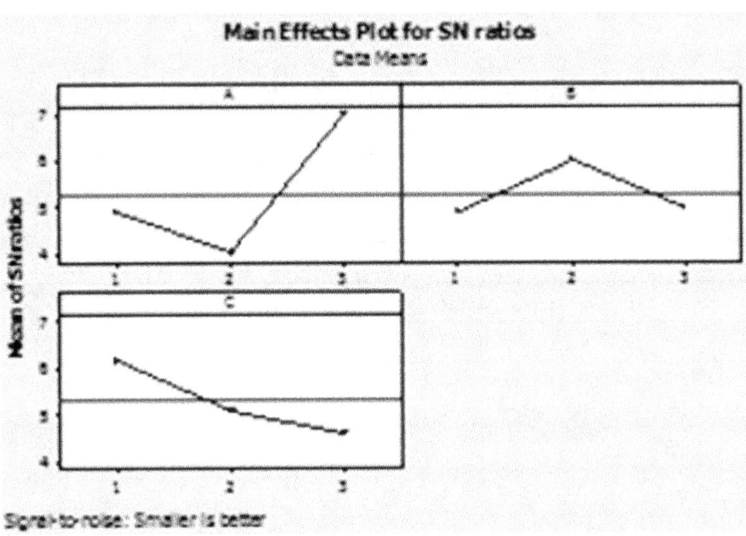

Table 11. S/N ratio Response Table of grey relational grade

Level	A	B
1	-6.022	-6.767
2	-8.475	-5.653
3	-6.008	-3.695
4	-5.568	
5	-4.108	
6	-2.048	
Delta	6.426	3.072
Rank	1	2

and percentage of involvement of factors and also influence of interaction on responses (Montgomery, 2001). The summation of squared deviation from mean S/N ratio gives the portion involvement of the controlling factors. By performing ANOVA, it can be provided in outcome that which factor is most influencing. ANOVA table for grey relational grade are presented in Table 12.

Confirmation Test of Grey Relational Grade for Tensile and Flexural Strength

Confirmation test validates the Grey Taguchi test results and to estimate the accuracy of test. A comparison of actual value and estimated value is given in table 13. The increment of S/N ratio for initial to optimal condition for grey relational grade is 5.23801 dB.

Table 12. Analysis of variance table for grey relational grade

Source	Degree of Freedom	Sequential Sum of Square	Adjusted Sum of Square	Adjusted Mean of Square	F
A	5	0.299	0.299	0.060	36.520
B	2	0.123	0.123	0.062	37.640
Error	10	0.016	0.016	0.002	
Total	17	0.439			

Table 13. Confirmation Table for grey relational grade

Parameter	Mean Parameter	Optimal Parameter	
		Predicted	Experimental
Combination of testing parameters	A3B2	A6B3	A6B3
S/N ratio for Tensile Strength	50.891		49.400
S/N ratio for Flexural Strength	58.142		53.920
S/N ratio for Grey relational grade	-6.383	-0.372	-1.145

Wear Test Result

The graphs of wear rate and coefficient of friction (COF) with various percentage of reinforcement are shown in figure 7, 8, 9 and figure 10, 11, 12 respectively. The test disclosed that both wear rate and coefficient of friction tends to decrease with increasing percentage of reinforcement, which confirms that silica gel addition is beneficial for reducing the wear rate of MMCs.

Figure 7. Wear Rate with reinforcement 2%

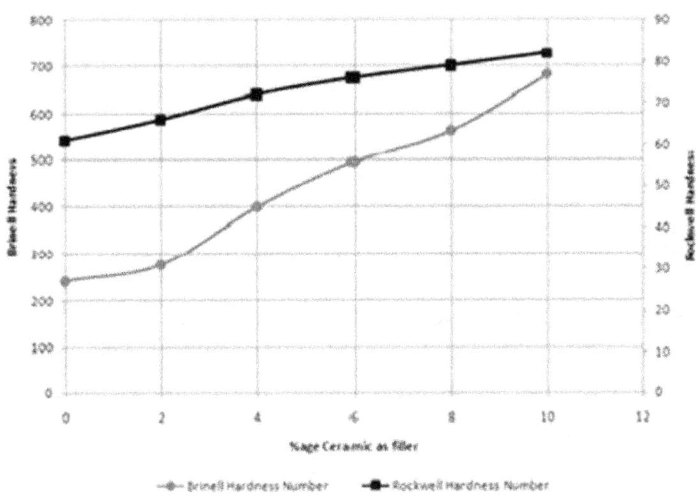

Figure 8. Wear Rate with reinforcement 6%

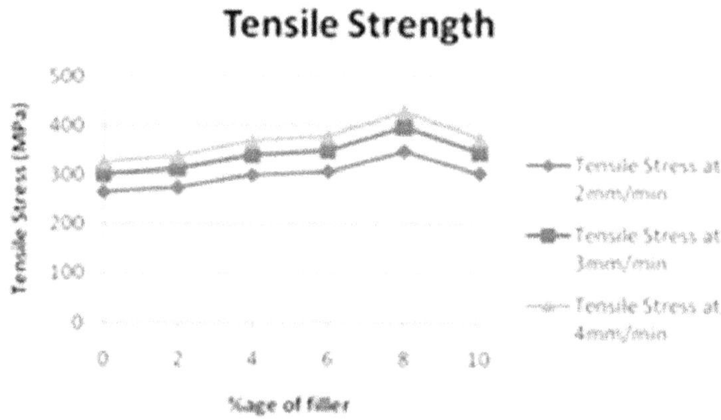

Figure 9. Wear Rate with reinforcement 10%

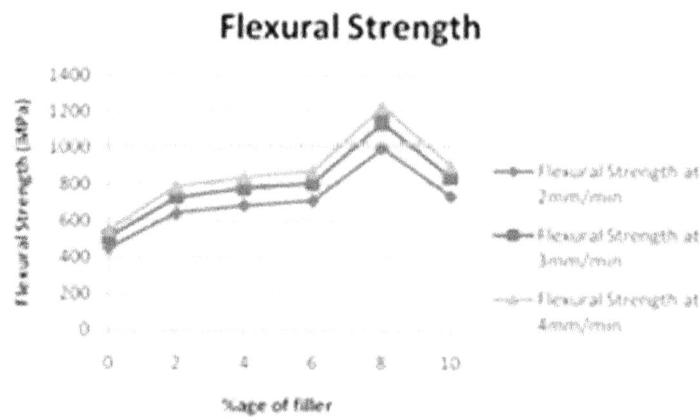

Wear Testing Data Analysis

The experimental plan was designed considering three factors which are percentage of reinforcement volume fraction (A), applied load (B), sliding speed (C) which gives L27 orthogonal array with three column and twenty seven rows. The L27 orthogonal array contained twenty six degree of freedom. For getting the optimal parameter setting for wear testing Taguchi method has been utilized. Further Grey-Taguchi approach for wear and coefficient of friction has also been employed. The level values of each process parameter corresponding to L27 orthogonal array with experimental results are shown in Table 14.

Figure 10. Coefficient of Friction with reinforcement 2%

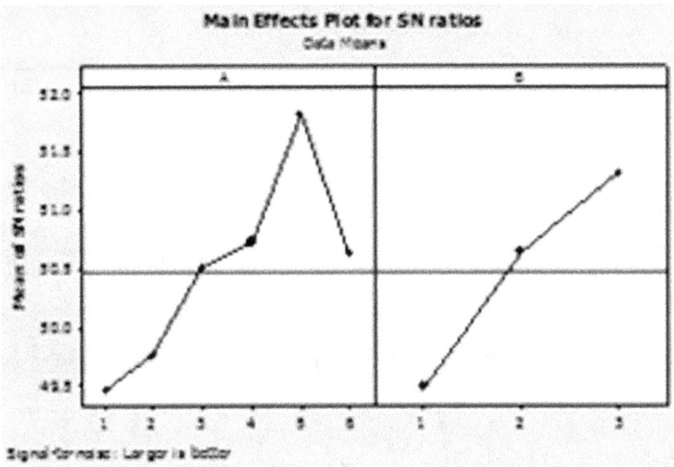

Figure 11. Coefficient of Friction with reinforcement 6%

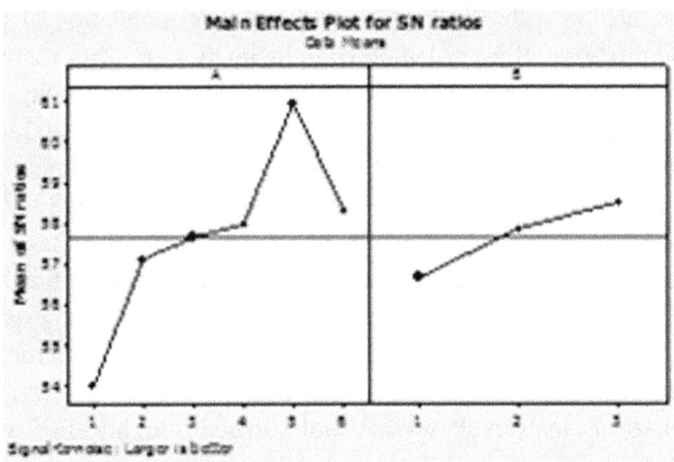

Taguchi Analysis for Wear

S/N ratio was estimated for combination of various factor are shown in this table 15. High value of S/N ratio leads to high quality, so optimum combination of design factors can be taken as A3B1C1.

As it can be observed from Table 16 that factor C i.e. sliding speed is the most dominating factor on wear followed by volume fraction and normal load (A & B).

Figure 12. Coefficient of Friction with reinforcement 10%

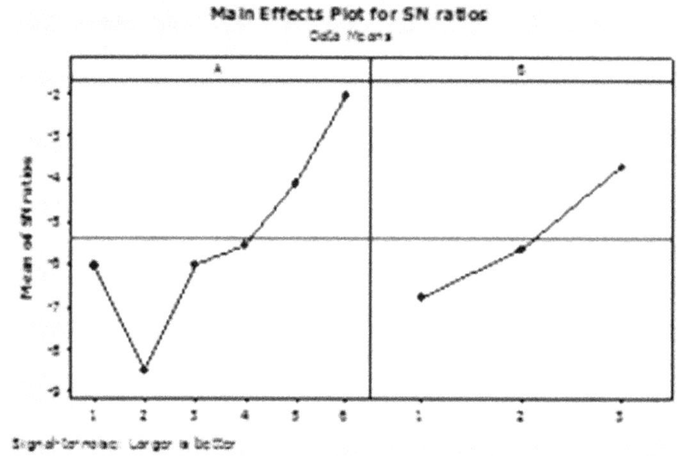

The main effect plot for S/N ratio for wear and interaction effect plot between design factors are presented in Figure 13 & 14. The inclination of main effect plot shows significance of every factor. High inclination of factor represents more significant. Main effect plot for S/N ratio (Figure 13) shows that, amongst the factors selected sliding speed (C) has the highest significance while normal load (B) the lowest. The interaction plot is analyzed on the basis of the extent of non-parallelism the various lines in the plot. If the lines of an interaction plot are not parallel, it represents that there is minimum interaction occurred and if the lines interact each other, then highly interaction occurred between the factors. Interaction plot for S/N ratio (Figure 14) shows that there is high interaction between sliding speed (C) and normal load (B), where moderate interaction occurred in between sliding speed (C) and volume fraction (A). The same is the case with volume fraction (A) and normal load (B). Hence this analysis shows that sliding speed is highly effective factor for wear characteristics of silica based aluminium composite.

Analysis of Variance of Wear

The test result has analyzed by analysis of variance (ANOVA), which is used to determine effect of various input factors, viz. volume fraction, normal load, sliding speed on wear. By conducting the analysis of variance, it can be decided which factor dominate over the other and the percentage contribution of that particular independent variable. Analysis of Variance (ANOVA) result for wear for three levels and three factor and interaction between them are presented in Table 17. Now a factor 'F', called variance ratio, and defined as the ratio of regression mean square

Table 14. Taguchi L27 OA with Wear and Coefficient of Friction results

Experiment Number	Volume Fraction	Applied Load	Sliding Speed	Wear (μm)	Coefficient of Friction
1	2	50	180	73.987	0.503
2	2	50	200	109.043	0.420
3	2	50	220	92.266	0.510
4	2	75	180	64.103	0.426
5	2	75	200	86.827	0.419
6	2	75	220	103.834	0.441
7	2	100	180	91.237	0.413
8	2	100	200	108.728	0.413
9	2	100	220	116.154	0.414
10	6	50	180	84.344	0.568
11	6	50	200	87.021	0.532
12	6	50	220	114.942	0.373
13	6	75	180	82.285	0.389
14	6	75	200	94.398	0.452
15	6	75	220	107.917	0.422
16	6	100	180	74.363	0.495
17	6	100	200	108.631	0.548
18	6	100	220	99.619	0.478
19	10	50	180	65.242	0.431
20	10	50	200	52.765	0.449
21	10	50	220	78.045	0.447
22	10	75	180	75.235	0.390
23	10	75	200	82.903	0.422
24	10	75	220	93.599	0.385
25	10	100	180	61.305	0.468
26	10	100	200	74.787	0.388
27	10	100	220	92.182	0.373

and the mean square error, is used as a yardstick for the level of significance of the factors and their interactions. A reference F value is selected in accordance with the confidence level (in this study, it is considered as 95 percent). Sources with higher F value than the reference were considered to have a statistically significant contribution to the performance measures.

Table 15. Wear and Coefficient of Friction S/N ratios

Experiment Number	Volume Fraction	Applied Load	Sliding Speed	S/N Ratio	
				Wear	COF
1	2	50	180	-37.383	5.968
2	2	50	200	-40.752	7.535
3	2	50	220	-39.300	5.848
4	2	75	180	-36.137	7.411
5	2	75	200	-38.773	7.555
6	2	75	220	-40.326	7.111
7	2	100	180	-39.203	7.681
8	2	100	200	-40.726	7.681
9	2	100	220	-41.300	7.659
10	6	50	180	-38.521	4.913
11	6	50	200	-38.792	5.481
12	6	50	220	-41.209	8.565
13	6	75	180	-38.306	8.201
14	6	75	200	-39.499	6.897
15	6	75	220	-40.661	7.493
16	6	100	180	-37.427	6.107
17	6	100	200	-40.719	5.224
18	6	100	220	-39.966	6.411
19	10	50	180	-36.290	7.310
20	10	50	200	-34.446	6.955
21	10	50	220	-37.846	6.993
22	10	75	180	-37.528	8.178
23	10	75	200	-38.371	7.493
24	10	75	220	-39.425	8.290
25	10	100	180	-35.749	6.595
26	10	100	200	-37.476	8.223
27	10	100	220	-39.292	8.565

It is obtained from Table 17 that sliding speed has highest influence on wear of silica gel reinforced aluminium metal composite followed by volume fraction and normal load respectively. The interaction terms has little or no influence on wear.

Table 16. Response table for S/N ratio for wear

Level	A	B	C
1	-39.320	-38.280	-37.390
2	-39.460	-38.780	-38.840
3	-37.380	-39.100	-39.930
Delta	2.070	0.810	2.530
Rank	2	3	1

Figure 13. Main effect plot for S/N ratio for wear

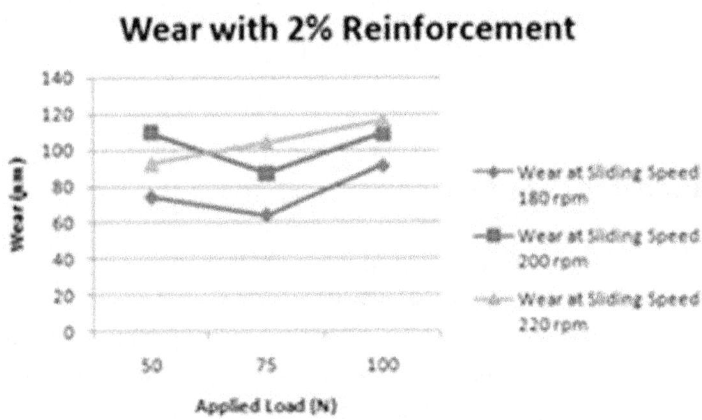

Figure 14. Interaction effect of design factor on wear

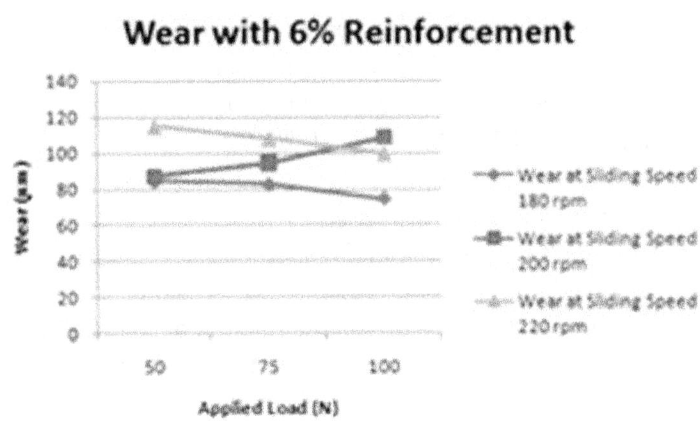

Table 17. ANOVA table for wear

Source	Degree of Freedom	Sequential Sum of Square	Adjusted Sum of Square	Adjusted Mean of Square	F	Percentage Contribution
A	2	2240.200	2240.200	1120.100	11.150	29.268
B	2	267.300	267.300	133.700	1.330	3.492
C	2	2878.000	2878.000	1439.000	14.330	37.601
A*B	4	906.300	906.300	226.600	2.260	11.841
A*C	4	396.700	396.700	99.200	0.990	5.183
B*C	4	162.100	162.100	40.500	0.400	2.117
Error	8	803.300	803.300	100.400		
Total	26	7653.900	7653.900			

Confirmation Test of Wear

Dry sliding test were carried on using unique set of factors to validate the statistical analysis. It is important to conduct verification test of optimum level of factors to estimate accuracy of test and validate experimental result. The confirmation test for wear is given in Table 18 which shows improvement of S/N ratio for wear from initial to optimum condition is 3.27088 dB.

Grey-Taguchi Analysis for Wear and Coefficient of Friction

The experimental results (wear rate, and coefficient of friction) are simplified in the range between zero and one in data preprocessing. Grey relational analysis depends on the quality performances of a data sequence. Different techniques of data preprocessing are available (Kumar et al. 2010).

The normalized S/N ratio was calculated based on quality performance. The higher normalized S/N ratio indicates higher performance.

The main effect plot for grey relation is based on the values of grey relational grade with respect to controlling factors and shown in Figure 15. The main effect plot provides optimum level at 10% of volume fraction of reinforcement, 75 N of applied load, and 180 rpm of sliding speed for minimum wear rate and coefficient of friction.

The controlling factors are very essential in S/N ratio and it can be observed (from Table 20) that volume fraction of reinforcement getting a rank 1, so obviously it is the highest dominating factor on grey relational grade.

Table 18. Confirmation table for wear

Parameter	Mean Parameter	Optimal Parameter	
	A2B2C2	A3B1C1	
		Predicted	Experimental
Wear	94.398		65.242
S/N Ratio	-39.499	-35.778	-36.290

Table 19. S/N ratio values, normalized S/N ratio values, deviation sequences, Grey relational coefficient, and grey relational grade

Exp. No	S/N Ratios		Normalised S/N Ratios		Deviation Sequences (Δoi)		Grey Relational Coefficient		Grey Relational Grade
	Wear Rate	Coefficient of Friction	Wear Rate	Coefficient of Friction	Wear Rate	Coefficient of Friction	Wear Rate	Coefficient of Friction	
1	-37.383	5.969	0.428	0.711	0.572	0.289	0.467	0.634	0.550
2	-40.752	7.535	0.920	0.282	0.080	0.718	0.862	0.411	0.636
3	-39.301	5.849	0.708	0.744	0.292	0.256	0.631	0.661	0.646
4	-36.138	7.412	0.247	0.316	0.753	0.684	0.399	0.422	0.411
5	-38.773	7.556	0.631	0.277	0.369	0.723	0.576	0.409	0.492
6	-40.327	7.111	0.858	0.398	0.142	0.602	0.779	0.454	0.616
7	-39.203	7.681	0.694	0.242	0.306	0.758	0.620	0.398	0.509
8	-40.727	7.681	0.916	0.242	0.084	0.758	0.857	0.398	0.627
9	-41.301	7.660	1.000	0.248	0.000	0.752	1.000	0.399	0.700
10	-38.521	4.913	0.594	1.000	0.406	0.000	0.552	1.000	0.776
11	-38.793	5.482	0.634	0.844	0.366	0.156	0.577	0.763	0.670
12	-41.210	8.566	0.987	0.000	0.013	1.000	0.974	0.333	0.654
13	-38.306	8.201	0.563	0.100	0.437	0.900	0.534	0.357	0.445
14	-39.499	6.897	0.737	0.457	0.263	0.543	0.655	0.479	0.567
15	-40.662	7.494	0.907	0.293	0.093	0.707	0.843	0.414	0.629
16	-37.427	6.108	0.435	0.673	0.565	0.327	0.469	0.605	0.537
17	-40.719	5.224	0.915	0.915	0.085	0.085	0.855	0.854	0.855
18	-39.967	6.411	0.805	0.590	0.195	0.410	0.720	0.549	0.635
19	-36.291	7.310	0.269	0.344	0.731	0.656	0.406	0.432	0.419
20	-34.447	6.955	0.000	0.441	1.000	0.559	0.333	0.472	0.403
21	-37.847	6.994	0.496	0.430	0.504	0.570	0.498	0.467	0.483
22	-37.528	8.179	0.450	0.106	0.550	0.894	0.476	0.359	0.417
23	-38.371	7.494	0.573	0.293	0.427	0.707	0.539	0.414	0.477
24	-39.425	8.291	0.726	0.075	0.274	0.925	0.646	0.351	0.499
25	-35.750	6.595	0.190	0.540	0.810	0.460	0.382	0.521	0.451
26	-37.477	8.223	0.442	0.094	0.558	0.906	0.473	0.356	0.414
27	-39.293	8.566	0.707	0.000	0.293	1.000	0.467	0.333	0.482

Figure 15. Main effect plot for S/N ratio for grey relational grade

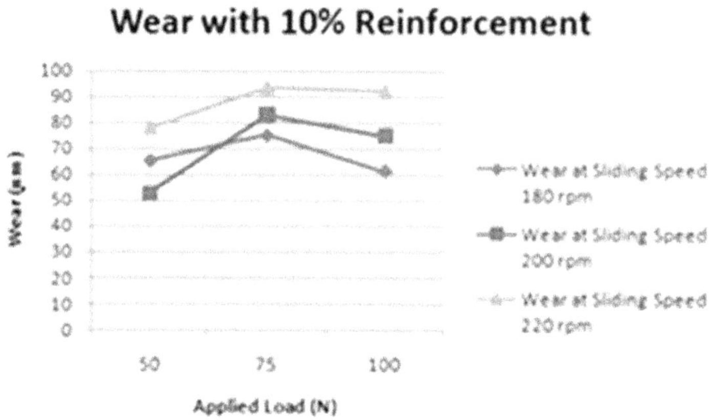

Table 20. S/N ratio Response Table of grey relational grade

Level	A	B	C
1	4.892	4.895	6.163
2	4.008	6.016	5.100
3	6.973	4.961	4.609
Delta	2.965	1.121	1.554
Rank	1	3	2

Analysis of Variance (ANOVA) of Grey Relational Grade for Wear and Coefficient of Friction

By performing ANOVA, it can be provided in outcome that which factor is most influencing. ANOVA table for grey relational grade are presented in Table 21. The ANOVA table 21 also confirms that the volume fraction of filler silica gel is most contributing factor on combined effect of wear rate and coefficient of friction.

Confirmation Test of Grey Relational Grade for Wear and Coefficient of Friction

Comparisons of experimental value and estimated value are presented in Table 22, which shows improvement of S/N ratio for grey relational grade of wear and coefficient of friction from initial to optimum condition is 2.66752 dB.

Table 21. Analysis of variance table for grey relational grade

Source	Degree of Freedom	Sequential Sum of Square	Adjusted Sum of Square	Adjusted Mean of Square	F	Percentage Contribution
A	2	0.171	0.171	0.085	16.300	46.307
B	2	0.033	0.033	0.017	3.180	9.032
C	2	0.041	0.041	0.021	3.940	11.198
A*B	4	0.030	0.030	0.008	1.460	8.266
B*C	4	0.026	0.026	0.007	1.260	7.140
A*C	4	0.025	0.025	0.006	1.180	6.697
Error	8	0.042	0.042	0.005		
Total	26	0.369				

Table 22. Confirmation Table for grey relational grade

Parameter	Mean Parameter	Optimal Parameter	
		Predicted	Experimental
Combination of testing parameters	A2B2C2	A3B2C2	A3B2C1
S/N ratio for Specific wear rate	-39.499		-37.528
S/N ratio for Coefficient of friction	6.897		8.179
S/N ratio for Grey relational grade	4.923	0.477	7.590

FUTURE RESEARCH DIRECTIONS

There are several possibilities to extend this research. Some of them are technical, quantitative in nature, others are managerial extensions. Some options are listed down for further research:

- Same metal matrix composites can be manufactured by using other manufacturing techniques like spray casting etc. and results can be compared with stir casting technique.
- In this study silica gel particle of size 120-200 μm has been used. This can further be extended by varying the particle size and then effect of particle size on the wear behaviour and mechanical properties of the composite can be studied.

- Heat treatment of the MMCs at different temperature range and quenching media like water, oil and brine solution etc. can be used to achieve better results.
- All the results of mechanical and wear properties can be optimized by some other techniques like ANN, Genetic Algorithm, Fuzzy etc.
- The optimized results of silica gel based aluminium composite can be compared with other result of different reinforcement.

CONCLUSION

Various conclusions are drawn from the present study. First is experimental conclusion of this investigation, which concludes that silica gel based aluminium metal matrix composites have been effectively casted by stir casting method. The results assured that stir formed Aluminium with Silica gel reinforced composites is obviously better to pure Aluminium in the comparison of mechanical properties such as hardness, tensile strength and flexural. Hardness of composite is continuously increasing with silica gel. Tensile strength, flexural strength is found to be maximum for 8% of silica gel. Dispersion of Silica gel particles in aluminium matrix improves the wear resistance properties of the composites. It is also found that wear rate and coefficient of friction tends to decrease with increasing wt. percentage (10%) of reinforcement, which verify that silica gel addition is beneficial for reducing the wear rate of MMCs. At 2% of filler reinforcement, minimum wear found with 50 N of applied load and 180 RPM of sliding speed and maximum wear found with 100 N of applied load and 220 RPM of sliding speed. At 6% of filler reinforcement, minimum wear found with 50 N of applied load and 180 RPM of sliding speed and maximum wear found with 100 N of applied load and 220 RPM of sliding speed. At 10% of filler reinforcement, minimum wear found with 50 N of applied load and 200 RPM of sliding speed and maximum wear found with 100 N of applied load and 220 RPM of sliding speed. At 2% of filler reinforcement, minimum COF found with 50 N of applied load and 200 RPM of sliding speed and maximum COF found with 50 N of applied load and 220 RPM of sliding speed. At 6% of filler reinforcement, minimum COF found with 50 N of applied load and 220 RPM of sliding speed and maximum COF found with 50 N of applied load and 180 RPM of sliding speed. At 10% of filler reinforcement, minimum COF found with 100 N of applied load and 220 RPM of sliding speed and maximum wear found with 100 N of applied load and 180 RPM of sliding speed.

The second conclusion is based on the optimization techniques. In this study the tensile and flexural strength of silica gel reinforced aluminium composite was optimized with the variation of percentage of reinforcement and speed by Taguchi technique. It can be validated by ANOVA which give percentage of reinforcement most affected the tensile as well as flexural strength at 95% confidence level. The percentage of reinforcement played a very crucial role in both type of strength. In case of Tensile strength the improvement of S/N Ratio was observed as 1.97637 dB whereas for Flexural the same was 3.94159 dB. Tensile strength and flexural strength of silica gel based aluminium composite were optimized by grey relational analysis. Optimum level for maximum tensile and flexural strength was attained at 10 percent of reinforcement and 4 mm/min of testing speed. The confirmation test represents the improvement of S/N ratio. Improvement of S/N ratio of grey relational grade is 6.01171. The wear characteristic of silica gel reinforced aluminium composite was optimized under the variation of volume fraction, normal load, and sliding speed with the help of Taguchi method and ANOVA. The experimental result confirmed that sliding speed has highest effect on wear followed by volume fraction and normal load for silica gel reinforced aluminium composite. The sliding speed Regression equation developed for silica gel base aluminium metal matrix composites can be used to forecast the wear of above composite for different input conditions with reasonable accuracy. The confirmation test showed that improvement of S/N ratio of the proposed combination of the input parameters to the mean values for the same is 8.89%. Wear rate and coefficient of friction of silica gel based aluminium composite were optimized by grey relational analysis. Optimum level for minimum wear rate and coefficient of friction were attained at 10% of volume fraction of reinforcement, 75 N of applied load, and 180 rpm of sliding speed. From ANOVA, for both wear rate and coefficient of friction, volume fraction of filler silica gel (46.3072%) followed by sliding speed (11.1979%) and applied load (9.0321%) can be ranked as per the extent of significance. Interactions between volume fraction of filler silica gel, applied load, and sliding velocity on the dry sliding are of a little fraction, so they are measured statistically insignificant.

At the end, third or last conclusion is defined the application of silica gel reinforced aluminium composite into the manufacturing of piston of IC engine. The overall investigation shows that tensile and flexural strength as well as hardness of silica gel reinforced composite is tremendously high as compare to aluminium alloy, which is beneficial for sustaining long life of engine piston under the cycling loading. Silica gel reinforced composite has a low wear rate and coefficient of friction (COF). Less COF provides the smooth functioning of piston in the engine cylinder and low wear rate prevents from the surface abrasion and provides better life of the piston. So piston made of silica gel reinforced aluminium composite piston is better option rather than aluminium alloy piston.

REFERENCES

Abdel-Azim, A. N., Shash, Y., Mostafa, S. F., & Younan, A. (1995). Casting of 2024-A1 alloy reinforced with A12O3 particles. *Journal of Materials Processing Technology*, *55*(3-4), 199–205. doi:10.1016/0924-0136(95)01954-5

Arsha, A. G., Jayakumar, E., Rajan, T. P. D., Antony, V., & Pai, B. C. (2015). Design and fabrication of functionally graded in-situ aluminium composites for automotive pistons. *Materials & Design*, *88*, 1201–1209. doi:10.1016/j.matdes.2015.09.099

Chiang, K. T., & Chang, F. P. (2006). Application of grey-fuzzy logic on the optimal design of an injection mould part with a thin shell feature. *International Communications in Heat and Mass Transfer*, *33*(1), 94–101. doi:10.1016/j.icheatmasstransfer.2005.08.006

Deng, J. (1982a). Control problems of grey systems. *Systems & Control Letters*, (5): 288–294.

Deng, J. (1982b). Grey Control System. *Journal of Huazhong University of Science and Technology*, *3*, 9–18.

Deng, J. (1989). Introduction to Grey System. *Journal of Grey System*, *1*(1), 1–24.

Deng, J. (1990). *Theory of Grey System*. Wuhan: Huazhong University of Science and Technology Press.

Diler, E. A., & Ipek, R. (2012). An experimental and statistical study of interaction effects of matrix particle size, reinforcement particle size and volume fraction on the flexural strength of Al-SiCp composites by P/M using central composite design. *Materials Science and Engineering A*, *548*, 43–55. doi:10.1016/j.msea.2012.03.066

Edwin, B. D. (1992). Taguchi approach to design optimization for quality and cost: An overview. *Annual Conference of the International Society of Parametric Analysis*.

Feng, Y. C., Geng, L., Zheng, P. Q., Zheng, Z. Z., & Wang, G. S. (2008). Fabrication and characteristic of Al-based hybrid composite reinforced with tungsten oxide particle and aluminum borate whisker by squeeze casting. *Materials & Design*, *29*(10), 2023–2026. doi:10.1016/j.matdes.2008.04.006

Fisher, R. A. (1951). *Design of Experiments*. Edinburgh, UK: Oliver & Boyd.

Haq, A. N., Marimuthu, P., & Jeyapaul, R. (2008). Multiresponse optimization of machining parameters of drilling Al/SiC metal matrix composite using grey relational analysis in the Taguchi method. *International Journal of Advanced Manufacturing Technology*, *37*(3-4), 250–257. doi:10.1007/s00170-007-0981-4

Haque, M. M., & Sharif, A. (2001). Study on wear properties of aluminium–silicon piston alloy. *Journal of Materials Processing Technology*, *118*(1-3), 69–73. doi:10.1016/S0924-0136(01)00869-X

Himanshu, K., Mer, K. K. S., & Sandeep, K. (2014). A Review on Mechanical and Tribological Behaviors of Stir Cast Aluminum Matrix Composites. *Procedia Materials Science*, *6*, 1951–1960. doi:10.1016/j.mspro.2014.07.229

Horng, J. T., & Chiang, K. T. (2008). A grey and fuzzy algorithms integrated approach to the optimization of turning Hadfield steel with Al2O3/TiC mixed ceramic tool. *Journal of Materials Processing Technology*, *207*(1-3), 89–97. doi:10.1016/j.jmatprotec.2007.12.060

Huang, X., Liu, C., Lv, X., Liu, G., & Li, F. (2011). Aluminum alloy pistons reinforced with SiC fabricated by centrifugal casting. *Journal of Materials Processing Technology*, *211*(9), 1540–1546. doi:10.1016/j.jmatprotec.2011.04.006

Jha, A. K., Dan, T. K., Prasad, S. V., & Rohatgi, K. (1986). Aluminum alloy solid lubricant talc particle composites. *Journal of Materials Science*, *21*(10), 3681–3685. doi:10.1007/BF02403021

Kala, H., Mer, K. K. S., & Kumar, S. (2014). A review on mechanical and tribological behaviors of stir cast aluminum matrix composites. *Procedia Materials Science*, *6*, 1951–1960. doi:10.1016/j.mspro.2014.07.229

Kumar, S., & Balasubramanian, V. (2010). Effect of reinforcement size and volume fraction on abrasive wear behavior of AA7075 Al/SiCp P/M composites-A statistical analysis. *Tribology International*, *43*(1-2), 414–422. doi:10.1016/j.triboint.2009.07.003

Kuo, Y., Yang, T., & Huang, G. W. (2008). The use of grey relational analysis in solving multiple attribute decision-making problems. *Computers & Industrial Engineering*, *55*(1), 80–93. doi:10.1016/j.cie.2007.12.002

Mahapatra, S. S., & Patnaik, A. (2007). Optimization of wire electrical discharge machining (WEDM) process parameters using Taguchi method. *International Journal of Advanced Manufacturing Technology*, *34*(9-10), 911–925. doi:10.1007/s00170-006-0672-6

Modi, O. P., Yadav, R. P., Mondal, D. P., Dasgupta, R., Das, S., & Yegneswaran, A. H. (2001). Abrasive wear behaviour of zinc-aluminium alloy 10% alumina composite through factorial design of experiment. *Journal of Materials Science*, *36*(7), 1601–1607. doi:10.1023/A:1017523214073

Montgomery, D. C. (2001). *Design and analysis of experiments*. New York: John Wiley.

Nicolettoa, G., Rivaa, E., & Di Filippo, A. (2014). High Temperature Fatigue Behavior of Eutectic Al-Si-Alloys used for Piston Production. *Procedia Engineering*, *74*, 157–160. doi:10.1016/j.proeng.2014.06.241

Park, J. O., Kim, K. J., Kang, D. Y., Lee, Y. S., & Kim, Y. H. (2001). An experimental study on the optimization of powder forging process parameter for an aluminium alloy piston. *Journal of Materials Processing Technology*, *113*(1-3), 486–492. doi:10.1016/S0924-0136(01)00663-X

Rahul, K., Kaushik, K., & Sumit, B. (2014). Optimisation of mechanical properties of epoxy based wood dust reinforced green composite using Taguchi method. *Procedia Material Science, 5*, 688-696.

Ravindra, S. R., Rajesh, P., Anil, S., & Saraswati, R. (2014). Optimization of Wear Performance of Aa 5083/10 Wt. % Sic Composites Using Taguchi Method. *Procedia Materials Science*, *6*, 503–511. doi:10.1016/j.mspro.2014.07.064

Rebba, B., & Ramanaiah, N. (2014). Evaluation of mechanical properties of aluminium alloy (Al-2024) reinforced with molybdenum disulphide (MOS2) metal matrix composites. *Procedia Materials Science*, *6*, 1161–1169. doi:10.1016/j.mspro.2014.07.189

Reddy, A. C., & Zitoun, E. (2010). Matrix Al-alloys for silicon carbide particle reinforced metal matrix composites. *Indian Journal of Science and Technology*, *3*(12), 1184–1187.

Rohatgi, P. (1991). Cast aluminum-matrix composites for automotive applications. *Journal of Materials*, *43*(4), 10–15.

Roy, R. K. (1990). A primer on the Taguchi method. Society of Manufacturing Engineers.

Sahin, Y. (2005). The prediction of wear resistance model for the metal matrix composites. *Wear*, *258*(11-12), 1717–1722. doi:10.1016/j.wear.2004.11.024

Sannino, A. P., & Rack, H. J. (1995). Dry sliding wear of discontinuously reinforced aluminium composites: Review and discussion. *Wear*, *189*(1-2), 1–19. doi:10.1016/0043-1648(95)06657-8

Sonawane, P. M., & Karnik, M. G. (2017). Experimental Investigation of Optimum level of Reinforcements with in 6061Al Alloy for better Tribological and Mechanical Properties. *Materials Today: Proceedings*, *4*(2), 3290–3299. doi:10.1016/j.matpr.2017.02.215

Soner, B., & Serdar, O. Y. (2007). Relationship between thermal and sliding wear behavior, of Al6061/Al2O3 metal matrix composites. *Journal of Springer Science*, *42*, 4485–4493.

Srivathsan, S. (2014). Comparative study of aluminium alloy pistons manufactured by casting and powder forging. *International Journal of Scientific & Engineering Research*, *5*(6), 662–664.

Subbaya, K. M., Suresha, B., Rajendra, N., & Varadarajan, Y. S. (2012). Grey-based Taguchi approach for wear assessment of SiC filled carbon–epoxy Composites. *Materials & Design*, *41*, 124–130. doi:10.1016/j.matdes.2012.04.051

Sudeepan, J., Kumar, K., Barman, T. K., & Sahoo, P. (2014). Study of friction and wear properties of ABS/Kaolin polymer composites using grey relational technique. *Procedia Technology*, *14*, 196–203. doi:10.1016/j.protcy.2014.08.026

Sulima, I., Jaworska, L., Wyzga, P., & Perek-Nowak, M. (2010). The influence of reinforcing particles on mechanical and tribological properties and microstructure of the steel-TiB2 composites. *Journals of Achievements in Materials and Manufacturing Engineering*, *48*, 52–57.

Surappa, M. K. (2003). Aluminium matrix composites: Challenges and opportunitie. *Sadhana*, *28*(1), 319–334. doi:10.1007/BF02717141

Surappa, M. K., Prasad, S. V., & Rohatgi, P. K. (1982). Wear and abrasion of cast Al-alumina particles composites. *Wear*, *77*(3), 295–302. doi:10.1016/0043-1648(82)90055-2

Taguchi, G. (1990). *Introduction to quality engineering*. Asian Productivity Organization.

Taguchi, G. (1993). *Taguchi on Robust Technology Development: Bringing Quality Engineering Upstream*. New York, NY: ASME press. doi:10.1115/1.800288

Tjong, S. C., & Tam, K. F. (2006). Mechanical and thermal expansion behavior of hipped aluminum-TiB2 composites. *Materials Chemistry and Physics*, *97*(1), 91–97. doi:10.1016/j.matchemphys.2005.07.075

Tosun, N. (2006). Determination of Optimum Parameters for Multi-Performance Characteristics in Drilling by Using Grey Relational Analysis. *International Journal of Advanced Manufacturing Technology*, *28*(5-6), 450–455. doi:10.1007/s00170-004-2386-y

Tsui, K. L. (2007). An overview of Taguchi method and newly developed statistical methods for Robust Design. *IIE Transactions*, *24*(5), 44–56. doi:10.1080/07408179208964244

Venkatprasad, S., & Subramanian, R. (2011). Influence of parameters on the dry sliding wear behavior of aluminium/flyash/graphite hybrid metal matrix composite. *European Journal of Scientific Research*, *53*(2), 280–290.

Vishal, S., & Sanjeev, D. (2007). *Synthesis and Interfacial Characterization of Al-4.5wt%Cu/ Zircon Sand/ Silicon Carbide Hybrid Composite* (Master's thesis). Department of Physics and Materials Sciences, Thapar University, Patiala, India.

Zahi, S., & Daud, A. R. (2010). Fly ash characterization and application in Al–based Mg alloys. *Journal of Elsevier-. Materials & Design*, *32*(3), 1–10.

Zhang, Z., Zhang, L., & Mai, Y. (1995). Wear of ceramic particle-reinforced metal-matrix composites'. *Journal of Materials Science*, *30*(8), 1967–1971. doi:10.1007/BF00353019

Chapter 5
Development of Aerospace Composite Structures Through Vacuum-Enhanced Resin Transfer Moulding Technology (VERTMTy):
Vacuum-Enhanced Resin Transfer Moulding

Raghu Raja Pandiyan Kuppusamy
Birla Institute of Technology, India

ABSTRACT

Quality products with low cost manufacturing routes are the major objectives for the product development in any application. The current statement is evident for polymer-matrix composites, particularly in high end applications such as aerospace and mass transit structures. These applications require advanced composite materials tailored to meet the property demands posted by dynamic load conditions, and hence, the use of wide spectrum of constituents and architectures are vital to cater the needs. Consequently, the development of novel composite materials with the permutations of ingredients leads to the innovative processing techniques. To address the gap in the manufacturing with economical processing routes of thick sectioned advanced composite parts showing superior properties at different wall sections, an innovative composite manufacturing technology coupling resin transfer moulding (RTM) processing and vacuum applications, namely vacuum enhanced resin transfer moulding technology (VERTMTy), is conceptualized, proposed, and developed.

DOI: 10.4018/978-1-5225-5216-1.ch005

INTRODUCTION

Quality products with low cost manufacturing routes are the major objectives for the product development in any application. The current statement is evident for polymer-matrix composites, particularly in high end applications such as aerospace & mass transit structures. These applications require advanced composite materials tailored to meet the property demands posted by dynamic load conditions and hence, the use of wide spectrum of constituents and architectures are vital to cater the needs. Consequently, the development of novel composite materials with the permutations of ingredients leads to the innovative processing techniques. Importantly, part thickness also plays a major role in bearing dynamic mechanical load at the application conditions. Most of the thick sectioned composite parts for an application are manufactured by primitive processing routes such as hand layup method or costly autoclave processing. However, the manufacturing processes fail to prepare economical thick sectioned composite parts that meet the demands such as uniform resin-fibre volume fractions, uniform product thickness, reduced porosity, surface properties, dynamic loading etc. posted by higher end applications. Notably, most composite moulding processes fail to manufacture thick sectioned composite parts owing to problems that rose due to ununiform mould filling and resin curing across the part section. To address the gap in the manufacturing with economical processing routes of thick sectioned advanced composite parts showing superior properties at different wall sections, an innovative composite manufacturing technology coupling resin transfer moulding (RTM) processing and vacuum applications namely "Vacuum Enhanced Resin Transfer Moulding Technology (VERTMTy)" is conceptualized, proposed and developed.

Review of Status of Research and Development in VERTMTy

To the best knowledge of the investigator there are no reports in literature that directly deals with "Vacuum Enhanced Resin Transfer Moulding Technology (VERTMTy)", since it is newly conceptualized. However, researchers reported work based on thick sectioned composite development methods and their related issues.

C. Soutis (2005) has made a complete review on the application of fibre reinforced composites in aircraft construction. He detailed the use of different raw materials and processing routes in manufacturing composite materials for aerospace applications: resin ranging from epoxy to unsaturated polyester, reinforcements ranging from carbon to glass, reinforcement architectures ranging from unidirectional to woven roven and process routes ranging from hand-lay up to autoclave moulding.

Yijun Jiang & Suong V. Hoa (2006) developed a hand-layup novel technique to manufacture thick sectioned composites. Pre-catalyzing fabric method was the newly developed technique applied in manufacturing thick composites with hand layup process. By applying a peroxide catalyst to the fabric instead of mixing it into the bulk resin system, this technique can slow down the polymerization reaction rate and subsequently reduce the internal temperature of thick composites.

Dennis J. Michaud (2010) submitted his PhD dissertation in Delware University, USA on thick sectioned composite laminates manufactured through Resin Transfer Moulding (RTM) process. He has shown how processing thick-sectioned composites (>1/2") can be difficult due to the exothermic nature of the resin and the low thermal conductivity of the composite. This was particularly true for the RTM process examined here. Also, he shown that the temperature profile used to polymerize the resin, otherwise known as a "cure cycle," must be carefully chosen to reduce thermal gradients within the composite while ensuring satisfactory processing times. Instead of trial and error methods that are expensive, time consuming, and non-optimal, he proposed a knowledge-based optimization strategy to predict the proper cure cycle to get uniform cure throughout the part cross section

James K. Sutter et al. (2010), NASA, has shown the manufacturing and testing methodology to develop composite laminates for heavy lift launch architectures. Autoclaved and hand-laid epoxy-carbon fibre composites were fabricated and the results highlighted differences in processing conditions, laminate quality, as well as initial room temperature thermal and mechanical performance. Laminate cure, glass transition temperature. tackiness, voidage, fracture toughness were measured and compared between autoclave and hand laid composite structures

Daniel Walczyk and Jaron Kuppers (2012) had shown that the thermal press curing of advanced composite laminate is alternate to the autoclaved composite parts. They prepared carbon-epoxy composite laminates using both autoclave and hot press technique and laminate properties were compared. The results had shown that the hot pressed composite.

Raghu Raja Pandiyan et al. (2010 & 2014) showed the development of a new composite product through resin transfer molding (RTM) process requires the identification of an effective injection strategy and optimization of the process variables and raw material parameters of the manufacturing process. They developed an effective manufacturing technology for a cab front model using RTM isothermal mold-filling simulations & mould scaling down strategy and verified simulated results by conducting isothermal mould filling experiments at the scaled down mould.

A.R.Upadhya et al. (2011) detailed the application of autoclaves for developing aerospace structures with its issues and challenges at Indian scenario. The complete control and automation of composite laminates autoclave processing were explained.

Vijaya Kumar et al. (2013) demonstrated the development of carbon-epoxy composite laminates through vacuum bagging process. They analyzed the root causes for the vacuum failures during curing process. Laminates made of Hexcel 913 carbon/ epoxy prepeg are prepared and cured in the autoclave by varying the vacuum levels at different salient points of cure process during curing. The Experimental results has shown that ILSS may not vary much if the vacuum fails after (Tg) Glass Transition point, hence the vacuum can be maintained up to the Tg point and thereby it can be slowly reduced to zero. This showed that there is a scope to optimize the manufacturers recommended cure cycle so as to reduce the curing cost and time. This research work helps the designers in developing the tools and optimum cure cycles for carbon epoxy composite structures with autoclave moulding.

Kundan Kumar Verma et al. (2013) have developed a novel infusion technique named Vacuum Enhanced Resin Infusion Technology (VERITy) to develop carbon-epoxy composites for aircraft structures.

Importance of the Proposed Manufacturing Technique in the Context of Current Status

From the literature review, it can be seen that the manufacturing of advanced composite materials are made viable through either costly autoclave moulding or primitive hand-layup method. Though autoclave moulding process produces quality composite products for aerospace structures, the costing factor and cumbersome process & operational development at the autoclave for quality products makes it harder to make it a choice especially for thick sectioned composite laminates. The primitive hand-layup primarily fails to meet voidage constraint as well as uniformity in fibre and resin volume fractions even in thin sectioned composites and hence, it is never been a choice of manufacturing method for higher end applications. Though, other out-of-autoclave processing of composite moulding processes such as Hot Press Curing (HPC), RTM, VARTM etc. is utilized to develop advanced composite laminates but it has not been satisfactorily resolved for thick sectioned composite parts. Notably, none of the aforementioned processing routes ensures a composite part that bears both costing factor and reduced processing difficulties. To address the gap of developing thick sectioned advanced composite parts through economical out-of-autoclave processing, an innovative "Vacuum Enhanced Resin Transfer Moulding Technology (VERTMTy)" was conceptualized, proposed and developed. Hence, the main objective of this book chapter is to demonstrate a manufacturing technique for thick sectioned advanced composite parts for higher end applications through economical processing routes. The manufacturing of the composite laminates should consent to specific raw material and parameters to meet the demands of higher end applications such as aerospace structures:

- Composite panel of specific polymer matrix and reinforcement material to the specific fabric architecture and lay-up

- Specific geometry

VERTMTy Process

The work phase to manufacture the composite panels was initiated with the development of mould design of specified product dimension. The mould was designed considering the mechanical, thermal and flow aspects of the RTM process. The mould was designed to the dimensions of as 600 mm x 300 mm with a varying part thickness from 2 mm to 8 mm. The mould assembly of proposed VERTMTy process is shown in Figure 1. Controller based electrical cartridges and water circulation cooling will be used for curing resin at elevated temperature and system cooling during VERTMTy processing.

Isothermal Mould Filling Simulations

Preliminary isothermal mould filling simulations using the proposed product geometry were performed with idealized raw material properties and process parameters to identify the effective injection strategy. The preliminary simulations predicted an effective injection strategy of single point injection port at the centre and two point vents at the sides. Moreover, the injection port and vent diameter of 4 mm delivers an acceptable mould fill time with the possible variations in the raw material and process parameters.

A model isothermal mould filling simulation using the under mentioned raw material and process parameters was performed. The simulation results showing the flow front positions and mould fill times are shown in Figure 2. Simulation inputs: Single point injection of 4 mm diameter, injection pressure 2 Bars (200000 Pa), vacuum 10000Pa, resin viscosity 0.1 Pa-sec, reinforcement permeability (isotropic)

Figure 1. Vacuum Enhanced Resin Transfer Moulding Technology (VERTMTy) Process

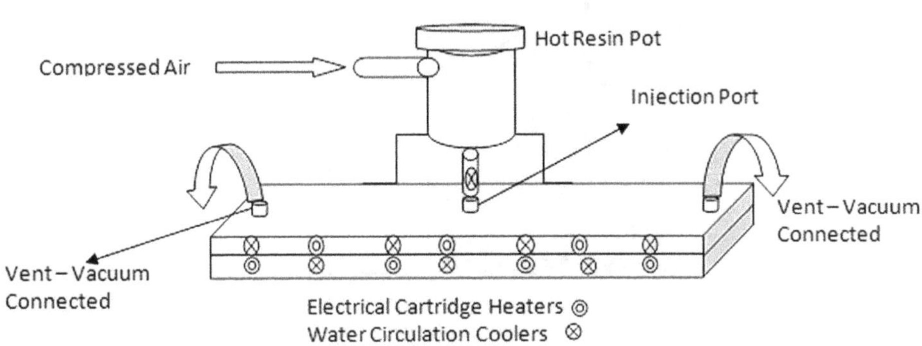

10-10 m2, laminate thickness 2 mm, porosity 0.5 Simulation outputs: Flow front position as a function of time, mould fill time

Mould Senstivity Analysis

A mould sensitivity analysis was performed using isothermal mould filling simulations to evaluate the effect of raw material and process parameters on the mould fill time. The possible variations in the raw material properties and parameters are listed below:

1. Resin injection pressure was varied from 1 Bar to 5 Bars in the order of 1 Bar, since RTM process is practiced mostly within this pressure limit.
2. Resin viscosity was varied from 0.1 Pa-sec to 0.5 Pa-sec in the order of 0.1 Pa-sec, since most resin grades fall into this boundary at the room temperatures. At higher temperatures, the resin melts causing the drop in viscosity.

Figure 2. Isothermal Mould Filling Simulations

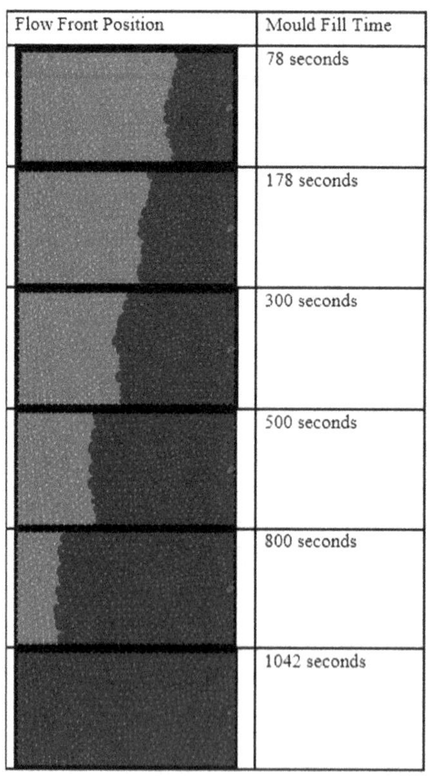

**For a more accurate representation see the electronic version.*

3. Reinforcement isotropic permeability was varied from 10^{-10} m^2 to 10^{-9} m^2 in the order of 2*10^{-10} m^2, since most reinforcement mats falls into this permeability range.
4. Laminate thickness was varied from 2 mm to 8 mm in the order of 2 mm

An injection strategy of single point injection port and single point vent of diameter 4 mm was kept constant for the mould sensitivity analysis. There were 600 possible combinations and the simulation results of few combinations were tabulated in Table 1.

Mould Heat Calculation and Cartridge Heater Power Design

The heating of mould to controlled temperatures are designed through application of electrical resistance cartridges. The requirement of minimum number of cartridges to elevate the mould required temperature is calculated using below given heat load calculations.

Table 1. Mould Sensitivity Analysis

Resin Viscosity, Pa-sec.	Injection Pressure, Bars	Permeability, (m^2)	Laminate Thickness, mm	Mould Fill Time, Seconds
0.1	1	2*10^{-10}	2	1295
0.2	2	2*10^{-10}	2	2015
0.3	3	2*10^{-10}	2	2756
0.4	4	2*10^{-10}	2	3632
0.5	5	2*10^{-10}	2	4540
0.1	1	4*10^{-10}	4	1223
0.2	2	4*10^{-10}	4	1908
0.3	3	4*10^{-10}	4	2745
0.4	4	4*10^{-10}	4	3648
0.5	5	4*10^{-10}	4	4560
0.1	1	6*10^{-10}	6	2550
0.2	2	6*10^{-10}	6	5002
0.3	3	6*10^{-10}	6	7503
0.4	4	6*10^{-10}	6	10004
0.5	5	6*10^{-10}	6	12505

Heat required to raise the temperature of the mould, Q = {mould mass * specific heat of the mould material * temperature}

Volume of the mould = 0.010019904 m^3
Density of carbon steel = 7850 Kg/ m^3
Mass of the bottom half of the mould = 78.65 Kg
Specific heat of carbon steel = 490 J/Kg. K
Heat required to raise the mould temperature to 200°C = 78.65*490*200 = 770.8307 KJ

By trial and error, we have to obtain the cartridge heater power which takes minimum time to reach the set temperature, i.e 200°C. Let us select the cartridge heater having capacity of 1000 watts and check the time required to reach the set temperature, i.e 200°C. Mould is designed with 13 heater holes and total power is 13*1000.

Time required by the heaters (13 Nos.*1000 watts) to raise the mould to 200°C assuming 10% heat loss due to the free surface = [(7708308)+(0.1*7708308)]/(13*1000)] = 11 minutes, which is acceptable. Hence, 13 cartridge heaters with a capacity of 1000 watts can heat the designed mould to 200°C within 12 minutes

Mould Assembly

A complete mould lay-out was prepared as per the design guidelines and Figure 3 provides the complete mould dimensions. The mould material was adjudged as the carbon steel. Bottom half thickness of 40 mm and top half thickness of 23 mm were selected to carry the mechanical aspects of the mould.

Cartridge heaters with 13 nos. of power 500 watts were used to heat the mould up to 200°C. A resin pre-heating cylinder of capacity 2 litres with heating capacity to 150°C was designed and fabricated. 7 heaters were placed at the mould bottom half and 6 heaters were placed at the mould top half to aid the temperature cycle to cure the resin. Provisions were made for the water circulation at the both halves of the mould. It was also tested the mould cooling capability of the circulating water and found that the present cooling set-up is enough to meet the cooling demands. The heaters and coolers points were placed in such a way that the heating and cooling load is uniformly distributed over the mould. Respective temperatures sensors were placed at top and bottom halves of the mould and the resin pre-heater. All these temperatures were controlled through the respective sensors and the data acquisition system was installed to capture temperature variations as a function of time. The complete assembly of mould with heating and cooling elements is shown in Figure 4.

Figure 3. Mould Lay-out

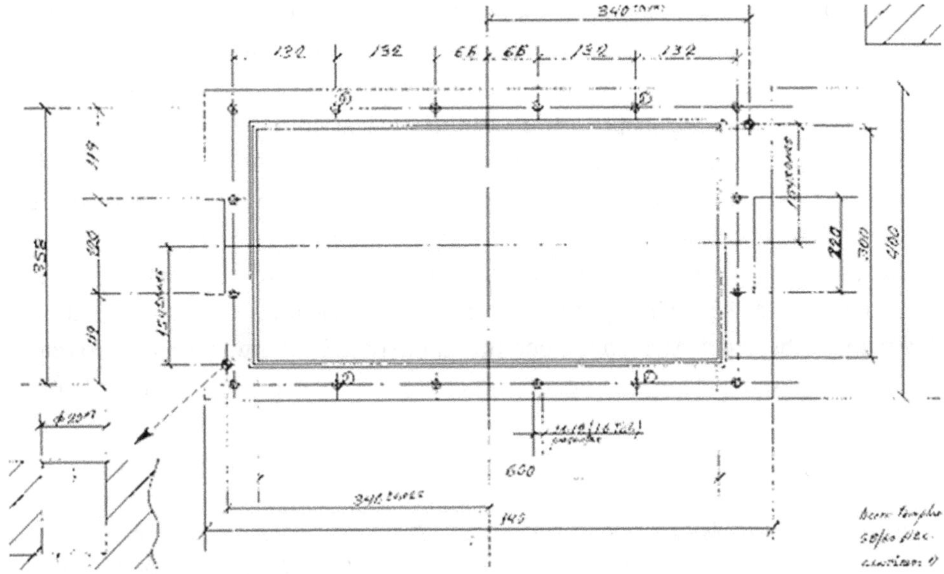

Figure 4. Mould Assembly

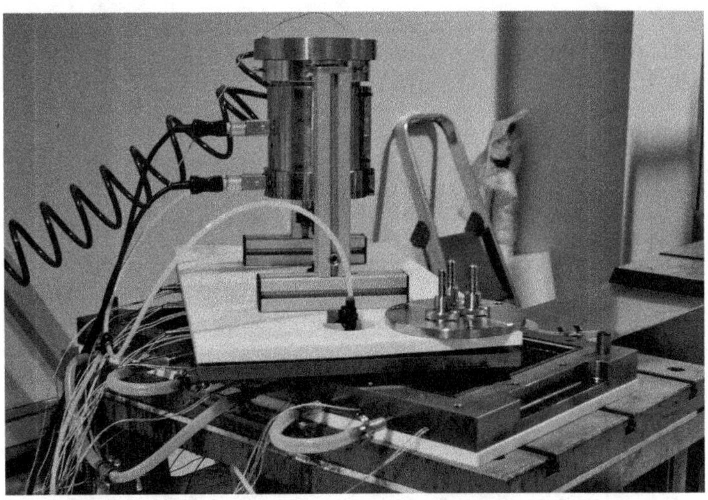

Fully Cured Composite Panels from VERTMTy Process

In this work, the Hexcel RTM6 Epoxy resin and G0926 5H satin weave carbon fabric are the polymer matrix and reinforcement fibres proposed for the development of advanced semi cured composite panels. Typically, 6 layers of carbon fibre woven fabric of architecture 90° / 90° / 90° / 90° / 90° / 90° are proposed to constitute the lay-up for the semi cured composite laminates to the geometry 30 cm x 30 cm. Two trials were performed to manufacture the fully cured composite panel. Initially, the resin was preheated to 80°C to make cold stored resin to the low viscous flow able resin. The resin inlet was kept closed and the vents at both the sides of the mould were connected to vacuum. After sufficient vacuum application, the resin inlet valve was opened and the preheated resin was pressurized to flow into the mould through the reinforcement medium. The mould temperature was kept at 120°C during mould filling. In both the cases, 6 layers of carbon woven roven reinforcement mats at 45° architecture were used. The injection was held until a clear resin free of air bubbles flown out of the vents. Then, a temperature ramp of 3°C/min was initiated to reach the cure temperature 180°C. As per the proposed cure cycle, the cure temperature

Figure 5. (a) VERTMTy Fully Cured Panel -1, Cure Cycle: 180°C for 120mins., Degree of Cure: 99%, Panel Thickness: 2.62 mm (b) VERTMTy Fully Cured Panel -2, Cure Cycle: 180°C for 120mins., Degree of Cure: 99%, Panel Thickness: 2.55 mm

Figure 5(a) VERTMTy Fully Cured Panel -1, Cure Cycle: 180°C for 120mins., Degree of Cure: 99%, Panel Thickness: 2.62 mm

Figure 5(b) VERTMTy Fully Cured Panel -2, Cure Cycle: 180°C for 120mins., Degree of Cure: 99%, Panel Thickness: 2.55 mm

Figure 6. (a) RTM Semi Cured Panel -1, Cure Cycle: 140°C for 30mins., Degree of Cure: 15%, Panel Thickness: 2.6 mm (b) RTM Semi Cured Panel -2, Cure Cycle: 140°C for 60mins., Degree of Cure: 21%, Panel Thickness: 2.55 mm (c) RTM Semi Cured Panel -3, Cure Cycle: 140°C for 90mins., Degree of Cure: 30%, Panel Thickness: 2.65 mm (d) RTM Semi Cured Panel -4, Cure Cycle: 140°C for 120mins., Degree of Cure: 38% Panel Thickness: 2.6 mm

Figure 6(a) RTM Semi Cured Panel -1, Cure Cycle: 140°C for 30mins., Degree of Cure: 15%, Panel Thickness: 2.6 mm

Figure 6(b) RTM Semi Cured Panel -2 Cure Cycle: 140°C for 60mins. Degree of Cure: 21%, Panel Thickness: 2.55 mm

Figure 6(c) RTM Semi Cured Panel -3 Cure Cycle: 140°C for 90mins. Degree of Cure: 30%, Panel Thickness: 2.65 mm

Figure 6(d) RTM Semi Cured Panel -4 Cure Cycle: 140°C for 120mins. Degree of Cure: 38%, Panel Thickness: 2.6 mm

Figure 7. VERTMTy Cure Evolution

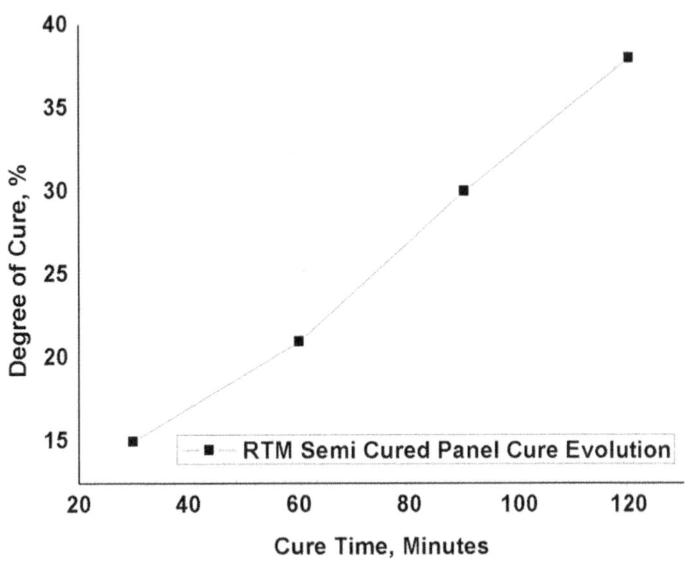

of 180°C was applied to the interval of 2 hours. The results of the fully cured VERTMTy composite panels are given in Figure 5.

Semi Cured Composite Panel From VERTMTy Process

Four trials were performed using 140°C with the time starting from 30 minutes to 2 hours with a time interval of 30 minutes. The main objective of these trials was to find out the resin cure evolution to the applied temperatures. The results of the developed VERTMTy semi cured panels are given in Figure 6 and the VERTMTy mould cure evolution is shown in Figure 7.

Practical Challenges During VERTMTy Processing

- It is required to cure the composite panel to a minimum degree of cure, say to the least of 30% curing. Otherwise, the extraction of panel from the mould becomes difficult. Extraction of under-cured panels using forcible means can lead to the panel damage and alteration n panel dimensions.
- Proper mould sealing and sealing thickness should be ensured. Bubbles get entrapped due to proper sealing and product thickness increase with increase sealing thickness.
- The conditions of resin such as storage, preheating should be standardized to each trial. The panel cure may differ to the trials of same cure cycle, if the resin preheating & mould filling time. Hence, care should be taken to ensure the idealistic resin preheating and mould filling time at each trial.

CONCLUSION

1. VERTMTy fully cured composite panels of dimension 600 mm x 300 mm were manufactured using the cure cycle 180°C for 120 Minutes
2. VERTMTy semi cured composite panels of dimension 600 mm x 300 mm to the several degrees of cure were manufactured using different cure cycles and the effect of cure cycle on the degrees of cure was measured.

ACKNOWLEDGMENT

This research is supported by EMR Scheme of SERB, DST-India [EMR/2016/005562].

REFERENCES

Michaud. (2010). *Simulation-based Design Optimization and Control of Thick Composite Laminates Manufactured by Resin Transfer Molding* (Doctoral dissertation). Department of Chemical Engineering, University of Delaware.

Raghu Raja Pandiyan, K., Kundu, G., Neogi, S., & Patel, J. (2010). Development of Manufacturing Technology for Cab Front using Resin Transfer Molding Process. *Journal of Composite Materials*, *44*(18), 2217–2231. doi:10.1177/0021998310364168

Raghu Raja Pandiyan, K., & Neogi, S. (2014). Development of Resin Transfer Molding Process Using Scaling Down Strategy. *Polymer Composites*, *35*(9), 1683–1639. doi:10.1002/pc.22821

Soutis, C. (2005). Fibre reinforced composites in aircraft construction. *Progress in Aerospace Sciences*, *41*(2), 143–151. doi:10.1016/j.paerosci.2005.02.004

Sutter, Kenner, Pelham, Miller, Polis, Nailadi, … Fikes. (2010). *Comparison of Autoclave and Out-of-Autoclave Composites*. NASA Report. Retrieved from ntrs.nasa.gov/archive/nasa/casi.ntrs.nasa.gov/20110016095.pdf

Upadhya, A. R., Dayananda, G. N., Kamalakannan, G. M., Ramaswamy Setty, J., & Christopher Daniel, J. (2011). Autoclaves for Aerospace Applications: Issues and Challenges. International Journal of Aerospace Engineering, Article ID 985871.

Verma, Dinesh, Singh, Gaddikeri, Srinivasa, Kumar, & Sundaram. (2013). Development of Vacuum Enhanced Resin Infusion Technology (VERITy) Process for Manufacturing of Primary Aircraft Structures. *Journal of the Indian Institute of Science*, *93*(4), 621–633.

Vijaya Kumar, , Safiulla, & Ahmed. (2013). Analysis of Vacuum Failures during Curing of CFRP Composites. *International Journal of Scientific & Technology Research*, *2*(5), 220–225.

Walczyk, D., & Kuppers, J. (2012). Thermal Press Curing of advanced Thermoset Composite Laminate Parts. *Composites. Part A, Applied Science and Manufacturing*, *43*(4), 635–646. doi:10.1016/j.compositesa.2011.12.008

Chapter 6
A Decision-Making Approach for Material Selection of Polymeric Composite Bumper Beam

Divya Zindani
National Institute of Technology Silchar, India

Saikat Ranjan Maity
National Institute of Technology Silchar, India

Sumit Bhowmik
National Institute of Technology Silchar, India

ABSTRACT

Selection of material is a crucial part of the voyage of obtaining the optimal product. The task is challenging for the designer who has to select the optimal material from a list of thousands of materials that satisfy the tangible and intangible criteria. The problem of identification of the optimum material can be solved using different MCDM (multi-criteria decision-making) tools. This chapter proposes to select the optimal material for polymeric composite bumper beam using TODIM (tomada de decisao interativa multicriterio, which means in Portuguese "interactive and multi-criteria decision-making") method. Six different composite materials were considered. Glass fiber epoxy turns out to be the optimal material for the bumper beam. Further, finite element analysis was carried out to study the behavior of the candidate materials under specific loading conditions.

DOI: 10.4018/978-1-5225-5216-1.ch006

INTRODUCTION

Bumper system is one of the most important components of a vehicle. This is because in case of any accident it is the bumper system which absorbs the energy of the impact. To a great extent, it protects the passengers since it is the first to accept the collision. The bumper system comprises of three main components: energy absorber, fascia, and the bumper beam.

However, in the past decade the role of design engineers has been to produce the vehicular model that has the minimum weight. Materials often play a crucial role in achieving this objective. A number of polymeric composites are now in use to manufacture bumper beam.

Selection of a suitable material for the bumper beam is crucial as a poor material can lead to damage and failure of the bumper beam. A large number of parameters must be considered in making an optimal selection. Thus, one of the difficult tasks that a design engineer faces is with the selection of an optimal material from a vast list of materials, with new materials being added to the list, under the consideration of conflicting tangible and non-tangible criteria.

The development of MCDM (Multi-Criteria Decision Making) tools has made the task a tad easier for the design engineers. Some of the prominent MCDM tools used in the automotive material selection include VIKOR, Artificial Neural Network, Analytic Hierarchy Process (AHP), Fuzzy Logic, etc. However, as far as material selection of bumper beam is concerned, the reports on MCDM tools are scarce.

The present work proposes to select polymeric composites for the bumper beam, using an MCDM tool in the form of TODIM and relatively simple to implement in comparison to the other reported MCDM tools. The polymeric composites make the list of alternatives for selection of optimal material for the bumper beam. Further, FEM analysis was carried out with the candidate materials to study their behavior under specific loading conditions.

Background

A vehicular bumper system comprises usually of a bumper beam, structural components for absorbing energy, stays and cover. The main structural component in the bumper system is the bumper beam. The bumper beam absorbs the energy due to bending resistance. Further, by collision it dissipates the high impact energy (Davoodi et al, 2008). Therefore, as appropriately described by Cheon et al. (Cheon, Choi and Lee, 1995), a bumper beam is the main backbone of the entire bumper system.

The bumper beam must be deformable to the extent that it absorbs most of the impact energy and thereby leading to the reduced risk of causing injury to the vulnerable road users. At the same time the bumper beam should have adequate strength to provide protection to the nearby vehicular components. Therefore a bumper beam must overcome both the scenarios of low-impact condition and crashworthiness.

A suitable material during the preliminary design process of the bumper beam is vital so that the bumper beam can perform its intended function and meet the service requirement. Composite materials make suitable alternatives for a suitable material selection. This is because of their enhanced physical and mechanical properties. It is a common practice to generally find a suitable material for the bumper beam and then manufacture the same which otherwise would have been a costly affair.

Further, a number of factors need to be considered for a systematic material selection process. Mallick (Malick, 2008) and Edwards (Edwards, 2004) have reported the factors that are required to be considered while making a suitable material selection for the bumper beam. Besides the number of criteria to be considered, the nature of the criteria also makes the selection arduous process. The criteria may be conflicting and non-commensurable. Therefore, a systematic approach is therefore necessary to arrive at the best material. MCDM techniques resolve and provide for a systematic approach to material selection.

However, as far as the bumper beam is considered, the work reported in comparison to that reported on the materials for bumper beam and its analysis is scarce. For instance, weighted objective method was employed by Sapuan et al (Sapuan et al., 2005). For the selection of material for the bumper beam. Materials for bumper beam were classified on the basis of cost and impact loading using LS-DYNA ANSYS 5.7 by Hosseinzadeh et al. (Hosseinzadeh, Shokrieh and Lessard, 2005). Suddin et al. (Suddin et al., 2007) utilized the knowledge base system (KBS) for making a systematic material selection for bumper beam. The analytic hierarchy process (AHP) was used by Hambali et al. (Hambali et al., 2010).

On the other hand a number of works on materials and analysis of bumper beams have been reported. For instance, Davoodi et al. (Davoodi, Sapuan and Yunus, 2010) developed a hybrid material using kenaf and synthetic glass fibre as reinforcement. A review on development process was presented by Davoodi et al. (Davoodi et al., 2012). Beyene et al. (Beyene et al., 2014) reported the issues in the design and fabrication of the lightweight solution for the bumper. The design and crashworthiness analysis for a bumper made of functionally graded foam-filled bumper beam was carried out by Xiao et al. (Xiao et al., 2015). Kim et al. (Kim, Kim and Kim, 2015) used glass/carbon fibre reinforced composites for the bumper beam. Alternative lightweight materials for the bumper were reported by Belingardi et al. (Belingardi et al., 2015) in their technical report. The modified particle swarm optimizer was utilized by Liu et al. (Liu, Lu and Zhu, 2016) for designing the automotive composite beam. Finite element analysis of energy absorption for bumper beams of different cross sections was carried out by Nasiruddin and Rosidah (Nasiruddin and Rosidah,

2017) using ANSYS LS DYNA. The best design concept was selected based on the criteria of energy absorption. Design and analysis of automotive bumper covers were undertaken by Liu et al. (Liu et al., 2016) using the commercial FEA solver. The beam was under transient loading conditions. The study proved to be a cost effective method to predict the performance of automotive bumper cover. Design and optimization for a composite bumper beam in situations of high-speed frontal crashes was done using the commercial FEA solver and the Fruit Fly Optimization Algorithm (FFO) by Zheng et al. (Zheng et al., 2016). Percentage reduction of about 56% was reported for the head injury criteria. Further, the new composite beam was reported to be 4.84% lighter than the steel counterpart.

MAIN FOCUS OF THE CHAPTER

Issues, Controversies, Problems

A much research has been carried out on the materials for bumper beam and the design, optimization and manufacturing of the bumper beam. It is well established fact that the design engineers cannot neglect the role of materials in the designing of a component. Therefore, the systematic process of selection of suitable material for a component is vital in the preliminary design phase of any component. The different MCDM tools provide for the systematic process of selection of a suitable material. But, the number of research reports on the MCDM tools for selection of suitable material for bumper beam is scarce and very complex to be implemented. Therefore, keeping the theme of the book in mind, the present work focuses to select a suitable material for polymeric composite bumper beam using an MCDM tool in the form of TODIM.

The remaining part of the main focus of the chapter is organized as follows: section 1 describes in brief the TODIM method which is followed by section 2 discussing the materials available for the polymeric composite bumper beam and the factors to be considered while making the material selection for the polymeric composite beam. In section 3 the TODIM method for the case in hand is implemented.

TODIM Method

TODIM is a discrete MCDM method based on the concept of prospect theory (Gomes and Lima). It is observed that all the existing MCDM methods are based on the principle that the decision maker always looks for a solution which corresponds to the maximum of some global measure of value. On the other hand, TODIM method is based on the adaptation of a global measurement of value which is calculated while implementing the paradigm of prospect theory. This method is thus based on a description, empirically proved, of how people effectively formulate decisions in face of risk.

Consider a decision making problem consisting of a set of n alternative options to be evaluated based on m quantitative/qualitative criteria and also assume that among those criteria, one is to be treated as the reference criterion. For each of the criteria c, the contribution of each alternative i to the objective associated with the criterion is then estimated. The TODIM method is also quite suitable for dealing with both quantitative and qualitative criteria. Both types of the data are subsequently normalized. The relative measure of dominance of one alternative over another is determined for each pair of alternatives. It is calculated as the sum over all the criteria of both relative gain/loss values for the considered alternatives. The components in this sum may be either gains, losses or zeros, depending on the performance of each alternative with respect to every criterion. After determination of the criteria weights and their normalization, the partial matrices of dominance and the final matrix of dominance are developed. In this stage, a criterion r is to be chosen as the reference criterion for the calculations according to the relative priority assigned to each criterion. Usually, the criterion having the highest weight of importance is identified as the reference criterion. Thus, w_{cr} becomes the weight of criterion c divided by the weight of the reference criterion r. Using w_{cr} allows all pairs of differences between the performance measurements to be directed towards the reference criterion. Based on the prospect theory, the dominance of alternative A_i over alternative A_j is now measured. Finally, the global measures obtained permit the complete rank ordering of all the alternatives.

Thus, the key idea of TODIM method lies in measuring the dominance degree of each alternative over the remaining ones using the prospect value function. The ranking of the candidate materials is done on the basis of this dominance degree.

The procedural steps of TODIM method are presented as below (Adali, Işik and Kundakci, 2016; Gomes and Lima, 1992; Zindani et al., 2017):

- **Step 1:** As the first step for the implementation of TODIM method, the corresponding decision matrix X needs to be developed. In this matrix, there are n and m numbers of alternatives and evaluation criteria respectively.

$$X = [x_{ic}]_{n\times m} = \begin{bmatrix} x_{11} & x_{12} & \cdots & x_{1m} \\ x_{21} & x_{22} & \cdots & x_{2m} \\ \cdots & \cdots & \cdots & \cdots \\ x_{n1} & x_{n2} & \cdots & x_{nm} \end{bmatrix} \quad (i=1,2,...,n; c=1,2,...,m) \tag{1}$$

where, x_{ij} denotes the performance of ith alternative with respect to j^{th} criterion.

- **Step 2:** In order to make this decision matrix dimensionless and all of its elements comparable, it is then normalized. For beneficial criteria requiring higher values, Equation (2) can be adopted. On the other hand, for non-beneficial criteria preferring lower values, Equation (3) can be utilized.

$$P_{ij} = \frac{x_{ic}}{\sum_{i=1}^{n} x_{ic}} \tag{2}$$

$$P_{ij} = \frac{1/x_{ic}}{\sum_{i=1}^{n} 1/x_{ic}} \tag{3}$$

where, P_{ij} is the normalized value of x_{ij}.

- **Step 3:** Using AHP or Shannon's entropy method, the priority weights (relative importance) (w_j) of all the considered criterion are determined. The relative weight (w_{cr}) of criterion C_c (c = 1,2,…,m) with respect to the reference criterion C_r is then computed using Equation (4).

$$w_{cr} = w_c \tag{4}$$

where, w_r is the weight of the reference criterion. Usually, the reference criterion is selected as that criterion having the maximum weight.

- **Step 4:** Next, the dominance degree is calculated. The dominance degree shows the dominance of alternative A_i over alternative A_j and is calculated employing the following equation:

$$\delta(A_i, A_j) = \sum_{c=1}^{m} \varphi_c(A_i, A_j) \qquad \forall(i,j) \tag{5}$$

In the above equation, the dominance degree of alternative A_i over alternative A_j, i.e. $\phi_c(A_i,A_j)$, concerning criterion C_c ($c = 1,2,\ldots,m$), is evaluated using Equation (6).

$$\varphi_c(A_i, A_j) = \begin{cases} \sqrt{\dfrac{w_{cr}(P_{ic} - P_{jc})}{\sum_{c=1}^{m} w_{cr}}} & \text{if } (P_{ic} - P_{jc}) > 0 \\ 0 & \text{if } (P_{ic} - P_{jc}) = 0 \\ \dfrac{-1}{\theta}\sqrt{\dfrac{(\sum_{c=1}^{m} w_{cr})(P_{ic} - P_{jc})}{w_{cr}}} & \text{if } (P_{ic} - P_{jc}) < 0 \end{cases} \tag{6}$$

where, $(P_{ic} - P_{jc}) > 0$ denotes the gain and $(P_{ic} - P_{jc}) < 0$ represents the loss. The gain and loss is calculated for the ith alternative over j^{th} alternative. θ is the attenuation factor of the losses.

- **Step 5:** The overall dominance degree of alternative A_i (ζ_i) is determined applying Equation (7).

$$\xi_i = \frac{\sum_{j=1}^{n} \delta(A_i, A_j) - \min \sum_{j=1}^{n} \delta(A_i, A_j)}{\max \sum_{j=1}^{n} \delta(A_i, A_j) - \min \sum_{j=1}^{n} \delta(A_i, A_j)} \tag{7}$$

- **Step 6:** The alternatives are finally ranked based on the descending order of their dominance scores and the alternative having the maximum dominance score becomes obviously the best choice.

Material and Factors to be Considered for Material Selection for the Polymeric Composite Bumper Beam

The main motive of the automotive designers is to reduce the weight of the vehicle. Materials play a very vital role in achieving this task. The component manufactured using a particular component should also meet the functionality and the performance requirements. Composites have replaced steel and aluminum for the manufacturing of the bumper beam. Many technical reports have been published on the application of polymeric composites for the automotive bumper beam. The use of structural reinforced injection moulding composite for the bumper beam was reported by Mohan (Mohan, 1987). Clark et al. (Clark, Bals and Layson, 1991) made use of glass fibre reinforced composites and analyzed the performance using ABAQUS software. Glass fibre epoxy composites were used for developing the bumper beam by Cheon et al. (Cheon, Choi and Lee, 1995). A report on the analysis of performance for the bumper beam made using glass-mat thermoplastics were presented by Kumar and Johnston (Kumar and Jhonston, 1998). Kelman and Nelson (Kelman and Nelson, 1998) also have reported the use of composite for the bumper beam.

The materials considered for the bumper beam material selection have been tabulated in the table 1 below.

The criteria under which the selection of material for the polymeric composite bumper beam was made: Impact toughness (J/cm), flexural strength (MPa), flexural modulus (GPa), Cost (USD/Kg), Density (kg/m^3), water absorption (%) and recyclability. Impact toughness, flexural strength, flexural modulus, and recyclability are beneficial criteria (higher the better), whereas cost, density and water absorption are non-beneficial criteria (lower the better).

Implementation of TODIM Method

The different steps that were followed while making the material selection for the polymeric composite beam:

- **Step 1:** The decision matrix is shown in the table 2 below
- **Step 2:** The normalized matrix is shown in the table 3 below:
- **Step 3:** The weights for different criteria were obtained using Shannon's Entropy method and have been tabulated in the table 4 below. Further, the recyclability has the highest criteria weight and is selected to be the reference criterion. The relative weights (w_{cr}) (c=1,2..7) for the different criteria as obtained have also been tabulated in the table 4.

Table 1. Candidate materials for the material selection of the bumper beam (Hambali et al., 2010)

S. No.	Composite Materials
A1	Glass fibre reinforced epoxy
A2	Carbon fibre reinforced epoxy
A3	Carbon fibre reinforced polypropylene
A4	Glass fibre reinforced polypropylene
A5	Glass fibre reinforced polyester
A6	Glass fibre vinylester (SMC)

Table 2. Decision matrix for the polymeric composite beam material selection problem (Hambali et al., 2010)

Alternative	Criteria						
	Impact Toughness (J/cm) (C_1)	Flexural Strength (MPa) (C_2)	Flexural Modulus (GPa) (C3)	Cost (USD/Kg) (C_4)	Density (Kg/m³) (C_5)	Water Absorption (%) (C6)	Recyclability (C7)
Glass fibre reinforced epoxy (A1)	21.20	483.0	20.7	4	1400	0.10	0
Carbon fibre reinforced epoxy (A2)	10.60	656.0	34.5	6	1600	0.10	0
Carbon fibre reinforced polypropylene (A3)	3.20	75.8	13.8	5	1110	0.01	1
Glass fibre reinforced polypropylene (A4)	7.52	294	11.4	1	1560	0.07	1
Glass fibre reinforced polyester (A5)	8.54	179	11.0	2	1850	0.25	1
Glass fibre vinylester (SMC) (A6)	12.80	427	17.0	3	1900	0.05	0

Note 1: cost of the material is from high (6) to low (1)
Note 2: For no recyclability it is 0 and for recyclability it is 1.

Table 3. Normalized decision matrix for polymeric composite bumper beam

Material	C_1	C_2	C_3	C_4	C_5	C_6	C_7
A_1	0.332	0.228	0.189	0.102	0.181	0.063	0
A_2	0.166	0.311	0.316	0.068	0.158	0.063	0
A_3	0.051	0.036	0.126	0.082	0.228	0.632	0.33
A_4	0.118	0.139	0.104	0.408	0.163	0.090	0.33
A_5	0.134	0.085	0.101	0.204	0.137	0.025	0.33
A_6	0.201	0.202	0.156	0.136	0.134	0.126	0

Table 4. Weights and relative weights for different criteria

Criteria	Weights (w_c)	Relative Weights (w_{cr})
C_1	0.096	0.337
C_2	0.107	0.375
C_3	0.084	0.295
C_4	0.123	0.432
C_5	0.058	0.204
C_6	0.248	0.871
C_7	0.285	1

- **Step 4:** Equation 6 is used for the calculation of dominance degree. The dominance degree shows the dominance of a particular alternative over the other alternatives. The dominance degree for alternative material A1 over other alternatives has been tabulated in the table 5 below. The value of attenuation factor θ has been kept 1.
- **Step 5:** The overall dominance degree of alternative A_i (ζ_i) is determined using Equation 7. Table 6 shows the overall dominance degree for different alternatives.

Finite Element Analysis

Finite element analysis (FEA) tools are often used by the design engineers to analyze how a particular material behaves when subjected to a specific loading condition. Whenever a product underperforms, the material used for its design and development is replaced with a superior one to help the product to perform its intended function for the entire service life. The FEA divides product geometry into various sub-domains or number of elements.

Table 5. Dominance degrees of the first material alternative over the others with respect to each criterion

Material	C_1	C_2	C_3	C_4	C_5	C_6	C_7	$\delta(A_i,A_j)$
A_2	0.126	-0.882	-1.23	0.065	0.036	0	0	-1.885
A_3	0.164	0.143	0.073	0.049	0.899	-1.515	-1.07	-1.257
A_4	0.143	0.097	0.085	-1.58	0.032	-0.33	-1.07	-2.623
A_5	0.138	0.123	0.086	-0.83	0.051	0.097	-1.07	-1.415
A_6	0.112	0.052	0.053	-0.526	0.052	-0.564	0	-0.761

Table 6. Overall dominance degrees of the material alternatives

Material	$\sum_{j=1}^{n}\delta(A_i, A_j)$	ζ_i
A_1	-7.941	-0.275
A_2	-11.545	-0.714
A_3	-14.303	-0.867
A_4	-9.801	-0.517
A_5	-16.547	-1.151
A_6	-13.571	-1.030

The differential equation governing the geometry is then solved for each of these elements. A large system of equations is thus formulated by compounding the differential equations of various elements. A boundary condition is also imposed into the equation and the entire system is analyzed to obtain a feasible solution. The number of elements in the geometry usually determines the accuracy of the solution. Furthermore, the interpolation functions adopted in the analysis determines the precision of the derived results. One of the most commonly used FEA tool is ANSYS. In this paper, ANSYS (Version 14.0) is applied to predict the performance of the bumper beam for different candidate materials under specific loading conditions.

In order to analyze the product geometry, the corresponding 3D model of the bumper beam is first developed in the design modeler of ANSYS, as shown in Figure 1. However, the designer can export the external CAD file in either IGES or STEP format from any of the 3D modeling software. For FEA, defining the boundary condition is one of the important elements for the analysis. The fixed support for the bumper beam, as shown in Figure 2, is treated as the boundary condition.

The bumper beam under analysis is assumed to be under load of 0.126 N/mm^2 and it is chosen as the reference load for comparing the Von-Mises stress developed in the bumper beam. The loading condition is exhibited in Figure 3. The bumper beam is fixed at the upper end, i.e., the displacement is considered to be zero for the upper end of the beam. The loading and the displacement together condition comprise of the boundary conditions. In order to analyze the system, the geometry of the bumper beam is meshed. The mesh size depends on the relevance center, transition, smoothing and span angle center. For the present analysis, the relevance center is set to fine, transition to slow, smoothing to high and span angle center to coarse. The element considered is Beam type element for the analysis of the polymeric bumper beam. The entire geometry is discretized into a number of beam elements. A mesh convergence test was conducted to select the best mesh size and it was found to be

Figure 1. 3D model of bumper beam

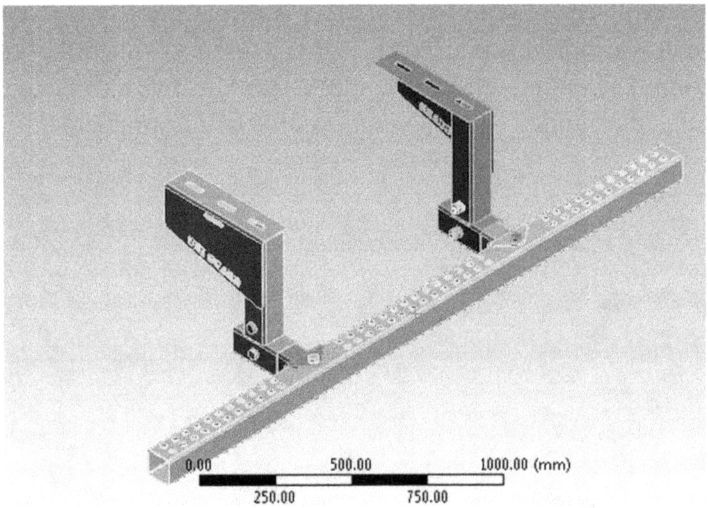

Figure 2. The fixed support for the bumper beam

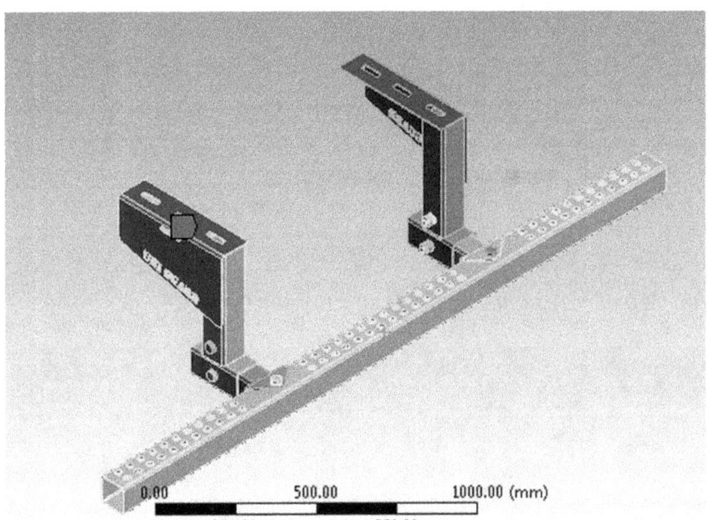

2.3 mm. The bumper beam is analyzed for the distribution of the Von-Mises stress for the candidate materials. Figure 4 shows the distribution of the von-Mises stress for glass fibre reinforced epoxy. The contour plot in the figure shows the variation of the Von-Mises stress in the bumper beam. The maximum stress corresponds to the red color of the plot, whereas the minimum stress is denoted by the bluish shade with least color shade.

The commercial FEA solver i.e., ANSYS thus predicts the stress to which the bumper beam will be subjected when under maximum loading condition. The material of construction which suffices the maximum stress will be the best material of construction. Therefore, ANSYS provides a cost-effective method to predict the performance of the bumper with different material of construction.

Figure 3. Loading condition on bumper beam

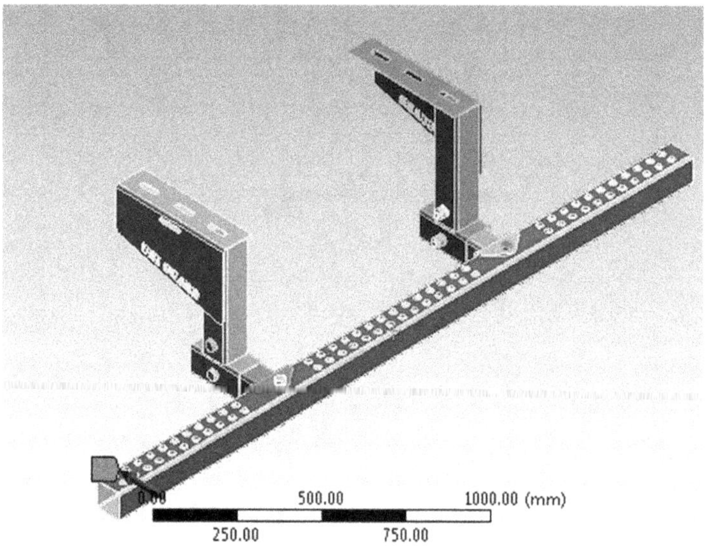

Figure 4. Von-Mises stress distribution of bumper beam

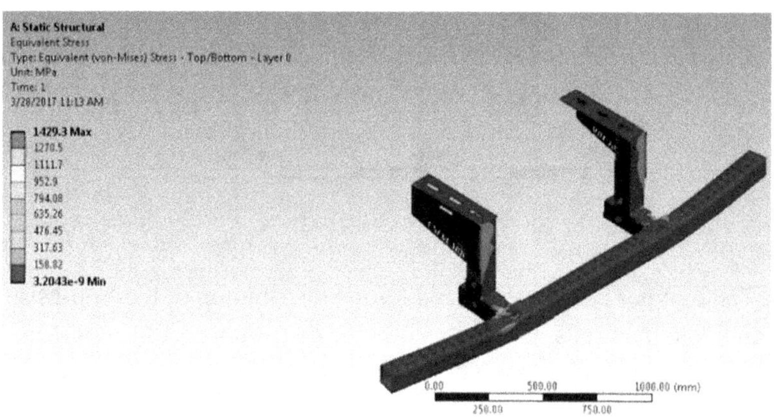

**For a more accurate representation see the electronic version.*

SOLUTIONS AND RECOMMENDATIONS

On the basis of the dominance degree, alternative material A_1 i.e., glass fibre reinforced epoxy with the highest dominance degree turns out to be the best candidate material for the polymeric composite beam. This also corroborates with the ranking established by Hambali et al. (Hambali et al., 2010) it was also seen that for different values of the attenuation factor the alternative material A_1 remains to be the best, although there was a slight variation in the intermediate rankings.

Further the FEM results also prove the robustness of the TODIM method. This is obvious from Table 7 which clearly shows that glass fibre reinforced epoxy can be selected as the most appropriate material for the polymeric composite bumper beam.

FUTURE RESEARCH DIRECTIONS

Bumper beam is one of the most important components of an automobile. It is the first of all the components to bear the impact in case of vehicular collision and therefore is required to absorb most of the impact energy. The bumper beam therefore decreases the amount of impact energy transferred to other components of the vehicle. For the bumper beam to serve its intended function, a suitable material for its design is of utmost importance. Till date, knowledge base system has been proposed for the same. The authors in this work have presented another MCDM tool in the form of TODIM for the material selection of polymeric composite beam. In the near future, other tools could be developed for the material selection of the bumper beam which could also consider the recently developed materials for the bumper beam. The material selection for bumper beam under fuzzy environment is another future research direction.

Table 7. Number of elements for convergence and the von-Mises stress

Material	Number of Elements	von-Mises Stress (MPa)
Glass fibre reinforced epoxy (A1)	948663	1429.3
Carbon fibre reinforced epoxy (A2)	1039841	1320.8
Carbon fibre reinforced polypropylene (A3)	1217211	1293.2
Glass fibre reinforced polypropylene (A4)	1457293	1218.5
Glass fibre reinforced polyester (A5)	1395425	1177.9

CONCLUSION

The present work has presented a relatively simpler tool in the form of TODIM for the selection of material for the polymeric composite beam. The results obtained using TODIM corroborates with the past researchers. FEM analysis also proves the robustness of the method. According to the TODIM method and the FEM analysis, Glass fibre reinforced epoxy turns out to be the best material amongst the considered alternative materials. However, with the development of new composite materials for the bumper beam, the new materials should also be considered in the near future while making selection of material for the bumper beam.

REFERENCES

Belingardi, G., Beyene, A. T., Koricho, E. G., & Martorana, B. (2015). Alternative lightweight materials and component manufacturing technologies for vehicle frontal bumper beam. *Composite Structures*, *120*, 483–495. doi:10.1016/j.compstruct.2014.10.007

Beyenea, A. T., Korichob, E. G., Belingardib, G., & Martoranac, B. (2014). Design and manufacturing issues in the development of lightweight solution for a vehicle frontal bumper. *Procedia Engineering*, *88*, 77–84. doi:10.1016/j.proeng.2014.11.129

Cheon, S., Choi, J. H., & Lee, D. G. (1995). Development of the composite bumper beam for passenger cars. *Composite Structures*, *32*(1-4), 491–499. doi:10.1016/0263-8223(95)00078-X

Clark, G. L., Bals, C. K., & Layson, M. A. (1991). Effects of fibre and property orientation on 'C' shaped cross sections. *SAE Technical Paper*, 910049.

Davoodi, M. M., Sapuan, S. M., Ahmad, D., Ali, A., Khalina, A., & Jonoobi, M. (2010). Mechanical properties of hybrid kenaf/glass reinforced epoxy composite for passenger car bumper beam. *Materials & Design*, *31*(10), 4927–4932. doi:10.1016/j.matdes.2010.05.021

Davoodi, M. M., Sapuan, S. M., Aidy, A., Osman, N. A. A., Oshkour, A. A., & Wan Abas, W. A. B. W. (2012). Development process of new bumper beam for passenger car: A review. *Materials & Design*, *40*, 304–313. doi:10.1016/j.matdes.2012.03.060

Davoodi, M. M., Sapuan, S. M., & Yunus, R. (2008). Conceptual design of a polymer composite automotive bumper energy absorber. *J Mater Des.*, *29*(7), 1447–1452. doi:10.1016/j.matdes.2007.07.011

Edwards, K. (2004). Strategic substitution of new materials for old: Applications in automotive product development. *J Mater Des*, *25*(6), 529–533. doi:10.1016/j.matdes.2003.12.008

Gomes, L. F. A. M., & Lima, M. M. P. P. (1992). *Foundations of Computing and Decision Sciences*. Academic Press.

Hambali, A., Sapuan, S. M., Ismail, N., & Nukman, Y. (2010). Material selection of polymeric composite automotive bumper beam using analytical hierarchy process. *J. Cent. South Univ. Technol.*, *17*(2), 244–256. doi:10.1007/s11771-010-0038-y

Hosseinzadeh, R., Shokrieh, M. M., & Lessard, L. B. (2005). Parametric study of automotive composite bumper beams subjected to low-velocity impacts. *Journal of Composite Structures.*, *68*(4), 419–427. doi:10.1016/j.compstruct.2004.04.008

Kelman, J., & Nelson, G. V. (1998). Composite motor vehicle bumper beam. Academic Press.

Kim, D. H., Kim, H. G., & Kim, H. K. (2015). Design optimization and manufacture of hybrid glass/carbon fiber reinforced composite bumper beam for automobile vehicle. *Composite Structures*, *131*, 742–752. doi:10.1016/j.compstruct.2015.06.028

Kumar A, Johnston C (1998). Comparative performance of a compression: Molded I-section bumper beam with integrated mounting stays versus other GMT bumper design. *SAE Technical Papers,* 980111.

Liu, C. H., Huang, Y. C., Chiu, C. H., Lai, Y. C., & Pai, T. Y. (2016). Design and Analysis of Automotive Bumper Covers in Transient Loading Conditions. *Key Engineering Materials*, *715*, 174–179. doi:10.4028/www.scientific.net/KEM.715.174

Liu, Z., Lu, J., & Zhu, P. (2016). Lightweight design of automotive composite bumper system using modified particle swarm optimizer. *Composite Structures*, *140*, 630–643. doi:10.1016/j.compstruct.2015.12.031

Mallick, P. K. (2008). *Fiber-reinforced composites: Materials, manufacturing, and design*. Boca Raton, FL: CRC Press.

Mohan, R. (1987). Composites for automotive structural applications. *Proceedings of the Third Annual Conference on Advanced Composites,* 57–62.

Nasiruddin, S. M., Hambali, A., & Rosidah, J. (2017). Energy absorption analysis of different cross section bumper beam using finite element analysis. *Proceedings of Innovative Research and Industrial Dialogue*, *2016*(1), 97–98.

Sapuan, S., Maleque, M., Hameedullah, M., Suddin, M., & Ismail, N. (2005). A note on the conceptual design of polymeric composite automotive bumper system. *Journal of Materials Processing Technology*, *159*(2), 145–151. doi:10.1016/j.jmatprotec.2004.01.063

Suddin, M. N., Harun, M. R., Hamzah, A. T., & Anuar, S. (2007). A prototype of KBS for material selection in bumper beam design. *Suranaree Journal Science Technology.*, *14*, 215–222.

Xiao, Z., Fang, J., Sun, G., & Li, Q. (2015). Crashworthiness design for functionally graded foam-filled bumper beam. *Advances in Engineering Software*, *85*, 81–95. doi:10.1016/j.advengsoft.2015.03.005

Zeng, F., Xie, H., Liu, Q., Li, F., & Tan, W. (2016). Design and optimization of a new composite bumper beam in high-speed frontal crashes. *Structural and Multidisciplinary Optimization*, *53*(1), 115–122. doi:10.1007/s00158-015-1312-2

Zindani, D., Maity, S. R., Bhowmik, S., & Chakraborty, S. (2017). A material selection approach using the TODIM (TOmada de Decisao Interativa Multicriterio) method and its analysis. *International Journal of Materials Research*, *108*(5), 345–354. doi:10.3139/146.111489

Chapter 7
Textile-Reinforced Composites for the Automotive Industry

Hande Sezgin
Istanbul Technical University, Turkey

Omer Berk Berkalp
Istanbul Technical University, Turkey

ABSTRACT

In the last decades, automotive industry has especially focused on developing and applying new materials and technologies for enhancing the comfort and security levels in the vehicles, but on the other hand for reducing the weight of the means of transportation in order to decrease the energy consumption. In this concept, textile-reinforced composite structures come in possession of one of the most favored materials in the automotive industry by satisfying these demands. In this chapter, usage of textile-reinforced composites in the automotive industry has been elucidated under three main sections: textile reinforced composites for 1) automobiles, 2) mass transportation vehicles, and 3) trucks. The aim of this chapter is to discuss the subject in detail by giving technical information about particular vehicle parts and composite structures utilized in the automotive industry and academia.

DOI: 10.4018/978-1-5225-5216-1.ch007

INTRODUCTION

The composites are gathering of two or more different materials that enable to acquire a material whose properties are greater than that of the constituent components (Sivakandhan & Prabhu, 2014). Composite structures can be classified into three categories according to their matrix system. These are; metal matrix composites, ceramic matrix composites and polymer matrix composites (Tucker & Lindsey, 2002). Polymer matrix composites are mostly reinforced with textile structures. These structures can be in different forms such as; discontinuous chopped fibers, continuous filament yarns, simple fabrics (2D) or advanced fabrics (3-D) (Misnon, Islam, Epaarachchi, & Lau, 2013). The first example of a fiber reinforced composite was developed after World War II by embedding glass fibers in polymeric resin for petrochemical industry (Bakis et al., 2002).

By using low-cost, light-weight polymers with high strength and high modulus fibers provides good combination of mechanical and technological properties to composites and these properties enables them to find numerous application areas in many sectors such as; automotive, marine, aerospace industry and so on (Bakis et al., 2002; Bindal, Singh, Batra, & Khanna, 2013; Murugan, Ramesh, Padmanabhan, Jeyaraam, & Krishna, 2014; Sezgin & Berkalp, 2016; Shalin, 2012).

Figure 1 shows the relative effect of matrix and reinforcement material on several properties of the composites. It is seen that while the properties such as corrosion, temperature and chemical resistance of composite materials are directly related to the properties of the matrix material, strength and stiffness of the composite is mostly associated with the reinforcement material. However, most of the mechanical properties of the composites are governed by both the reinforcement and the matrix material. By raising the amount of work that is needed to fracture the composite, the synergy of the matrix and the reinforcement material strengthen and toughen the composite material (Friedrich & Almajid, 2013).

The growth of the automotive industry has begun by means of the economic growth that occurred in the early 20th century (Kamath, Bhat, Parikh, & Mueller, 2005). Ever-rising population in the world is the key factor that drives the increment of the global automotive production year by year. The number of passenger cars and commercial vehicles in use worldwide is given in Table 1 (Statista, 2017). On the other hand, while the number of passenger cars and commercial vehicles that were manufactured in the year 2000 was 41.22 and 17.17 million respectively, it has increased to 72.11 and 22.53 million in 2016 (Statista, 2016).

This great expansion in the automotive industry has led the engineers to hunt for new materials (Kamath et al., 2005). All vehicles are subjected to loads which result in stress, vibration and noise and the components of a vehicle should possess required properties to withstand these loads (Harshit, Kumar, & Verma, 2016). In the

Figure 1. Relative effect of matrix and reinforcement material on properties of composites

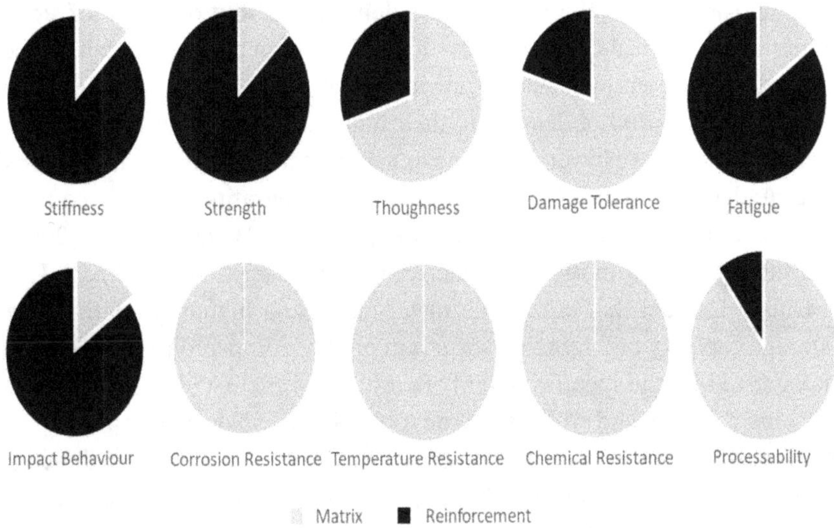

Table 1. Number of vehicles in use worldwide (Statista, 2017)

	Commercial Vehicles (Million)	Passenger cars (Million)
2006	247.488	679.154
2007	258.538	701.690
2008	264.970	727.405
2009	271.957	747.899
2010	280.127	775.573
2011	288.618	808.401
2012	306.721	834.916
2013	315.845	869.083
2014	327.825	907.062
2015	335.190	947.080

automotive industry, composite materials are taking place of the metal automotive parts since the application of the early composite structures to a 1953 Corvette car (Das, 2001; Shokrieh & Rezaei, 2003). The most important factors that an automotive composite structure must be captured, can be defined by three R's, these are; rapid, robust and reasonable cost (Chapman, 1999). Among all metals, steel and aluminum are the leading automotive materials due to their cheapness, strongness and stiffness (Brylawski & Lovins, 1998). When compared with steel or aluminum, polymer matrix composite structures mostly have a density of about one-fifth of steel and half of aluminum, and additionally, have a yield strength of more than ten times of them. Besides, these composites ensure improved noise, vibration and harshness performance to the automobiles due to their high viscoelastic damping properties. The composite parts have lower tooling costs and they are more resistant to corrosion. Moreover, under favor of their high potential for part integration, polymer composites have less manufacturing costs (Das, 2001; Lee, Lee, Lee, Hwang, & Kim, 2004; Lu, Pilla, & Society of Automotive Engineers, 2014; Sedlacek, Bernardin, & Lasova, 2016). Due to the fact that composite materials have a higher strength-to-weight ratio, storage capacity and more elastic strain energy compared to steel, composite materials can reduce the specific weight of a tool of a vehicle without a decrement on load carrying capacity and stiffness of the tool (Mehul, Shah, & Bhojawala, 2012).

Automotive industry manufacturers are exposed to strict government directives to reduce the emissions by decreasing the amount of fuel consumption according to the Kyoto Protocol. The easiest and the sanest thing to meet the emission requirements is reducing the weight of the vehicles. It is proved that fuel consumption can be lessened by 8.4% by reducing the weight of the vehicle 20% (Koncherry, Potluri, & Fernando, 2017). Moreover, it is known that due to the essential balanced ratio of front-rear weight distribution, light weighting is more important at the front parts of the vehicles (Akhshik, Panthapulakkal, Tjong, & Sain, 2017). Vehicle weight reduction can be obtained by selecting better material, design optimization and manufacturing techniques (Mehul et al., 2012).

In the automotive sector, glass and carbon fibers are the most common reinforcement materials for polymer matrix composites. It is considered that instead of metals, by using glass reinforced composites a 20-35% reduction can be achieved in the weight of the vehicle, while this ratio increases up to 40-65% when carbon reinforced composites are used (Das, 2001; Lu et al., 2014).

Carbon fiber has gained importance in the automotive industry due to its advantageous properties like higher specific stiffness, specific strength and damping capability compared to metallic materials (D. H. Kim, Choi, & Kim, 2014). Carbon fiber is mostly preferred where lightness and high stiffness are required (Pandya, Veerraju, & Naik, 2011). However, carbon fiber is 15 times more expensive than steel which is the most preferred material in the automotive industry (Koncherry et al., 2017). E-glass fiber reinforced composites are used in the automotive industry mostly to reduce the weight of the vehicles. E-glass fiber is one of the most favored reinforcement materials in the composite industry, thanks to its properties such as high tensile strength, high chemical resistance, and high dimensional stability (Bhagat, Biswas, & Dehury, 2014). The strength of any material is determined by the presence of flaws of critical dimension. E-glass fiber is a convenient reinforcement material that increases the surface to volume ratio by reducing the number of flaws of critical dimension and thus leading to increased strength (Hearle, 2001). However, their potential health hazards, poor recycling properties and relatively higher fiber densities put a question mark in the minds (Hu, Hu, & Yang, 2016).

There is an ever-increasing consciousness about environmental protection in the world, which is nowadays forcing the automotive industry to use more eco-friendly natural fibers in the vehicles. Utilization of natural fibers offers many advantages and superior properties for the industry such as low cost and density, high strength and modulus, renewability and low heat conductivity. Natural fiber reinforced composites are mostly used as dashboards, seat cushions, cabin linings, door panels and backrests in the automotive industry (AL-Oqla, Sapuan, Ishak, & Nuraini, 2016; Ayrilmis, Jarusombuti, Fueangvivat, Bauchongkol, & White, 2011; Hu et al., 2012, 2016). By means of its good acoustic properties, nontoxicity, high hardness, hard-wearing quality, and resistant to moisture and microbial and fungi degradation, coir fiber is one of the most preferred natural fibers in the automotive industry (Ayrilmis et al., 2011).

Composite structures can be manufactured by many different production methods such as hand lay-up, spray-up, injection moulding, resin injection moulding, filament winding, compression moulding, sheet moulding, bulk moulding, vacuum bagging and autoclave bonding according to their desired properties and area of utilization (Astrom, 1997). Resin transfer molding, injection molding, pultrusion and compression molding methods are the most favored composite manufacturing methods for automotive composite parts (Lee et al., 2004). While injection moulding technique is mostly preferred for low-weight reinforcements, short fiber reinforcements and glass felt reinforcements, compression moulding is mostly used at production of long fiber reinforced composites (Turkish Composites Manufacturers Association, 2014).

The automotive industry has an important duty not only in the transportation of individual people but also for the transportation of groups of people and goods (Kamath et al., 2005). According to Encyclopedia Britannica, automotive industry has principal and secondary products. While passenger automobiles and light trucks (pickups, vans and sport utility vehicles) are defined as primary products, commercial vehicles are defined as secondary products (Binder & Rae, n.d.). In this chapter textile reinforced composites that are used in automotive industry will be analyzed under three main headings: textile reinforced composites for automobiles, textile reinforced composites for mass transportation vehicles and textile reinforced composites for trucks.

FIELDS OF APPLICATION

Textile Reinforced Composites for Automobiles

Composites are utilized in many parts of an automobile structure. Together with body panels, they are used in bumper beams, leaf springs, wheel covers, drive shafts, valve covers and compressed natural gas fuel tanks (Das, 2001). The chassis frame is the fundamental assembly of an automobile and components of an automobile such as; power plant, clutch gearbox, propeller shaft, rear axle, wheels, suspension and controlling system are all mounted on the chassis frame (Figure 2). It forms the backbone of a vehicle. Chassis frame holds the vehicle rigid, stiff and provides less noise, vibration and harshness (Begum & Murthy, 2016).

Textile reinforced composite automobile parts will be examined in three main parts: body structures, powertrain components and suspension components.

Body Structures

Composite structures are generally preferred in the body panels of automobiles by means of giving high-quality surface finish, options of processing and styling details (Vivekanandan & Prakash, 2014). Moreover, car body light weighting and crashworthiness are the two important objectives of car body design and improvement (Li, Lin, Jiang, & Chen, 2003). Due to the fact that, car body structures (floor parts, roof, hood, deck lid and etc.) constitute a great amount of a vehicle mass, weight reduction becomes an important factor for body structures (Cole & Sherman, 1995;

Figure 2. Chassis frame of a vehicle ("Car chassis frame | shutterstock.com," n.d.)

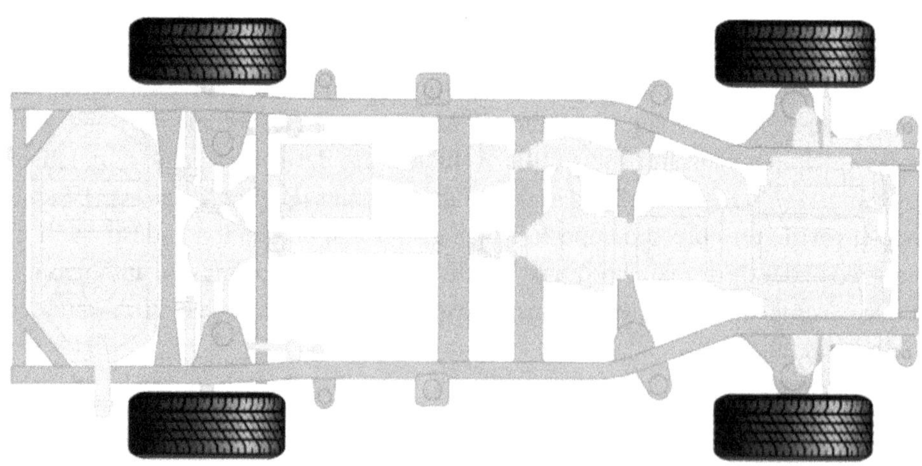

Meschut, Janzen, & Olfermann, 2014). Vehicle weight is one of the most important factors, which affects the fuel consumption of the vehicle. Roughly, 75% of the fuel consumption directly associated with the weight of the vehicle and approximately 13% fuel saving could be achieved with a 20% decrement of the weight (Ning, Pillay, & Vaidya, 2009).

Crashworthiness, which is the response of the vehicle when it undergoes an impact, is a critical factor for the passengers' safety. Bumper beam is one of the most important parts of the automobile which protects the passenger and frontal part of the car while an impact is applied (Muhammad Nasiruddin, Hambali, Rosidah, Widodo, & Ahmad, 2017). Polymeric materials are mostly used in bumper beams to fulfill the required performance and the weight (Sapuan, Maleque, Hameedullah, Suddin, & Ismail, 2005).

In the literature, there are a large number of studies about composite body structures. In one of these studies, glass reinforced composite roof panels were manufactured with various volume fractions of resin/fiber (50/50, 55/45 and 60/40) for TATA SUMO vehicle. Epoxy was preferred as the matrix material and composite structures were produced by hand lay-up process with the help of a wooden die. It was indicated that highest impact strength (25.6 N/mm^2) was achieved with the sample which has a resin/fiber volume fraction of 55/45 and it was decided to use this sample for manufacturing the roof panel (Vivekanandan & Prakash, 2014). Liu et al. (2013) developed a carbon fabric reinforced composite body structure for an electric vehicle using a multiscale approach. Material properties were predicted by finite element based on the homogenization method and compared with the uniaxial tensile and three point bending test results. It was observed that a weight saving of 28% and a highly increased amount of crashworthiness were obtained with the carbon fabric reinforced sample compared to its predecessor made of glass fiber reinforced plastic. In another study, Duan et al. (2017) manufactured a long glass fiber reinforced polymer composite bumper beam and investigated its energy absorbing property and weight reduction. The experimental results revealed that specific energy absorption of the composite bumper beam was considerably higher than those of conventional bumper beams. Besides, a 51-58% weight reduction was observed by using long glass fiber reinforced polymer composite. Zeng et al. (2016) applied Fruit Fly Optimization Algorithm to the frontal bumper beam of a car to enhance its energy absorption ability and to reduce the weight of the bumper beam. The design optimization of the bumper beam was done by using a composite structure. It was concluded that the weight of the bumper beam was reduced from 4.813 kg to 4.58 kg while the value of head injury criterion was reduced by 6.37% compared to steel one. In addition, it was noted that composite bumper beam has an advanced efficiency in collision energy absorption.

Natural fibers are also used as composite automotive parts. Daimler Chrysler developed a banana (abaca) fiber reinforced composite structure for under-floor protection of passenger cars. Rieter Automotive undertook the compression molding manufacturing process. It was stated that banana is the first natural fiber which provides the required properties of components used on the exterior of the road vehicles. (Samal, Mohanty, & Nayak, 2009). Ayrilmis et al. (2011) studied the physical, mechanical and flammability properties of coir fiber reinforced composites that were produced for automotive interior applications (door trim and cabin linings). They used polypropylene as a matrix material and four different amounts of coir fiber (40, 50, 60, and 70% wt.) as a reinforcement material. They also mixed polypropylene powder and maleic anhydride grafted polypropylene to the blend. Results showed that, while flexural and tensile strength of composite structures increased at a rate of 26% and 35% by reinforcing polypropylene matrix with 60 wt % coir fiber, the modulus of elasticity, Janka hardness, and the flame retardancy of samples increased at a rate of 32%, 69% and 7% by reinforcing polypropylene matrix with 70% coir fiber, respectively. They concluded that the optimal configuration of a composite structure for automotive interior applications that was achieved in this study was the sample which was reinforced by 60% coir fiber, 37% polypropylene powder and 3% maleic anhydride grafted polypropylene. It was also emphasized that coir fiber could be used instead of costlier and heavier glass fiber at composite manufacturing. In one of the studies, Hu et al. (2016) manufactured three different composites (glass/polypropylene, jute/polypropylene and jute/polylactide) to use in the car interior. The aspects and microstructures (loose and porous microstructures) of all three composite materials found similar. The tensile strength results indicated that strengths of jute reinforced composite samples were a bit higher than traditionally used glass reinforced sample.

Powertrain Components

The main element of a powertrain component is a drive shaft. The mission of an automotive propeller shaft or drive shaft is to transmit the power from the engine to the differential gears of the rear wheel drive in the vehicles (Kumar, Jain, & Patil, 2016; Lee et al., 2004). Shapes of the shafts are mostly solid or hollow tubes (Figure 3). Aluminum and steel are commonly preferred in the manufacturing of drive shafts but there are some limitations about their weight and vibration characteristics. Studies revealed that composite drive shafts could bring a novel solution to overcome these problems. One of the problems is that when manufacturing a steel drive shaft, which has a length of more than 1.5 m, it should be made in two pieces to increase the fundamental natural frequency that is proportional to the square root of the specific modulus. Two-piece drive shafts consist of some universal joints with jaw coupling and this makes drive shafts heavier and non-fuel-efficient. Since composite materials have a higher specific elastic modulus, drive shafts can be manufactured in one piece and this provides higher safety. Moreover, by using composite drive shafts, higher cabin comfort, reduction of wear on drive train components and vibration damping can be achieved (Abu Talib, Ali, Badie, Azida Che Lah, & Golestaneh, 2010; Sivakandhan & Prabhu, 2014). Kevlar, graphite, glass and carbon are the most preferred materials in composite drive shaft manufacturing due to their high specific strength and modulus (Patil, Shinde, & Sawant, 2016).

In 1985, Spicer U-Joint Division of Dana Corporation developed the first composite propeller shaft for the Ford Econoline van models (Abu Talib et al., 2010; Kumar et al., 2016; Lee et al., 2004). In a study, Abu Talib et. al (2010) designed composite drive shafts. Carbon/epoxy, glass/epoxy and a hybrid of these two fabrics were wrapped around aluminum tubes. Fabrics were used in [0°/90°] and [±45°] direction lay-ups. From the results, it was observed that stacking sequence of fabric layers had an effect on buckling strength of composite drive shafts. While the highest buckling strength (2303.1 Nm) was achieved with the sample which has a stacking sequence of [45°/-45°/0°/90°], the lowest (1242 Nm) was achieved with [0°/90°/-45°/45°] sample. Also, it was indicated that the natural frequency of the drive shaft

Figure 3. Powertrain components of a vehicle ("front-wheel drive | engineering | Britannica.com," 2008)

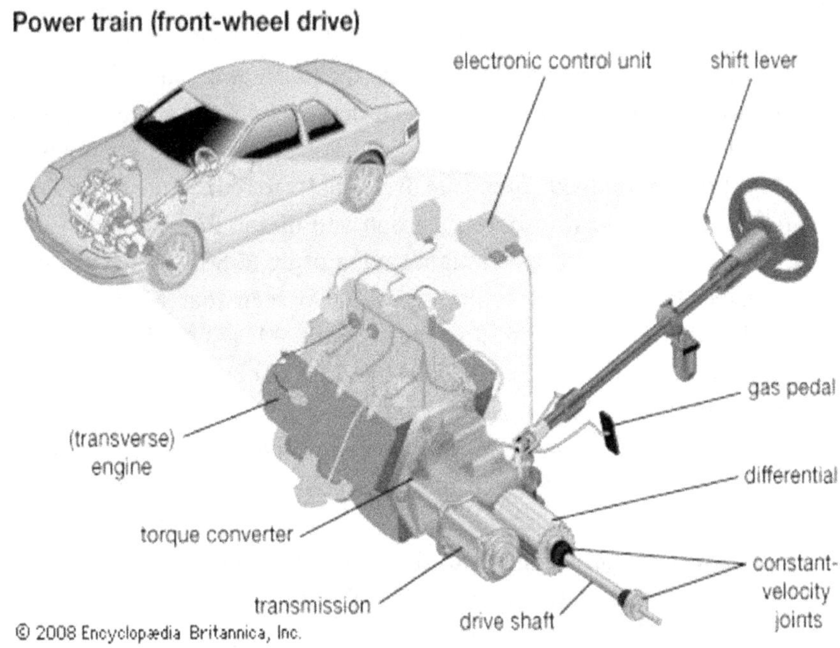

was lessened at the rate of 44.5% by changing carbon fibers winding angle from 0° to 90°. Sivakandhan and Prabhu (2014) developed single-piece composite drive shafts by reinforcing epoxy with E-glass, E-glass/carbon and E-glass/carbon/S-glass with different fiber orientations. According to the mechanical test results, it was observed that tensile properties, yield stress, breaking load and torque transmitting capability of the hybrid composite drive shaft, which was reinforced with E-glass/carbon/S-glass fibers, were superior to those of other composite drive shafts and they concluded that it is a favorable material to replace the metallic drive shafts of light vehicles. In another study, Bhajantri et al. (2014) produced three different (E-glass/epoxy, high strength carbon/epoxy and high modulus carbon/epoxy) single-piece composite drive shafts for automobile shafts. A weight saving of 50% was obtained with carbon fiber reinforced composite and results revealed that orientation of fibers has a huge effect on the composite shafts static characteristics. Babu and Sangeethkumar (2015) fabricated single-piece composite drive shafts for rear wheel drive with glass and carbon fibers and compared their properties with steel drive shafts. In one of the configurations, silicon carbide was also added to glass fiber reinforced composite. It was realized from ANSYS results that, glass fiber reinforced with silicon carbide composite sample has had the best mechanical properties due to the improvement in its morphological property, which was achieved by addition of silicon carbide. Moreover, when the high cost of carbon fiber was taken into consideration, glass fiber reinforced composite with low cost and better mechanical properties, became a promising material for drive shafts. Patil et al. (2016) designed a carbon/epoxy composite drive shaft to substitute steel drive shafts. Results indicated that a weight reduction of 81% was obtained by using carbon/epoxy material instead of steel. In addition, natural bending frequency and critical speed of composite drive shaft were higher than those of steel drive shafts, which demonstrate the lower vibration of the system and prevention of whirling of the shaft, respectively. In another study, Maheta et al. (2015) developed composite drive shafts to reduce the weight of steel drive shafts. They used Kevlar, boron, carbon and aluminum boron to reinforce the epoxy resin. It was noticed that weight saving of up to 80% was obtained with composite materials. From the results, it was also seen that Kevlar reinforced composite drive shaft has the highest strength, while 90° angle of fibers provided the best fundamental frequency. Badie and Hamouda (2011) investigated the effects of stacking sequence and fiber orientation angles on the frequency and mechanical properties of composite drive shafts. They used glass and carbon fibers to reinforce the epoxy resin. Results showed that although carbon/epoxy and E-glass/epoxy samples showed catastrophic failure modes, this was eliminated by developing E-glass/carbon/epoxy hybrid composite drive shafts. Moreover, it was observed that composite drive shafts, which have ±45° fiber orientation angles, had higher torsional stiffness and load carrying capacity.

Suspension Components

The essential objectives of a suspension system are to avoid the road shocks, to protect the passengers from road shocks and to ensure the stability of the vehicle in pitting or rolling while travelling (Lakshmi & Satyanarayana, 2012). Springs and the dampers are the main components of a suspension system (Lakshmi & Satyanarayana, 2012). Springs are used for absorbing, storing and releasing the energy (Al-Qureshi, 2001; Dhoshi, Ingole, & Gulhane, 2011). There are different types of springs. The simplest form of a spring is leaf spring and it is mostly used for suspension in wheeled vehicles (Figure 4(a)). They are long, narrow plates and are joined to the trailer frame (Lakshmi & Satyanarayana, 2012). To permit the suspension movement, leaf springs bend and slide on each other (Mehul et al., 2012). One of the most important parameters of designing a spring is the strain energy of the material. Materials which have lower modulus and density will have a greater strain energy and this makes composites favorable materials for this application (Al-Qureshi, 2001). Composite springs have superior corrosion resistance and their fatigue lives are five times more than steel springs'. Besides, using a composite spring instead of a steel spring could decrease the weight of the suspension system at an amount of 60% and it is known that leaf springs account for 10-20% of the unsprung weight of the vehicle (Ambare & Mishra, 2016; J. P. Hou, Cherruault, Nairne, Jeronimidis, & Mayer, 2007).

While performing their duties, springs oscillate and dampers, which are also known as shock absorbers, restrict these oscillations to an acceptable level (Figure 4(b)) (Lakshmi & Satyanarayana, 2012). The damper is a mechanical apparatus, which decreases the effect of travelling over rough surfaces by absorbing and dissipating the energy. Also, it increases the comfort by reducing the amplitude of disturbance (Chavhan, Burande, & Dhole, 2014). Lower arm, which is located at the front of the passenger compartment, is also one of the suspension units and props a cross member and a knuckle component. A lower arm should have high strength, high stiffness and damping capability because it has issues of reducing the vibration coming from the ground during travelling and providing the reliability of wheel alignment (D. H. Kim et al., 2014).

Studies about composite suspension components are summarized in this section. In one of these studies, Suprith et al. (2013) generated a model for a multi leaf spring using ANSYS software program. Analysis was done for steel, glass fiber reinforced composite and hybrid (steel and composite) composite leaf springs. Results pointed out that stresses in leaf springs were least at the composite leaf spring, while it was highest at steel spring. Although it was found out that glass fiber reinforced composite leaf spring was much more costly than the steel spring, the weight of the composite sample (19.461 kgs) was approximately three times less than the weight of the

Figure 4. (a) leaf spring (Etaphop photo, n.d.) and (b) damper of an automobile ("damper | shutterstock.com," n.d.)

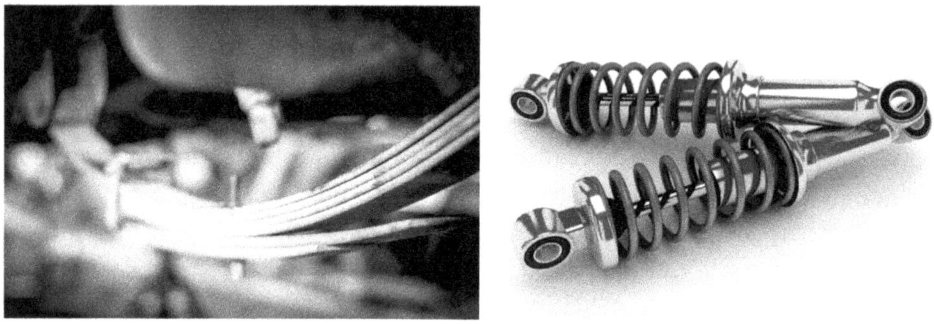

steel sample (58.757). Mehndiratta et al. (2015) fabricated a glass fiber reinforced composite leaf spring and compared its load bearing capacity, stiffness and weight with a steel leaf spring by using theoretical, experimental and numerical analysis methods. Eventually, it was obtained that both leaf springs had approximately same bending stress and deflection. However, 86% lighter and 56% cheaper leaf springs were achieved with composite one. Kim et al. (2014) designed a carbon/epoxy composite lower arm with finite element analysis and designed an optimization to maximize the buckling load capability. They achieved a weight reduction of 50% compared to steel lower arm. A micro-genetic algorithm was used to optimize the stacking sequence of the composite layers. Eventually, they found that failure load and stiffness of composite lower arm (3383 kgf and 70.9 kgf/mm) was approximately two times higher than those of steel one (1500 kgf and 32.75 kgf/mm) with a weight saving of 50%. In an another study, Gopalakrishnan et al. (2017) developed a composite leaf spring by vacuum bagging technique. They reinforced epoxy resin with E-glass, carbon and graphite fibers with different stacking sequences of fabric layers. They obtained that better mechanical properties were achieved with the sample which has a stacking sequence of $[\{0/90/0\}^{gr}\{90/0/90\}^{e\text{-}gl}\{0/90/0\}^{ca}\{90/0/90\}^{gr}\{0/90/0\}^{e\text{-}gl}\{0/90/0\}^{ca}\{90/0/90\}^{gr}]$ and stiffness of the composite leaf spring was increased with the addition of carbon and graphite fibers. Dubey et al. (2016) designed an E-glass/epoxy composite leaf spring as an alternative material to steel leaf spring. The analysis was done by finite element method and results showed that composite leaf spring (79.164 kg) was considerably lighter than steel leaf spring (287.64 kg). Moreover, E-glass/epoxy composite leaf spring had less strain, deflection, stress and higher strength than conventional steel spring. Katake et al. (2016) used glass fiber and carbon fiber to manufacture composite leaf springs for a light-weight vehicle. They developed a glass fiber reinforced composite and a hybrid composite from glass and carbon fiber. Numerical and experimental strength analysis of samples were done and it was seen that hybrid sample has less deformation, less stress and higher strength than glass fiber reinforced and steel leaf springs. Also, weight reduction of about 70% was obtained with composite structures compared to steel one. In a study, Rady (2016) performed a computational analysis of a glass reinforced composite multi leaf spring and a conventional steel spring. Deformation and stress observed in the composite leaf spring were less than those ones of steel spring. Approximately 44% weight saving was also acquired with glass reinforced multi leaf spring. Kassie et al. (2014) designed a single E-glass/epoxy leaf spring for light-weight vehicles to reduce the weight of leaf spring and so the vehicle. They specified that by using composite structure not only a weight reduction of 68.14% was achieved but also sufficient fatigue life and design stresses that satisfy the acceptable limits were obtained. Chavhan et al. (2014) compared the strength capabilities, damping factors and shock absorbing capacities of carbon fiber reinforced composite, spring steel and beryllium copper materials in ANSYS software program to determine the most available material for a shock absorber. It was attained that, although carbon fiber reinforced composite has the lowest breaking strength, it has the greatest shock absorbing property owing to its high deflection (Chavhan et al., 2014)

Textile Reinforced Composites for Mass Transportation Vehicles

With the increment of the private car usage, many problems like traffic congestion, air pollution and demand on non-renewable sources come in sight and these problems lead people to use mass transportation vehicles (Redman, Friman, Gärling, & Hartig, 2013). Composite materials are used in many parts of mass transportation vehicles including bus structures, non-structural interior panels, structural roof panels in high-speed railway coaches and front cabins of high-speed locomotives (Balcombe et al., 2004; Ning, Vaidya, Janowski, & Husman, 2007). Although automotive industry does not include railway vehicles, owing to the fact that railway vehicles comprise a great part of mass transportation, in this chapter, usage of composite materials in mass transportation are classified into two main sections. These are; buses and railway vehicles.

Buses

Buses are the securest transportation vehicle after airplanes. Over the last few years, there has been a public welfare in security advancement for buses and this brings out the use of composite materials in bus structures. Main advantages of using composite materials in buses are weight saving and by means of this low fuel consumption and higher speeds (Castejon, Miravete, & Cuartero, 2006).

Quite a few number of studies were conducted about utilization of composite structures in buses. One of them is Compo-bus, it was developed by Tillotson Pearson Inc in 2006. All parts of this bus were composite materials. Sandwich structures were used with E-glass fiber reinforced vinyl ester skins and balsa core. Compo-bus was 30% lighter and required 60% less power to run than the conventional 9-meter metallic bus. Moreover, it ensured improved sound/vibration damping and fuel efficiency (Ning et al., 2009; "TPI Composites Inc - Home," 2006; Vaidya et al., 2004). Ning et al. (2009) designed a thermoplastic composite air conditioning cover roof door for a mass transit bus. They used thermoplastic polyolefin (outer skin) with lofted glass mat thermoplastic sheets (inner layer). Thermoforming processing technology was used as the manufacturing technique. When the properties of the composite structure compared to a metallic bus door, it occurred that 39% weight saving and enhanced rigidity with 42% reduced free-standing deflection was achieved with the composite door. In another study, Lakshmi and Satyanarayana (2012), designed E-glass, S-class, and C-glass reinforced epoxy composite multi leaf springs for heavy vehicles and compared their load carrying capacity, stiffness and weight savings with mild-steel spring. Firstly, they compared 8 leafs E-glass/epoxy and mild-steel springs and they clarified that E-glass/epoxy had higher yield strength and lower weight than the mild-steel one. Afterward, they compared the 12 leafs springs (C-glass, E-glass, S-glass, mild-steel) and concluded that owing to its good structural stability, good efficiency and low production cost, S-glass reinforced epoxy composite sample was the foremost material to manufacture leaf springs for heavy vehicles (Lakshmi & Satyanarayana, 2012). Ning et al. (2007) manufactured a sandwich composite structure, which has a polypropylene honeycomb core and E-glass reinforced polypropylene composite face sheets, as a mass transit bus side body panel. They used single diaphragm forming process to manufacture composite sheets. The composite panel displayed a 55% weight saving compared to the conventional aluminum bus. In a study, Vaidya et al. (2004) designed and manufactured a hat-sine stiffened glass fabric reinforced polypropylene composite by hot-melt impregnation as a floor segment of a mass transit bus. Pro/Engineer, HyperMesh, and ANSYS 7.0 programs were used for the analysis and finite element analysis used for modeling the design. Generated composite floor segment compared with the traditional steel frame/plywood counterpart and results indicated that 22% and 41% weight savings on a unit cell basis and entire midsection of the floor were achieved, respectively. Furthermore, the weak points of the traditional steel/plywood (corrosion, high cost, and deterioration problems) were eliminated by using composite structure. Thattaiparthasarathy et al. (2008) manufactured a ribbed composite structure (E-glass fiber/polypropylene) as a battery box access door for a mass transit bus to replace the heavy metal door. Extrusion-compression molding process was used as the fabrication technique. From the analysis, it was observed that fiber distribution was uniform and there was not any significant fiber matrix separation throughout the structure. Moreover, 60% weight saving was achieved with E-glass reinforced composite door compared to a metal door. Ning et al. (2007) designed and fabricated a thin-walled carbon fiber reinforced polyphenylene sulfide composite frame segment (sub-element of a mass transit bus). They compared different cross-section configurations (rectangular, V-shape and rounded C-shape) and lamina stacking sequences ($[0/90]_{6,}$ $[\pm45/(0/90)_2]_s$ and $[\pm45/(0/90)]_3$ in their analysis and a rounded C-shape profile with $[0/90]_6$ stacking sequence was chosen for manufacturing. The compression testing results verified the design and the analysis. Primarily tensile dominated fracture behavior was obtained from flexure and low-velocity impact tests and high crystallinity was detected in the composite specimen.

Railway Vehicles

Railway vehicles have many advantages like higher capacity of transportation and energy efficiency compared to other mass transportations. Similar to other transportation vehicles, lightness is also an important factor for trains and it can result in escalated speed, less wear on the rails/wheels, load capacity and comfort of the ride. Also, yield capacities of lighter vehicles increase by providing a higher amount of transportation capacity and by decreasing the destructive effects of an increased service on the railway infrastructure (Jeon, Shin, & Kim, 2011; J. S. Kim, Lee, & Shin, 2007). Owing to their properties like lightness, durability and easily molding to aerodynamic shapes, composite materials have many potential usages in the railway vehicles. Moreover, composite materials can decrease the noise pollution and track force due to their intrinsic damping properties. They are mostly used at rail vehicle cabs, suspension systems, interior components and bogies (J. Hou & Jeronimidis, 2012; J. S. Kim et al., 2007).

One of the main parts of a railway structure is the bogie of the vehicle (Figure 5). It is located between the rail and the wheel and makes up the 37% of the total weight of the vehicle. Main issues of the bogies are to sustain the weight of the car body and passengers, to absorb vibration and to control the wheel sets. A bogie frame, which is mostly made of solid steel or welded structures, makes up the 20% of the bogie weight. It carries heavy static and dynamic loads, braking, accelerating and traction loads. Due to this fact, fatigue characteristics and life of composite structures, which are used as bogie frame, are the main factors that should be verified (Jeon et al., 2011; J. S. Kim & Yoon, 2011).

In one of the studies in the literature, Sedlacek et al. (2016) designed a composite (glass fiber/epoxy) leaf spring for railway vehicles. They created a geometric optimization to find the proper shape for the leaf spring and they obtained a great weight reduction by using a composite material. The weight of the leaf spring

Figure 5. Bogie of a railway vehicle ("Wheels and bogie | shutterstock.com," n.d.)

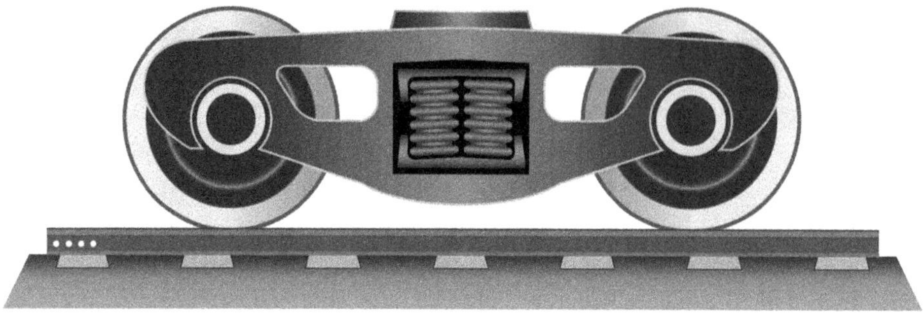

was decreased from 12.6 kg to 1.9 kg. Jeon et al. (2011) developed a composite bogie frame for urban subway trains. Glass fiber and epoxy resin were used as the reinforcement and the matrix, respectively. Fatigue test results showed that the structural integrity was enough to bear the various external loading conditions and it had adequate fatigue strength for a long-term service life of urban railway trains. Additionally, the weight of the composite bogie frame was less than the weight of the conventional metal bogie frame. Hou and Jeronimidis (2012) designed glass fiber reinforced composite bogie frames, predicted their performance by finite element analysis and compared their properties with those of steel bogie frames. Achieved results showed that the weight of the composite bogie frame was 1 tonne less than steel bogie frame and the cost of the bogie frame was reduced by using composite structure. In another study, Kim and Yoon (2011) designed and manufactured a glass fiber reinforced composite bogie frame for urban subway trains by autoclave curing method. Structural safety of the composite bogie frame was assessed by static tests and finite element analysis. It was observed that experimental and numerical analysis had similar results and composite bogie frame was within the safe region according to the Goodman diagram. In addition, the maximum stress took place at the strain gauge which was between the side and cross beam. Kim et al. (2007) manufactured a sandwich composite train car body (stainless steel under the frame and composite body shell) using a specially designed large autoclave. From the results, it was seen that the cutouts for the side windows and the side entrance doors are the stress concentration points of the composite structure. Moreover, the maximum stress of the composite body shell was of 12.2% of the strength of carbon/epoxy and the stiffness of the composite was enough for the necessities. Kim and Jeong (2006) assessed the numerical and experimental natural frequencies of a composite car body of the Korean tilting train. The car body was a sandwich structure which consists of an aluminum honeycomb core and carbon fabric reinforced epoxy composite face sheet. The composite car body had a first bending and torsional frequency of

11.67 Hz and 14.4 Hz from the finite element analysis and 10.25 Hz and 11 Hz from measurement analysis, respectively. Both results met the design requirements and it was concluded that composite car body could avoid the undesirable periodic motions during service.

Textile Reinforced Composites for Trucks

The truck is a heavy motor vehicle which is used for transporting loads (Rahman, Tamin, & Kurdi, 2008). Truck transportation is commonly used for distribution of nondurable packaged and high-value goods worldwide due to its short transport time and availability to reach regions where the railway and sea transports could not reach (Jarimopas, Singh, & Saengnil, 2005; Rissi, Singh, Burgess, & Singh, 2008). By virtue of the fact that, mostly high-value and delicate products are transported by trucks, it is important to compute and quantify the vibration that occurs during travelling (Singh, Singh, & Joneson, 2006).

In one of the studies, Vinoth et al. (2016) designed two composite structures for truck body application. One of them was reinforced with banana fiber/rice husk and the other one was reinforced with rice husk separately in fiber form. They realized tensile, impact, water absorption and machining tests. From the results, it was investigated that hybrid composite sample had higher tensile, impact and drilling strength but the rice husk composite had lower water absorption. Al-Qureshi (2001) manufactured glass fiber/epoxy composite leaf springs for light trucks (jeep) and compared it with an existing multi leaf steel spring. Hand lay-up vacuum process was used as the production method. Results showed that there was a quite big weight saving by using composite spring. Besides, it was observed that steel spring had better rigidity than the composite one. However, authors claimed that this weakness could be improved by using carbon fiber. Finally, field test results demonstrated that glass fiber reinforced springs were more flexible than steel springs and also noise and harshness were reduced substantially. Tomar and Singh (2016) performed modal analysis to find out the natural vibrating frequency and vibration characteristics of a chassis frame of a truck by using steel-52, carbon fiber and E-glass/epoxy composite. In addition, harmonic analysis was executed to analyze the von misses stress in the chassis frame. Results showed that although composite chassis frame was more costly than the steel chassis frame, composite material has very less density and also induced stress compared to steel one. Moreover, due to its low stiffness, E-glass/epoxy composite showed a higher deformation than the steel chassis. However, it was concluded that this problem could be solved by increasing the thickness of the composite structure. While in another case, Mehul et al. (2012) compared the properties (load carrying capacity, stiffness, and weight saving) of laminated carbon/epoxy and graphite/epoxy composite leaf springs with steel leaf spring of a truck.

Analysis of the model was conducted using ANSYS 11.0 and Hypermesh programs. They indicated that weight savings of 90.09% and 79.62% were achieved with the mono and multi-leaf composite springs, respectively. Kumar et al. (2016) studied the modal and structural analysis of single piece HS carbon/epoxy and E-glass/polyester composite drive shafts for trucks. They developed a single piece composite drive shaft, which has advantages such as higher specific strength, lower weight, longer life, and high critical speed than the conventional two-piece steel drive shaft. It occurred that according to maximum principal stress, shear stress, total deformation, strain energy, and critical speed results, carbon reinforced composite and conventional steel were the most suitable materials for single piece drive shaft. Begum and Murthy (2016) designed a vehicle chassis for a truck. They used dimensions of Mahindra Bolero vehicle for modelling and analyzing in Pro-E and ANSYS. Epoxy resin was reinforced with carbon, E-glass, and S-glass. By considering three different cross-sections (C, I and Box type cross-sections), they modeled three composite vehicle chassis frames. Structural analysis results demonstrated that I type cross-section had the least deformation. When the reinforcement types were compared together with cross-section, it was observed that carbon/epoxy composite vehicle chassis with I type cross-section had the highest strength, stiffness and weight saving. In another study, short staple jute and polylactide fibers were mixed, carded and needle punched to fabricate felt by Hu et al. (2012). Afterward, real size truck liners were manufactured by hot press process. Composite samples were reinforced with two different amounts of reinforcement (60% and 70%). Mechanical performance analysis results displayed that the strength of the sample with 70% fiber volume fraction (3.69 MPa) was two times more than the strength of the sample with 60% fiber volume fraction (1.9 MPa). In their study, Ambare and Mishra (2016) aimed to reduce the weight of the suspension system of a light vehicle having a weight of 2.5 to 5 ton. They designed and analyzed the performance characteristics of E-glass fiber reinforced composite leaf springs. Performance analysis results displayed that stiffness values of both samples were so similar to each other. However, it was found out that the ultimate tensile strength and corrosion resistance of the composite leaf spring were higher than that of steel leaf spring and almost 50% weight reduction was achieved with the composite sample. Harshit et al. (2016) studied the static and fatigue analysis of steel EN45 and composite (Kevlar/epoxy, E-glass/epoxy, carbon/epoxy) parabolic leaf springs of a TATA-ACE mini truck. Pro-E creo-2 and ANSYS-14 software programs were used for CAD modelling and analysis of the parabolic leaf springs. According to the results, compared to steel leaf spring weight savings of 64.03%, 58.13% and 62.63% were obtained with Kevlar, E-glass and carbon reinforced composites, respectively. Additionally, E-glass/epoxy sample showed higher fatigue life and less stress among all composite samples. Tataroglu et al. (2016) designed and manufactured a single piece composite drive shaft for heavy vehicles. Modelling of carbon/epoxy composite material was done by finite element method and manufacturing was performed by filament winding method. They indicated that a weight saving of 28 kg (approximately 35% of the total structural weight) was achieved and except natural frequency, design requirements were all fulfilled with the composite drive shaft.

CONCLUSION

Human population has almost reached 7.6 billion people in the world. While it continues to expand, our planet does not. On the other hand, human needs are also changing and expanding according to the growth in the global economy. More food, more constructions and more merchandises are needed for their everyday life. They need to use vehicles to produce more food, to make more buildings and to transport their commercial commodities to other regions. They are moving and travelling more than ever in or out of the cities for working, for having fun or for trying to find new opportunities to survive. All these activities depend on all kind of vehicles, which also includes “energy efficiency”, “safety” and “CO_2 emission” concerns of the modern humans in the contemporary era.

In this perspective, composite materials that continue to progress rapidly has reached the quality of being the material of the future. They support the sustainable development and growth of the core industries and play a key role in meeting their future demands. The main advantages that composite materials offer the automotive applications are especially reducing the weight of the vehicle without comprising the safety issues which means “less energy”, “same or better safety” and “less CO2emission”. Throughout this chapter, textile reinforced composite structures for automotive applications were reviewed for the readers. The main target of the automotive industry is to improve environmentally friendly, energy efficient and lighter vehicles, due to the rise in the fuel oil prices and growing awareness about carbon dioxide emission and global warming. Most of these problems have been overcome with the utilization of composite structures and also it is assumed that by developing innovative composite materials with enhanced properties will provide much more advantageous in the automotive industry.

REFERENCES

Abu Talib, A. R., Ali, A., Badie, M. A., Azida Che Lah, N., & Golestaneh, A. F. (2010). Developing a hybrid, carbon/glass fiber-reinforced, epoxy composite automotive drive shaft. *Materials & Design*, *31*(1), 514–521. doi:10.1016/j.matdes.2009.06.015

Akhshik, M., Panthapulakkal, S., Tjong, J., & Sain, M. (2017). Life cycle assessment and cost analysis of hybrid fiber-reinforced engine beauty cover in comparison with glass fiber-reinforced counterpart. *Environmental Impact Assessment Review*, *65*, 111–117. doi:10.1016/j.eiar.2017.04.005

AL-Oqla, F. M., Sapuan, S. M., Ishak, M. R., & Nuraini, A. A. (2016). A decision-making model for selecting the most appropriate natural fiber – Polypropylene-based composites for automotive applications. *Journal of Composite Materials*, *50*(4), 543–556. doi:10.1177/0021998315577233

Al-Qureshi, H. A. (2001). Automobile leaf springs from composite materials. *Journal of Materials Processing Technology*, *118*(1–3), 58–61. doi:10.1016/S0924-0136(01)00863-9

Ambare, R., & Mishra, H. (2016). Design and performance testing of master leaf for Mahindra Pick-up. *International Research Journal of Engineering and Technology*, *3*(7), 1140–1143.

Astrom, B. T. (1997). *Manufacturing of Polymer Composites*. CRC Press.

Ayrilmis, N., Jarusombuti, S., Fueangvivat, V., Bauchongkol, P., & White, R. H. (2011). Coir fiber reinforced polypropylene composite panel for automotive interior applications. *Fibers and Polymers*, *12*(7), 919–926. doi:10.1007/s12221-011-0919-1

Babu, N. R., & Sangeethkumar, E. (2015). Improving the mechanical properties of the propeller shaft by using composite materials. *Applied Mechanics and Materials*, *813–814*, 1090–1097. doi:10.4028/www.scientific.net/AMM.813-814.1090

Badie, M. A., Mahdi, E., & Hamouda, A. M. S. (2011). An investigation into hybrid carbon/glass fiber reinforced epoxy composite automotive drive shaft. *Materials & Design*, *32*(3), 1485–1500. doi:10.1016/j.matdes.2010.08.042

Bakis, C. E., Bank, L. C., Brown, V., Cosenza, E., Davalos, J. F., Lesko, J. J., ... Triantafillou, T. C. (2002). Fiber-reinforced polymer composites for construction—state of the art review. *Journal of Composites for Construction*, *6*(2), 73–87. doi:10.1061/(ASCE)1090-0268(2002)6:2(73)

Balaji Thattaiparthasarathy, K., Pillay, S., Ning, H., & Vaidya, U. K. (2008). Process simulation, design and manufacturing of a long fiber thermoplastic composite for mass transit application. *Composites. Part A, Applied Science and Manufacturing*, *39*(9), 1512–1521. doi:10.1016/j.compositesa.2008.05.017

Balcombe, R., Mackett, R., Paulley, N., Preston, J., Shires, J., Titheridge, H., … White, P. (2004). *The demand for public transport: a practical guide* (No. TRL593). Academic Press.

Begum, S. N., & Murthy, S. B. (2016). Modelling and structural analysis of vehicle chassis frame made of polymeric composite material. *International Research Journal of Engineering and Technology*, *3*(8), 574–582.

Bhagat, V. K., Biswas, S., & Dehury, J. (2014). Physical, mechanical, and water absorption behavior of coir/glass fiber reinforced epoxy based hybrid composites. *Polymer Composites*, *35*(5), 925–930. doi:10.1002/pc.22736

Bhajantri, V. S., Bajantri, S. C., Shindolkar, A. M., & Amarapure, S. S. (2014). Design and Analysis of Composite Drive Shaft. *International Journal of Research in Engineering and Technology*, *3*(15), 738–745. doi:10.15623/ijret.2014.0315138

Bindal, A., Singh, S., Batra, N. K., & Khanna, R. (2013). Development of glass/jute fibers reinforced polyester composite. *Indian Journal of Marine Sciences*, *2013*, 1–6.

Binder, A. K., & Rae, J. B. (n.d.). Automotive industry. In *Encyclopedia Britannica*. Retrieved from https://www.britannica.com/topic/automotive-industry

Brylawski, M. M., & Lovins, A. B. (1998). Advanced composites: The car is at the crossroads. *Society for the Advancement of Material and Process Engineering. Materials and Process Affordability: Keys to the Future.*, *43*, 1135–1148.

Car chassis frame. (n.d.). Retrieved August 4, 2017, from https://www.shutterstock.com/tr/image-vector/plan-car-chassis-showing-wheels-transmission-84399883

Castejon, L., Miravete, A., & Cuartero, J. (2006). Composite bus rollover simulation and testing. *International Journal of Heavy Vehicle Systems*, *13*(4), 281–297. doi:10.1504/IJHVS.2006.010584

Chapman, G. B. (1999). *A Thermoplastic Approach to a Composite Automotive Body*. SAE Technical Paper. Retrieved from http://papers.sae.org/1999-01-3222/

Chavhan, G. R., Burande, S. W., & Dhole, L. P. (2014). Analysis of shock absorber using different material of spring. *International Journal of Advances in Engineering and Technology*, *5*(4), 19–21.

Cole, G. S., & Sherman, A. M. (1995). Light weight materials for automotive applications. *Materials Characterization*, *35*(1), 3–9. doi:10.1016/1044-5803(95)00063-1

Damper. (n.d.). Retrieved August 4, 2017, from https://www.shutterstock.com/tr/image-illustration/shock-absorber-car-done-3d-isolated-78229843

Das, S. (2001). *The cost of automotive polymer composites: a review and assessment of DOE's lightweight materials composites research*. Oak Ridge National Laboratory. doi:10.2172/777656

Dhoshi, N. P., Ingole, N. K., & Gulhane, U. D. (2011). Analysis and modification of leaf spring of tractor trailer using analytical and finite element method. *International Journal of Modern Engineering Research*, *1*(2), 719–722.

Duan, S., Yang, X., Tao, Y., Mo, F., Xiao, Z., & Wei, K. (2017). Experimental and numerical investigation of Long Glass Fiber Reinforced Polypropylene composite and application in automobile components. *Transport*, *0*(0), 1–9. doi:10.3846/16484142.2017.1323231

Dubey, J., Pandit, P. P., & Das, R. K. (2016). Weight Reduction Analysis of Leaf Spring using Epoxy Material. *International Journal of Engineering Technology Science and Research*, *3*(5), 87–91.

Etaphop photo. (n.d.). *Shock absorber*. Retrieved August 4, 2017, from https://www.shutterstock.com/tr/image-photo/closeup-shock-absorber-suspension-pickup-truck-676655524

Friedrich, K., & Almajid, A. A. (2013). Manufacturing Aspects of Advanced Polymer Composites for Automotive Applications. *Applied Composite Materials*, *20*(2), 107–128. doi:10.1007/s10443-012-9258-7

front-wheel drive. (2008). Retrieved July 24, 2017, from https://www.britannica.com/technology/front-wheel-drive

Gopalakrishnan, T., Raja, M., Prakash, V. M. J., & Gnanavel, C. (2017). Design and Fabrication of E-Glass /carbon/graphite epoxy hybrid composite leaf spring. *IOP Conference Series. Materials Science and Engineering*, *183*(1), 1–7. doi:10.1088/1757-899X/183/1/012004

Harshit, Kumar, S., & Verma, A. (2016). Design And Simulation of Leaf Spring for TATA-ACE mini Loader Truck Using FEM. *Imperial Journal of Interdisciplinary Research*, *2*(10), 1421–1427.

Hearle, J. W. (2001). *High-performance fibres*. Elsevier.

Hou, J., & Jeronimidis, G. (2012). A novel bogie design made of glass fibre reinforced plastic. *Materials & Design, 37*, 1–7. doi:10.1016/j.matdes.2011.12.026

Hou, J. P., Cherruault, J. Y., Nairne, I., Jeronimidis, G., & Mayer, R. M. (2007). Evolution of the eye-end design of a composite leaf spring for heavy axle loads. *Composite Structures, 78*(3), 351–358. doi:10.1016/j.compstruct.2005.10.008

Hu, R. H., Hu, K. H., & Yang, G. H. (2016). *Studies on jute fiber/polypropylene, jute fiber/Polylactide and glass fiber/Polypropylene composites for automotive application*. Presented at the *3rd International Conference on Mechatronics and Information Technology (ICMIT 2016) 3rd International Conference on Mechatronics and Information Technology (ICMIT 2016)*. doi:10.2991/icmit-16.2016.125

Hu, R. H., Ma, Z. G., Zheng, S., Li, Y. N., Yang, G. H., Kim, H. K., & Lim, J. K. (2012). A fabrication process of high volume fraction of jute fiber/polylactide composites for truck liner. *International Journal of Precision Engineering and Manufacturing, 13*(7), 1243–1246. doi:10.1007/s12541-012-0165-5

Jarimopas, B., Singh, S. P., & Saengnil, W. (2005). Measurement and analysis of truck transport vibration levels and damage to packaged tangerines during transit. *Packaging Technology & Science, 18*(4), 179–188. doi:10.1002/pts.687

Jeon, K. W., Shin, K. B., & Kim, J. S. (2011). A study on fatiguelife and strength of a GFRP composite bogie frame for urban subway trains. *Procedia Engineering, 10*, 2405–2410. doi:10.1016/j.proeng.2011.04.396

Kamath, M. G., Bhat, G. S., Parikh, D. V., & Mueller, D. (2005). Cotton fiber nonwovens for automotive composites. *International Nonwovens Journal, 14*(1), 34–40.

Kassie, A. A., Kumar, R. R., Rao, A., & ... (2014). Design of single composite leaf spring for light weight vehicle. *International Journal of Mechanical Engineering and Robotics Research, 3*(1), 191–197.

Katake, K. A., Mankar, S. H., Kale, S. A., Dabeer, P. S., & Deshmukh, S. J. (2016). Numerical and Experimental Stress Analysis of a Composite Leaf Spring. *IACSIT International Journal of Engineering and Technology, 8*(5), 2098–2104. doi:10.21817/ijet/2016/v8i5/160805428

Kim, D. H., Choi, D. H., & Kim, H. S. (2014). Design optimization of a carbon fiber reinforced composite automotive lower arm. *Composites. Part B, Engineering, 58*, 400–407. doi:10.1016/j.compositesb.2013.10.067

Kim, J.-S., & Jeong, J.-C. (2006). Natural frequency evaluation of a composite train carbody with length of 23 m. *Composites Science and Technology*, *66*(13), 2272–2283. doi:10.1016/j.compscitech.2005.11.036

Kim, J. S., Lee, S. J., & Shin, K. B. (2007). Manufacturing and structural safety evaluation of a composite train carbody. *Composite Structures*, *78*(4), 468–476. doi:10.1016/j.compstruct.2005.11.006

Kim, J. S., & Yoon, H. J. (2011). Structural behaviors of a GFRP composite bogie frame for urban subway trains under critical load conditions. *Procedia Engineering*, *10*, 2375–2380. doi:10.1016/j.proeng.2011.04.391

Koncherry, V., Potluri, P., & Fernando, A. (2017). Multifunctional Carbon Fibre Tapes for Automotive Composites. *Applied Composite Materials*, *24*(2), 477–493. doi:10.1007/s10443-016-9550-z

Kumar, A., Jain, R., & Patil, P. P. (2016). Dynamic Analysis of Heavy Vehicle Medium Duty Drive Shaft Using Conventional and Composite Material. *IOP Conference Series. Materials Science and Engineering*, *149*(1), 012156. doi:10.1088/1757-899X/149/1/012156

Lakshmi, B. V., & Satyanarayana, I. (2012). Static and dynamic analysis on composite leaf spring in heavy vehicle. *International Journal of Advanced Engineering Research and Studies*, *2*(1), 80–84.

Lee, D. G., Lee, C. S., Lee, H. G., Hwang, H. Y., & Kim, J. W. (2004). Novel applications of composite structures to robots, machine tools and automobiles. *Composite Structures*, *66*(1–4), 17–39. doi:10.1016/j.compstruct.2004.04.044

Li, Y., Lin, Z., Jiang, A., & Chen, G. (2003). Use of high strength steel sheet for lightweight and crashworthy car body. *Materials & Design*, *24*(3), 177–182. doi:10.1016/S0261-3069(03)00021-9

Liu, Q., Lin, Y., Zong, Z., Sun, G., & Li, Q. (2013). Lightweight design of carbon twill weave fabric composite body structure for electric vehicle. *Composite Structures*, *97*, 231–238. doi:10.1016/j.compstruct.2012.09.052

Lu, Y. C., Pilla, S., & Society of Automotive Engineers (Eds.). (2014). *Design of automotive composites*. Warrendale, PA: Society of Automotive Engineers.

Maheta, V. V., Patel, A. B., Maheta, V. V., & Patel, A. B. (2015). Design, Analysis and optimization in Automobile Drive shaft. *International Journal (Toronto, Ont.)*, *1*, 432–439.

Mehndiratta, A., Singh, N. K., & Singh, K. K. (2015). Analysis of GFRP Leaf Spring. *International Journal Of Modern Engineering Research*, *5*(5), 22–27.

Mehul, S., Shah, D. B., & Bhojawala, V. (2012). Analysis of composite leaf spring using FEA for light vehicle mini truck. *Journal of Information. Knowledge and Research in Mechanical Engineering*, *2*(2), 424–428.

Meschut, G., Janzen, V., & Olfermann, T. (2014). Innovative and Highly Productive Joining Technologies for Multi-Material Lightweight Car Body Structures. *Journal of Materials Engineering and Performance*, *23*(5), 1515–1523. doi:10.1007/s11665-014-0962-3

Misnon, M. I., Islam, M. M., Epaarachchi, J. A., & Lau, K. T. (2013). Textile material forms for reinforcement materials - a review. In *Third Malaysian Postgraduate Conference (MPC) 2013* (p. 105). MPC.

Muhammad Nasiruddin, S., Hambali, A., Rosidah, J., Widodo, W. S., & Ahmad, M. N. (2017). A Review of Energy Absorption of Automotive Bumper Beam. *International Journal of Applied Engineering Research*, *12*(2), 238–245.

Murugan, R., Ramesh, R., Padmanabhan, K., Jeyaraam, R., & Krishna, S. (2014). Experimental investigation on static mechanical properties of glass/carbon hybrid woven fabric composite laminates. *Advanced Materials Research*, *903*, 96–101. doi:10.4028/www.scientific.net/AMR.903.96

Ning, H., Janowski, G. M., Vaidya, U. K., & Husman, G. (2007). Thermoplastic sandwich structure design and manufacturing for the body panel of mass transit vehicle. *Composite Structures*, *80*(1), 82–91. doi:10.1016/j.compstruct.2006.04.090

Ning, H., Pillay, S., & Vaidya, U. K. (2009). Design and development of thermoplastic composite roof door for mass transit bus. *Materials & Design*, *30*(4), 983–991. doi:10.1016/j.matdes.2008.06.066

Ning, H., Vaidya, U., Janowski, G. M., & Husman, G. (2007). Design, manufacture and analysis of a thermoplastic composite frame structure for mass transit. *Composite Structures*, *80*(1), 105–116. doi:10.1016/j.compstruct.2006.04.036

Pandya, K. S., Veerraju, C., & Naik, N. K. (2011). Hybrid composites made of carbon and glass woven fabrics under quasi-static loading. *Materials & Design*, *32*(7), 4094–4099. doi:10.1016/j.matdes.2011.03.003

Patil, D. D., Shinde, R. M., & Sawant, S. M. (2016). Design of Carbon/Epoxy Composite Drive Shaft for Low Passenger Vehicles. *International Journal of Current Engineering and Technology*, *6*(5), 1532–1536.

Prabu, L., Vinoth, A., & Senthilkumar, D. M. (2016). Experimental investigation on mechanical properties of hybrid and rice husk fiber reinforced polyester based composite for truck body application. *International Journal of Innovative Technology and Research*, *4*(2), 2857–2863.

Rady, M. H. (2016). Computational Stress Analysis of a Composite Multi Leaf Spring. *International Journal of Current Engineering and Technology*, *6*(6), 2239–2242.

Rahman, R. A., Tamin, M. N., & Kurdi, O. (2008). Stress analysis of heavy duty truck chassis as a preliminary data for its fatigue life prediction using FEM. *Jurnal Mekanikal*, *26*, 76–85.

Redman, L., Friman, M., Gärling, T., & Hartig, T. (2013). Quality attributes of public transport that attract car users: A research review. *Transport Policy*, *25*, 119–127. doi:10.1016/j.tranpol.2012.11.005

Rissi, G. O., Singh, S. P., Burgess, G., & Singh, J. (2008). Measurement and analysis of truck transport environment in Brazil. *Packaging Technology & Science*, *21*(4), 231–246. doi:10.1002/pts.797

Samal, S. K., Mohanty, S., & Nayak, S. K. (2009). Banana/Glass Fiber-Reinforced Polypropylene Hybrid Composites: Fabrication and Performance Evaluation. *Polymer-Plastics Technology and Engineering*, *48*(4), 397–414. doi:10.1080/03602550902725407

Sapuan, S. M., Maleque, M. A., Hameedullah, M., Suddin, M. N., & Ismail, N. (2005). A note on the conceptual design of polymeric composite automotive bumper system. *Journal of Materials Processing Technology*, *159*(2), 145–151. doi:10.1016/j.jmatprotec.2004.01.063

Sedlacek, F., Bernardin, P., & Lasova, V. (2016). *Design of a composite leaf spring for railway vehicles*. In *Proceedings of the 27th DAAAM International Symposium on Intelligent Manufacturing and Automation*. Vienna, Austria: DAAAAM International.

Sezgin, H., & Berkalp, O. B. (2016). The effect of hybridization on significant characteristics of jute/glass and jute/carbon-reinforced composites. *Journal of Industrial Textiles*. doi:10.1177/1528083716644290

Shalin, R. E. (2012). *Polymer Matrix Composites*. Springer Science & Business Media.

Shokrieh, M. M., & Rezaei, D. (2003). Analysis and optimization of a composite leaf spring. *Composite Structures*, *60*(3), 317–325. doi:10.1016/S0263-8223(02)00349-5

Singh, J., Singh, S. P., & Joneson, E. (2006). Measurement and analysis of US truck vibration for leaf spring and air ride suspensions, and development of tests to simulate these conditions. *Packaging Technology & Science*, *19*(6), 309–323. doi:10.1002/pts.732

Sivakandhan, C., & Prabhu, P. S. (2014). Investigation of Hybrid Composite Drive Shaft under Tensile, Bending and Torsion Testing. *International Journal of Vehicle Structures & Systems, 6*(4), 110–114.

Statista. (2016). *Worldwide car production through 2016*. Retrieved November 6, 2017, from https://www.statista.com/statistics/262747/worldwide-automobile-production-since-2000/

Statista. (2017). *Number of vehicles in use worldwide 2015*. Retrieved November 8, 2017, from https://www.statista.com/statistics/281134/number-of-vehicles-in-use-worldwide/

Suprith, N., Annamalai, K., Naiju, C. D., & Mahadevan, A. (2013). Design and Analysis of Automotive Multi-Leaf Springs Using Composite Materials. *Applied Mechanics and Materials*, *372*, 533–537. . doi:10.4028/www.scientific.net/AMM.372.533

Tataroğlu, S., Ok, E., Kaya, S., Şen, A., Doğan, V. Z., Mecitoğlu, Z., ... Türkmen, H. S. (2016). Design and Manufacturing of a Composite Drive Shaft. *Key Engineering Materials*, *706*, 73–77. doi:10.4028/www.scientific.net/KEM.706.73

Tomar, A., & Singh, D. (2016). Modelling and Analysis of a Chassis Frame by Using Carbon Fiber and E-Glass Epoxy as Composite Material: A Comparative Study'. *International Research Journal of Engineering and Technology*, *3*(04), 2395–0072.

TPI Composites Inc - Home. (2006). Retrieved July 22, 2017, from http://www.tpicomposites.com/English/home/default.aspx

Tucker, N., & Lindsey, K. (2002). *An introduction to automotive composites*. iSmithers Rapra Publishing.

Turkish Composites Manufacturers Association. (2014, September). Kompozitler Dünyası ve Otomotiv. *Composites Turkey*, (8), 28–37.

Vaidya, U. K., Samalot, F., Pillay, S., Janowski, G. M., Husman, G., & Gleich, K. (2004). Design and Manufacture of Woven Reinforced Glass/Polypropylene Composites for Mass Transit Floor Structure. *Journal of Composite Materials*, *38*(21), 1949–1971. doi:10.1177/0021998304048418

Vivekanandan, R., & Prakash, M. (2014). Selection of alternative material for car roof panel. In *Second International Conference on Current Trends In Engineering and Technology - ICCTET 2014* (pp. 32–40). Academic Press. doi:10.1109/ICCTET.2014.6966259

Wheels and bogie. (n.d.). Retrieved August 4, 2017, from https://www.shutterstock.com/tr/image-vector/wheels-bogie-part-wagonvector-illustration-eps-237431683

Zeng, F., Xie, H., Liu, Q., Li, F., & Tan, W. (2016). Design and optimization of a new composite bumper beam in high-speed frontal crashes. *Structural and Multidisciplinary Optimization*, *53*(1), 115–122. doi:10.1007/s00158-015-1312-2

KEY TERMS AND DEFINITIONS

Automotive Industry: All companies and activities involved in the manufacturing of motor vehicles like automobiles, buses, trucks, etc.

Composite Material: A type of material that is made by macroscopically combining two or more different materials together. The two materials work together—they do not dissolve or blend into each other—to give the composite unique properties.

Energy Efficiency: A fact that shows how efficiently energy is consumed, and it covers wide-ranging topics related to energy efficiency, energy savings, energy consumption, energy sufficiency, and energy transition in all sectors across the globe (including the automotive industry).

Global Warming: It is the existing rise in temperature of the Earth's surface (both land and water) as well as its atmosphere. The consequences of global warming are expected to create a great danger for the future of humankind.

Manmade Carbon Dioxide Emission: It is resulting from the burning of fossil fuels such as coal and oil so that humans can power various vehicles, machinery, keep warm, and create electricity. It is one of the most important sources of global warming.

Polymer Matrix: Main component (thermoplastic polymer, thermoset polymer, or rubber) of the composite material that holds the reinforcement materials together.

Textile Reinforcement: The textile component (can be in fiber, yarn, or fabric form) in a composite material that carries the load and increases the strength of the composite system.

Chapter 8
Fabrication of Orthotic Calipers With Epoxy-Based Green Composite

Nisha Kumari
Birla Institute of Technology, India

K. Kumar
Birla Institute of Technology, India

ABSTRACT

The concern towards environmental issues and the need for more polymer-based composites has extended its collection towards polymer composites strengthened with natural fillers. The main aim of the chapter is to make the side metallic sticks (called braces, where aluminum alloys are being used) of orthotic calipers with epoxy-based composite reinforced with coir dust as filler. Its mechanical behavior together with density, tensile and three-point bending are observed and compared with the existing aluminum-based calipers components after validating the experimental as well as virtual results. Virtual testing is done using 3D software CREO and analysis is done with the help of ANSYS workbench. It is seen that the strength and stiffness of epoxy-based composite with natural fillers is more than that of presently used aluminum alloy. The microstructure of the composites is considered to summarize the general filler distribution in the matrix using scanning electron microscope.

DOI: 10.4018/978-1-5225-5216-1.ch008

INTRODUCTION

Human race has been affected by disability of various forms such as vision impairment, mental disability, vision impairment, hearing impairment, speech disability, locomotive disability etc. Out of these the major disability is locomotive disability (related to ankle, foot, leg). The main cause of such disability in developing countries like India is polio. Polio is crippling and highly infectious disease caused by polio virus and the virus spreads from person to person due to which the amputees are unable to walk and move with their own. These amputees need a support with the help of which they could lead their lives easily. Researchers/scientists invented an appliance for those survivors which are known as braces/calipers. A brace or caliper helps in mobilizing or supporting a body part to assist the polio patients to undergo movement (Genet et al., 2010). Calipers or braces are categorized into three groups (Lower extremity, Upper extremity and spinal) taking into consideration the area affected and in which the patients require support. Calipers for lower body [Figure 1] rejuvenate weight bearing capabilities to the affected leg and hence allow the patients to walk without any risk of falling and contracture of joints (Hancox, 1982). It also provides support to the joints of affected organ and hence reduces joint pain. It also prevents further disfigurement of the joints and makes walking more effective by correct positioning of the calipers (Rao et al., 1985). Nowadays calipers, being designed by the designers, are effective and safer as they are being made considering patients work area, their age and weight. Currently it is required to be made as light and cheap as possible for increasing the ease of usage and also affordability to the general population (Sethi, 1990).

For creating a lighter caliper capable of taking the weight of affected patient, it is necessary to work with materials giving low weight to volume ratio. One of the most important and presently being used widely is composite materials. Composite materials provide low weight to volume ratio with strength when compared to many of the metals and alloys, hence, are being used extensively. Composite materials are made by combining of two or more unique materials having a distinct and identifiable interface segregating them. In tune to the major application, composites can be explained as those combinations of distinct materials that comprises of high strength/stiffness reinforcements embedded in a high performance matrix material (Thomason, 1995). The composites of interest in this work are materials possessing high strength and stiffness relative to weight comprising of natural fillers such as coconut coir dust. It is expected that the developed composites will have low weight to volume ratio, density, higher mechanical properties, specific tensile ratio (ratio of material strength to density), and higher versatility, less cost and can be machined very easily (Vignesh and Sengottaiyan (2014).

Figure 1. Commercially available Orthotic Braces / Calipers

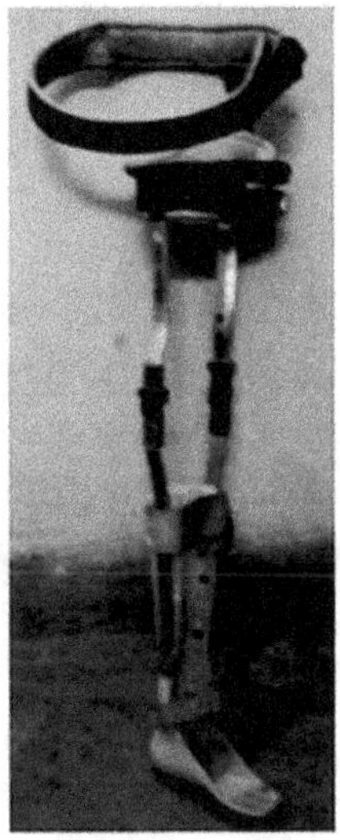

The natural filler composites, also known as green composites, have exhibited an increase of interest because of their disposal ease and abundant supply(Hossain, 2014).Inherent properties of natural filler composite, like light weight, low cost, renewable in nature, high specific strength and modulus, have enhanced their usage and offered a substantial reduction in the utilization of non-biodegradable polymers and non-recyclable resources(Biswas, 2012, Hossain, 2014, Ku et al., 2011, Lau et al., 2009, Rothon and DeArmitt, 2017, Verma et al., 2014).They are also offering much less abrasiveness than inorganic-mineral based composites to machines used for processing them and are naturally less harmful for the processing workers in case of exposure and above all they produce a final composite with lesser specific weight (in comparison to inorganic-mineral based composites)(Gnanasekar, 2015, Mohapatra et al., 2013).Additionally the natural filler based composite may be easily degraded at the end of their usage lifevia compositing or incineration in a furnace for reclamation of their calorific value which isn't feasible in inorganic-mineral based composites reinforced with fillers like glass (Kranthi and Sathapathy, 2010, Rathnakar and Shivanand, 2015).

Presently many natural types of filler are being utilized for composite product development (Puvanasvaran et al., 2011). Most prominent of all is Wood. Apart from wood derivatives, examples of alternative natural fillers are coconut dust and fibres, Date dust and Fibre, Jute dust and fibre, bamboo dust and fibre, sisal dust and fibres, polysaccharide, starch, banana, kenaf, agave, bagasse, pissava, etc (DeArmitt, 2017, Srinivasan et al., 2015).

Nearly all the normally available natural fillers are presently being utilized for reinforcement with polymeric non-biodegradable matrix materials like unsaturated polyester, epoxy resin, polyethylene and polypropylene etc. Amongst the above, epoxy resins are very flexible in nature(Jagannatha and Harish, 2015, Liu et al., 2014). They are one of the very important classes of thermosetting polymers which are exclusively used as matrices for particulate and fiber reinforced composite materials and also as commercial adhesives. Their structure is amorphous and being a cross-linked polymer offers numerous properties like higher tensile strength and modulus, ease in processing, fair thermal and chemical resistance, and above all dimensional stability(Ichetaonye et al., 2015, Roslon, 2015).

The prime attention of most of the researchers was on the experimental test of mechanical properties while working with natural filler composites (Kaewtatip and Thongmee (2012). The relationship between the mechanical properties and the composition of the composite along with the operating conditions is of utmost importance for fabricating proper composites to satisfy the required functionality(Gopinatha et al., 2014).

In the present work, the side braces of an orthotic calipers is being made up with epoxy based composite reinforced with Coconut Coir dust as filler. Its mechanical behaviour, including Density, Hardness, Tensile, 3 Point Bending etc. isobserved and compared with the existing aluminum based component.

WORK DEFINATION

The work can be subdivided in two parts:

1. Modeling and virtual estimation of mechanical properties
2. Fabrication of the natural composites and testing of the same.

A comparison between the virtual and experimental results were performed and also the effectiveness of composite based calipers over Aluminum ones were studied.

Virtual Testing

Modeling and Analysis

The braces were modeled using a solid modeling Software, CREO and the same was tested using ANSYS. The material properties used were of Aluminium Alloy (Currently used) and the proposed green composite.

Materials and Experimental Details

Materials

The standard epoxy matrix has a two component with bisphenol A, based resin, LY 556 and an amine-based hardener, HY 951.The standard epoxy resin was chosen as a matrix due to its excellent adhesion to different material, it has multidimensional stableness and is resistant to corrosion, chemical, environment friendly and provides excellent mechanical properties (Alajmi and Shalwan, 2015). Coconut coir dust was chosen as the reinforcement. Due to its higher strength and modulus property researchers/Scientists have considered it as a developing composite in the field of automotive industry (Bhaskar and Singh, 2013, Bhaskar and Singh, 2012). Coconut fillers offers good resistance, tensile strength, high hardness and are nonhazardous (Obele and Ishidi, 2015, Mulinari et al., 2011). The Coconut Coir Dust was collected from Coconut Shell grounded into flour particles through ball milling for 30 minutes by means of a motor driven ball mill. The coir dust fillers were then put onto a set of sieves in descending order of fineness (from 300 μm to 75 μm) to screen the particles less than 75 μm. The sieved particle fillers were screened with a magnet to remove any iron particle mixed during grinding process. A main challenging task was disintegration of green filler from the mixed soil particles, fill and tiny metal pieces. This was carried out through multiple water and acetone washing, then further by magnetic stirring process. The sieved particle fillers were washed thoroughly with distilled water multiple times and finally with acetone to remove dust or other contaminants and eventually dried in hot air oven at 60 °C for 8 hours.

CONDITIONING OF FILLER

Coir fillers obtained were dried at an optimal temperature of 80^0 centigrade for about 24 hours. After being dried they were examined to test the suitability of the filler and to know about the adhesion between the fillers and the matrix which is an important parameter in the preparation of the composites. The untreated coconut fillers are pre-treated with alkaline solution 1% and again re-treated in vacuum filters and the coconut fillers are washed with distilled water. After doing all this treatment they are dried in an oven for 24 hours at a temperature of 80^0 centigrade (Naghmouch et al., 2015, Panneerdhass et al., 2014. Roy et al., 2016).

COMPOSITE FABRICATION

The composite sample fabrication was initiated using hand layup technique. The mould of required dimension according to respective ASTM standard for different composite samples was prepared. The typical dimension of molds for tensile and 3 Point Bending specimen were 63.5 mm × 10 mm × 3.2 mm and 65 mm × 12.7 mm × 3.2 mm respectively with some tolerance according ASTM D-638 and ASTM D 790-03. Initially the resin and hardener were mixed in the ratio of 10:8 as recommended by manufacturer and mechanically stirred at 100 rpm for 8-10 minutes and then the dried coconut dust (10% by weight) (Pujari et al., 2014, Rowell et al., 1998, Rowell, 1998) was added to the mixture and mixed in a high speed mixer at 200 rpm for 10 – 15 minute. To minimize the void imperfections casting was done in vacuum casting chamber to eliminate air bubbles present if any in the specimen. Epoxy resin and corresponding hardener (in quantitative ratio), along with different natural fillers was amalgamated in the upper chamber of the vacuum casting (Figure 2). Rectangular specimen prepared was coated with glass paper for removal of the mold easily without any problem of sticking. The mould was kept in the chamber at room temperature for complete curing process for 6 hours. Afterward, the whole mould setup was placed inside the muffle furnace at 70 °C for 3 hours for post curing of samples. After the completion of the curing process the specimens (Figure 3) were removed and then were used for further examination and experimental work(Niyomwas, 2011).

Figure 2.Vacuum Casting Chamber (With inside details)
Source: http://www.igi-global.com/chapter/estimation-mechanical-tribological-properties-epoxy/156904(Roy et al. (2016))

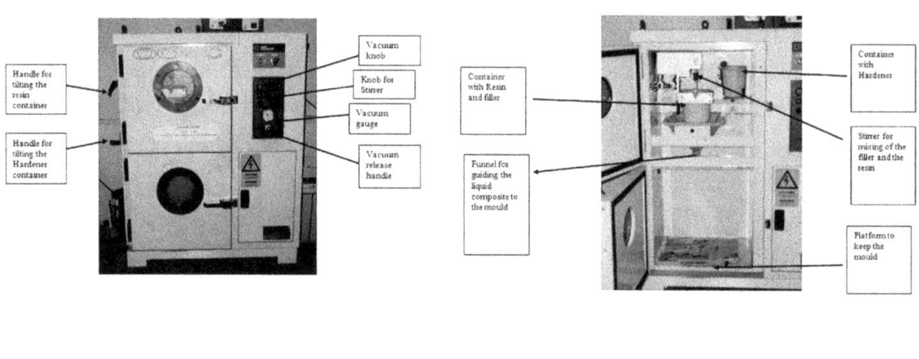

Figure 3. Samples A) Aluminium Alloy; B) Green Composite

Testing

Density and Void Content

The detrimental effect of the presence of manufacturing induced defect like voids makes it a critical issue in the applicability of composite material in engineering structures. This must be assessed to investigate the acceptability of developed material and find the allowable amount of void fraction. The density of polymeric based composites fundamentally reckons upon the mixture of the composite. The theoretical densities of composite material was calculated by using weight additive principle equation and the actual density of composite was determined experimentally by using simple water immersion method with the help of Mettler Toledo electronic balance (Figure 4). The void assessment test is carried out according to ASTM D 2734 – 94 which is the difference between the theoretical and the actual densities indicate the void content.

Figure 4. Mettler Toledo electronic balance

Tensile and 3 Point Bending Tests

Universal testing machine (UTM) was used to test the tensile and3 Point Bending behavior of composite materials. The mechanical tests were undertaken on Instron universal testing machine (Instron Ltd, UK) (Figure 5). The tensile test was conducted according ASTM D 638. The rectangular bar shaped specimens of size 100 mm long, 12 mm wide and 8 mm thickness were used. The 3 Point Bending Test was undertaken as per ASTM D 790 under the same condition as that of the Tensile test.

Morphological Study

It is important to study a composite material which is newly developed to ensure the structure; chemical composition, distribution etc. are exactly similar as desired. The microstructure characterization of a developed material helps one to quickly understand the reasons for variation of its properties, physical behavior, distribution of fillers into the matrix and performance under certain investigation, hence these

Figure 5. Instron universal testing machine

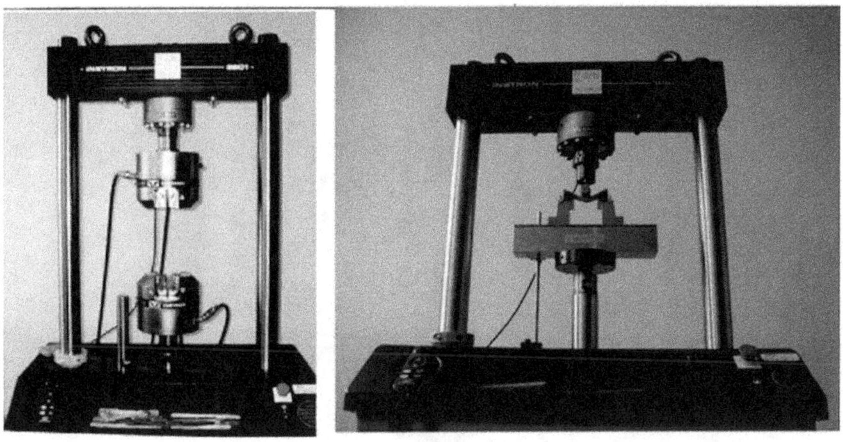

techniques are in demand to study and characterize materials. Microstructure characterization of a material is mostly studied by using Scanning Electron Microscopy (SEM) images and it is very useful in order to observe the microstructure in detail.

Scanning electron microscopy examinations was undertaken for the composite material to observe the filler distribution in the matrix. The surfaces of composite was coated with thin platinum film on the surface by sputtering and the images was observed in a JEOL (model JSM 6390LV, Japan) microscope.

RESULTS AND DISCUSSION

Modeling

In order to have a resemblance in virtual and actual experimentation results, a rectangular sample was modeled in a parametric software CREO with 100*20.22*6.03 mm^3 size. Figure 6 shows the modeled sample exported to ANSYS workbench for simulation.

Simulation

ANSYS Workbench was used for simulation purpose and then equivalent von-mises stress, strain and total deformation was calculated for both axial as well as transverse loading for coir dust. Material properties such as density, Young's modulus and Tensile Yield strength of the material was provided. After putting the boundary conditions such as meshing size geometry was imported and mesh size was considered as coarse

Figure 6. 3D model of sample imported to ANSYS

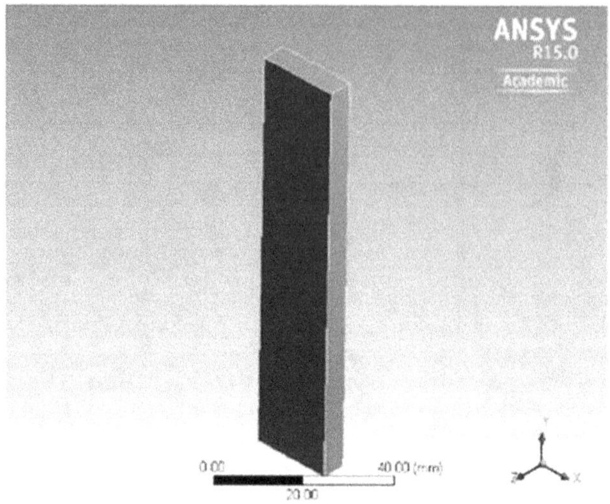

as given in the ANSYS Workbench. The maximum deflection for axial loading was found out to be 1.416mm and the maximum deflection for transverse loading was found out as 12.477mm. The equivalent von-mises stress for axial and transverse loading was 0.65506 MPa and 3.3609MPa respectively. The maximum strain under both loading conditions was observed to be 0.000655 mm/mm and 0.0036309 mm/mm. A sample picture can be viewed in Figure 7 showing equivalent von-mises stress for axial loading condition. All the other results are tabulated along with the experimental results to observe the validity of the outputs of the work.

Density and Void Content

Void content for the primarily based composites is examined by taking into account the experimental and theoretical information. It is seen that the experimental data is lesser than the theoretical showing existence of voids at the surface forming the common boundary within the composites. It can be assured from the table 1 that the density of Aluminum alloy (presently used orthotic calipers) is more than the epoxy (thermoset) composites. For all industrial and household products, void content is required to be less than 1 percent which was achieved as indicated in Table 1. Usually void presence in the composites is due to the entrapment of air bubbles during mixing and fabrication of composites. The extent of the same was reduced as fabrication was undertaken under vacuum conditions.

Figure 7. Von-Mises stress in Coir Dust reinforced Composite under Axial Loading Condition

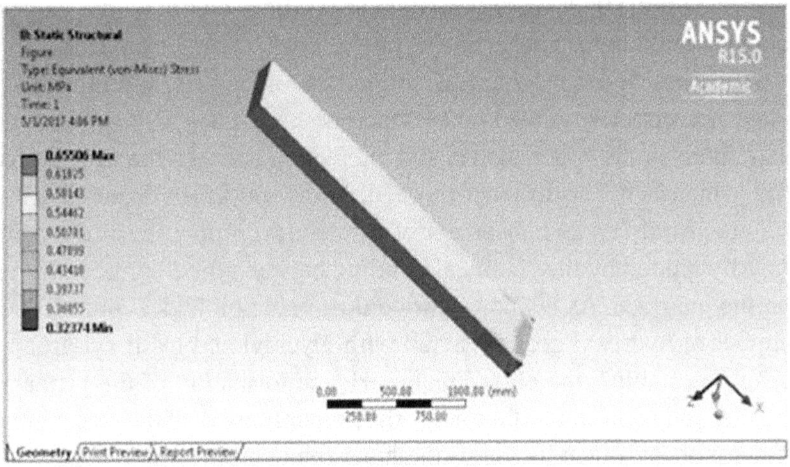

**For a more accurate representation see the electronic version.*

Table 1. Density and Void Content of Composite and Aluminum Alloy

Material of Construction	Density (g/cc)		Void fraction (%)
	Experimental	Theoretical	
Aluminium	2.705	--	--
Composite	1.137	1.144	0.616

Tensile and 3 Point Bending Tests

The tensile test was carried out in accordance to ASTM D 638 where the specimens are unit placed within the grips and are unit force till failure. The tensile tests are performed with the best possible load capacity of 1 kN with 50 mm gauge length and cross head speed of 2.5 mm / min. The information regarding the data of tensile modulus and tensile strength are entered on a computer being joined in close association with machine setup. The 3 point bending tests was executed as per ASTM D 790 where the test was started by enforcing the load on the specimen at the specified rate till the breaking of the specimen. The tests are carried out with a span length of 50 mm and cross head speed of 2.5 mm / min and the 3 Point Bending stress, 3 Point Bending strain and lateral deformation was recorded from the tests. According to the ASTM standard mechanical tests were performed on five samples for both aluminium and green composite average of the results were reported along with the standard deviation. The plots of the various values are depicted in the graphs. Figure 8 shows the tensile strength values for each of the materials along with their virtual values whereas Figure 9 illustrates variation in maximum strain for the same. Figure 10reveals maximum deformation undergone in axial loading condition.

The improvement of mechanical properties of coir dust reinforced epoxy composites are primarily guided by two factors: homogeneous distribution of the particulate fillers in the epoxy matrix and interaction between the filler surface and the matrix at interface. The distribution of filler was uniform as depicted in the SEM images. In this study the enhancement of mechanical properties of the composite was primarily guided by the chemical bonding between the reinforcement and the matrix at the interface. As is known, wood dust being biological product contains large number of hydroxyl group reacted with glycidyl and hydroxyl group of the epoxy resin to establish the ether linkage. Hence formation of the primary bond between filler and resin at interface helps the proper transfer of the load from matrix phase to filler phase which increases the mechanical properties.

The difference in the values of the virtual and experimental data can be attributed to the fact that the fabricated composites contains void as already found using density tests but the distribution of the same could not be determined. Hence, if the voids are not uniformly distributed throughout the product but gets clustered at a place, the results of the experimental data would get affected.

Figure 11 shows 3 Point Bending strength data whereas Figure 12 indicates the 3 Point Bending strain. The lateral deformation can be viewed in Figure 13. From all the graphs it is clear that the mechanical strength of green composite is more eminent than the existing aluminium alloy. The Equivalent Von-mises strain and Total Deformation is lesser for the composite. This can be imputed to same explanation as for the tensile properties. This shows that the Aluminium alloy made calipers can be fancied using coir reinforced composite. Moreover the similarity of the results between the virtual and experimental data confirms the validity of the experimentation.

As discussed above and depicted in the figures the maximum deformation and strain also varies in the similar trend as in case of Equivalent Von-mises stress both under axial and lateral loading condition.

Morphological Study

SEM images have been used to investigate the dispersion morphology and the interactions between matrix and filler for both thermoset based and thermoplast based composites. From the figures (Figure 14) it can be visualized that the dispersion and the interaction of filler with the matrix is evenly spread.

Figure 8. Comparison of Von mises Stress under axial loading with standard deviations

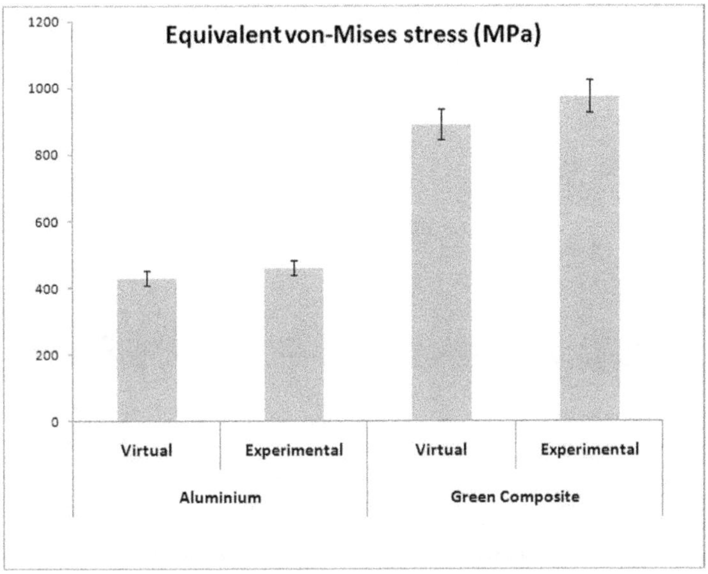

Figure 9. Comparison of Strain under axial loading with standard deviations

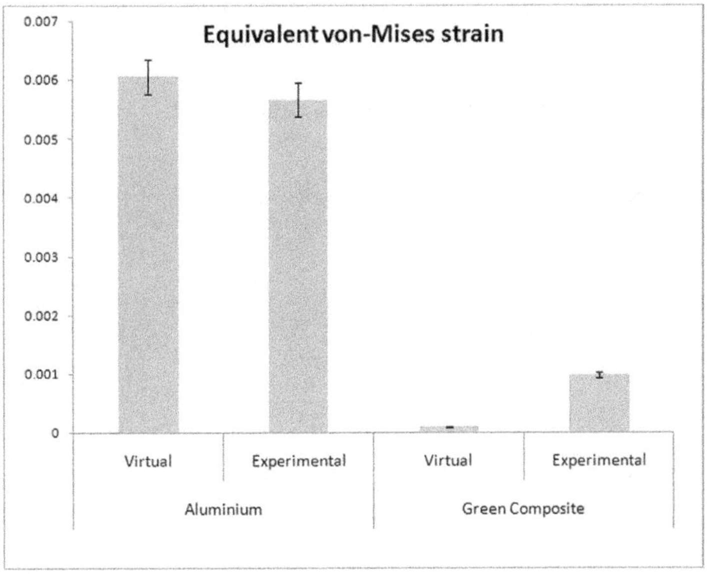

Figure 10. Comparison of Maximum Deformation under axial loading with standard deviations

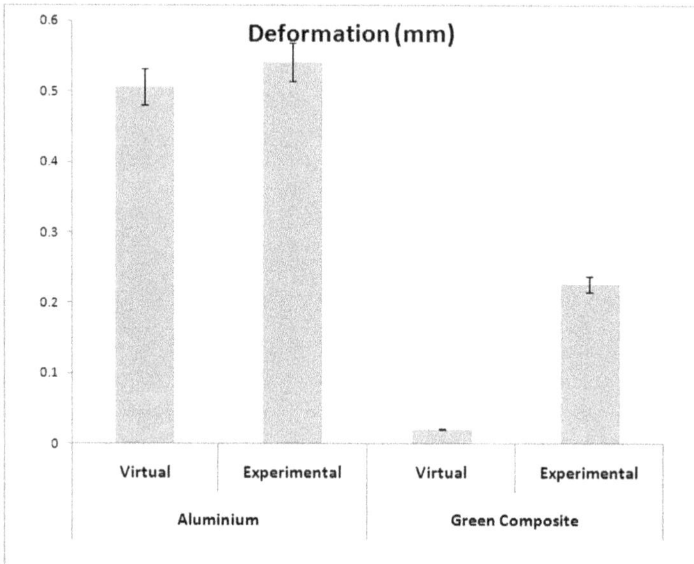

Figure 11. Comparison of Von mises Stress under 3 Point Bending with standard deviations

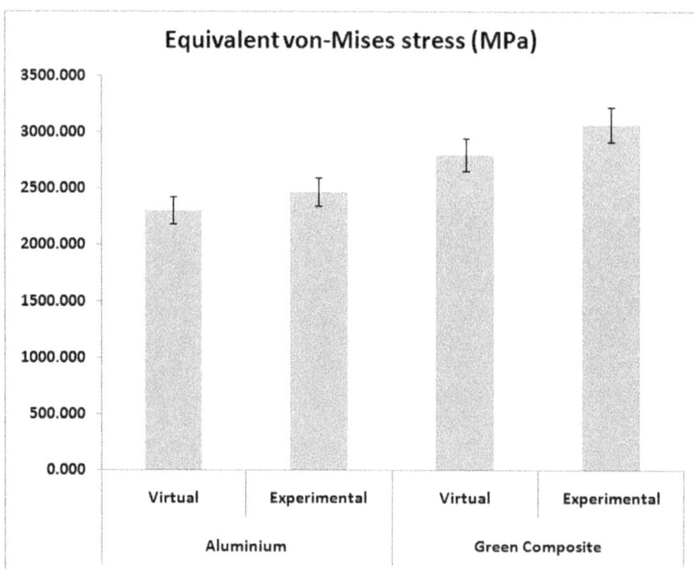

Figure 12. Comparison of Strain under 3 Point Bending with standard deviations

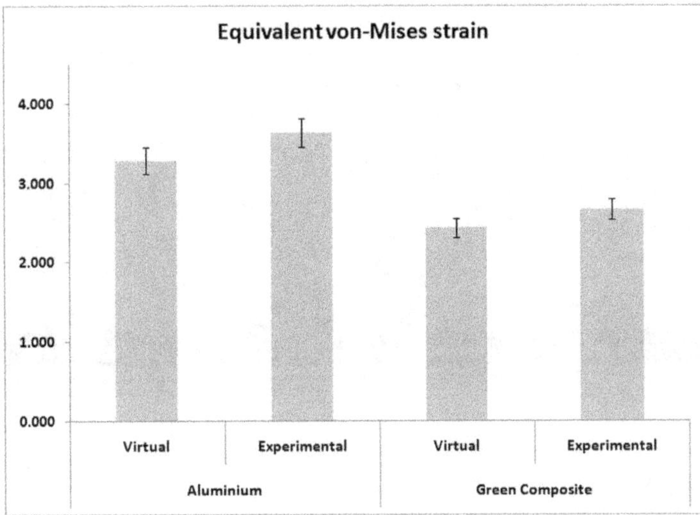

Figure 13. Comparison of deformation under 3 Point Bending with standard deviations

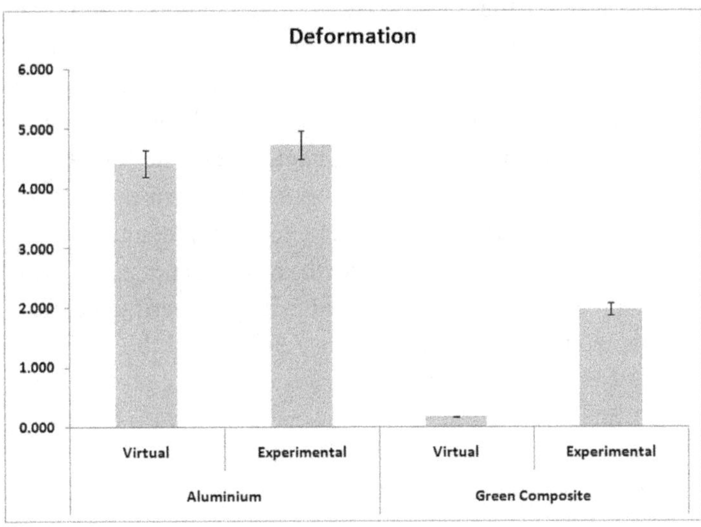

Figure 14. Scanning Electron Microscopy Images (a) Aluminium Alloy for an existing calliper (b): Green Composite

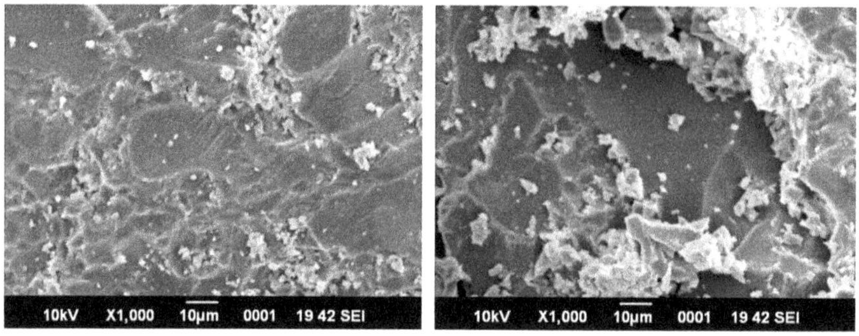

(a) (b)

CONCLUSION

An alternative material, if chosen, should provide results better than that of the presently used material (Aluminum alloy) at a lower cost for acceptability by the manufacturer and users of the lower body orthotic calipers. The literature review suggests that one of such material can be polymeric based green composite which apart from giving better strength would also provide lower weight to volume ratio. Hence, would decrease the self weight of the calipers. This would surely increase the ease of mobility of the polio infected patient. The carried out experimentation suggested that the proposed coir reinforced green composite provided such a solution. The results indicated that it would provides better strength and stiffness and lower weight to volume ration when compared with the results of presently used Aluminum based calipers. In order to validate the accuracy of the experimental results, the test pieces were subjected to virtual testing using ANSYS and the variation found was minimal.

FUTURE WORK

Since these supports undergo wear and tear and are always subjected to adverse climatic conditions hence Tribological testing and Dynamic Mechanical Testing would give a more insight about the material selection. Moreover, as the length of the braces is much more compared to the cross sectional dimensions, i.e. slenderness ratio is high, hence a buckling test would also provide a better mechanical behaviour of the alternative material. A complete lower body orthotic caliper is required to be fabricated and to be clinically tested under actual usage before replacing the aluminum alloy made calipers with proposed green composite ones.

ACKNOWLEDGMENT

The authors sincerely acknowledge the comments and suggestions of the reviewers that have been instrumental for improving and upgrading the paper in its final form. The authors also sincerely acknowledge the support of Central Instrumental Facility, Birla Institute of Technology, Mesra, India for mechanical testing.

REFERENCES

Alajmi, M., & Shalwan, A. (2015). Correlation between Mechanical Properties with Specific Wear Rate and the Coefficient of Friction of Graphite/Epoxy Composites. *Materials (Basel)*, *8*(7), 4162–4175. doi:10.3390/ma8074162 PMID:28793431

Banga, H., Singh, V. K., & Choudhary, S. K. (2015). *Fabrication and Study of Mechanical Properties of Bamboo Fibre Reinforced Bio-Composites*. Academic Press.

Bhaskar, J., & Singh, V. K. (2012). Physical and Mechanical Properties of Coconut Shell Particle Reinforced-Epoxy Composite. *Journal of Material Environmental Science, 4*(2), 227-232.

Bhaskar, J., & Singh, V. K. (2013). Water Absorption and Compressive Properties of Coconut Shell Particle Reinforced-Epoxy Composite. *Journal of Material Environmental Sciences (Ruse)*, *4*(1), 113–118.

Biswas, S. (2012). Mechanical properties of bamboo-epoxy composites a structural application. *Advanced Materials Research*, *1*(3), 221–231. doi:10.12989/amr.2012.1.3.221

DeArmitt, C. (2017). *23 - Functional Fillers for Plastics* (2nd ed.). Applied Plastics Engineering Handbook.

Genet, F., Schnitzler, A., Mathieu, S., Autret, K., Théfenne, L., Dizien, O., & Maldjian, A. (2010). Orthotic devices and gait in polio patients. *Annals of Physical and Rehabilitation Medicine*, *53*(1), 51–59. doi:10.1016/j.rehab.2009.11.005 PMID:20022835

Gnanasekar, R. (2015). Influence of Additives on the Mechanical Behaviors of Natural Fiber-Polyester. *Journal of Multidisciplinary Engineering Science and Technology*, *2*(3), 289–297.

Gopinatha, A., Kumar, S., & Elayaperumal, A. (2014). Experimental Investigations on Mechanical Properties of Jute Fiber Reinforced Composites with Polyester and Epoxy Resin Matrices. *Procedia Engineering*, *97*, 2052–2063. doi:10.1016/j.proeng.2014.12.448

Hancox, N. L. (1982). Developing a materials system for use in making orthopaedic calipers. *Composites*, *13*(2), 202–208. doi:10.1016/0010-4361(82)90058-1

Hossain, M. F., Islam, M. K., & Islam, M. A. (2014). Effect of chemical treatment on the mechanical and physical properties of wood saw dust particles reinforced polymer matrix composites. *Procedia Engineering*, *90*, 39–45. doi:10.1016/j.proeng.2014.11.811

Hossain, M. F., Shuvo, S. N., & Islam, M. A. (2014). Effect of types of wood on the thermal conductivities of wood saw dust particle reinforced composites. *Procedia Engineering*, *90*, 46–51. doi:10.1016/j.proeng.2014.11.812

Ichetaonye, S.I., Madufor, I.C., Yibowei, M.E., & Ichetaonye, D.N. (2015). Physico-Mechanical Properties of *Luffaaegyptiaca* Fiber Reinforced Polymer Matrix Composite. *Open Journal of Composite Materials, 5*(4), 110-117.

Jagannatha, T. D., & Harish, G. (2015). Mechanical properties of carbon/glass fiber reinforced epoxy hybrid polymer composites. *International Journal of Mechanical Engineering and Robotics Research, 4*(2), 131 – 137.

Kaewtatip, K., & Thongmee, J. (2012). Studies on the structure and properties of thermoplastic starch/luffa fiber composites. *Materials & Design*, *40*, 314–318. doi:10.1016/j.matdes.2012.03.053

Kranthi, G., & Satapathy, A. (2010). Evaluation and prediction of wear response of pine wood dust filled epoxy composites using neural computation. *Computational Materials Science*, *49*(3), 609–614. doi:10.1016/j.commatsci.2010.06.001

Ku, H., Wang, H., Pattarachaiyakoop, N., & Trada, M. (2011). A review on the tensile properties of natural fiber reinforced polymer composites. *Composites. Part B, Engineering*, *42*(4), 856–873. doi:10.1016/j.compositesb.2011.01.010

Lau, K. T., & Cheung, H. Y. (2009). Preface Natural fiber composites. *Composites. Part B, Engineering*, *40*(7), 591–593. doi:10.1016/j.compositesb.2009.04.002

Liu, R., Peng, Y., Cao, J., & Chen, Y. (2014). Comparison on properties of lignocellulosic flour/polymer composites by using wood, cellulose, and lignin flours as fillers. *Composites Science and Technology*, *103*, 1–7. doi:10.1016/j.compscitech.2014.08.005

Mohapatra, R.C., Mishra, A., & Choudhury, B. B. (2013). Investigations on tensile and flexural strength of wood dust and glass fibre filled epoxy hybrid composites. *International Journal of Mechanical Engineering and Technology, 4*(4), 180-187.

Mulinari, D. R., Baptista, C. A. R. P., Souza, J. V. C., & Voorwald, H. J. C. (2011). Mechanical Properties of Coconut Fibers Reinforced Polyester Composites. *Procedia Engineering*, *10*, 2074–2079. doi:10.1016/j.proeng.2011.04.343

Naghmouchi, I., Espinach, F. X., Mutjé, P., & Boufi, S. (2015). Polypropylene composites based on lignocellulosic fillers: How the filler morphology affects the composite properties. *Materials & Design*, *65*, 454–461. doi:10.1016/j.matdes.2014.09.047

Niyomwas, S. (2011). Synthesis and Characterization of TiC and TiC-Al2O3 Composite from Wood Dust by Self-Propagating High Temperature Synthesis. *Energy Procedia*, *9*, 522–531. doi:10.1016/j.egypro.2011.09.060

Obele, C., & Ishidi, E. (2015). Mechanical Properties of Coir Fiber Reinforced Epoxy Resin Composites for Helmet Shell. *Industrial Engineering Letters*, *5*(7), 67–74.

Panneerdhass, R., Gnanavelbabu, A., & Rajkumar, K. (2014). Mechanical Properties of Luffa Fiber and Ground nut Reinforced Epoxy Polymer Hybrid Composites. *Procedia Engineering*, *97*, 2042–2051. doi:10.1016/j.proeng.2014.12.447

Pujari, S., Ramakrishna, A., & Kumar, M. S. (2014). Comparison of Jute and Banana Fiber Composites: A Review. *International Journal of Current Engineering and Technology,* (2), 121-126.

Puvanasvaran, A. P., Hisham, S., & Kamil Sued, M. (2011). Investigation on The Mechanical Characteristics Of Sawdust And Chipwood Filled Epoxy. *Journal of Mechanical Engineering and Technology.*, *3*(1), 71–78.

Rao, D. L. V., Venkatappaiah, B., & Shanmugam, I. S. (1985). Two Dimensional Cinematographic Analysis-Handicapped Human Gait. *Biomedical Engineering*, *4*, 288–291.

Rathnakar, G., & Shivanand, H. K. (2015). A Review of Polymer Based Laminated Composites for Flexural and Shear Properties. *International Journal of Innovative Research in Science Engineering and Technology*, *4*(2), 40–43.

Roslan, S. A. H., Hassan, M. Z., Rasid, Z. A., Zaki, S. A., Daud, Y., Aziz, S., ... Ismail, Z. (2015). Mechanical properties of bamboo reinforced epoxy sandwich structure composites. *International Journal of Automotive and Mechanical Engineering*, *12*, 2882–2892. doi:10.15282/ijame.12.2015.7.0242

Rothon, R., & DeArmitt, C. (2017). *Fillers (Including Fiber Reinforcements)* (8th ed.). Brydson's Plastics Materials.

Rowell, R. M. (1998). *Economic opportunities in natural fiber-thermoplastic composites*. In P. N. Prasad, J. E. Mark, S. H. Kandil, & Z. H. Kafafi (Eds.), *Science and technology of polymers and advanced materials* (pp. 869–872). New York, NY: Plenum Publishing Corp. doi:10.1007/978-1-4899-0112-5_76

Rowell, R. M., Caulfield, D. F., Chen, G., Ellis, W. D., Jacobson, R. E., Lange, S. E., . . . Sain, M. (1998). Recent advances in agro-fiberhhermoplastic composites. Proc. Second International Symposium on Natural Polymers and Composites.

Roy, S., Davim, J. P., Bhowmik, S., & Kumar, K. (2016). *Estimation of mechanical and tribological properties of epoxy based green composites*. Academic Press.

Sakhtivel, M., Vijayakumar, S., & Ramesh, S. (2014). Production and Characterization of Luffa/coir Reinforced polyproplenecomposite. *Procedia Materials Science*, *5*, 739–745. doi:10.1016/j.mspro.2014.07.323

Sethi, P. K. (1990). Designing Aids for Physically Handicapped in Developing Countries. *Indian Journal of Physical Medicine and Rehabilitation*, *3*, 1–10.

Srinivasan, C., Sathish, S., & Vignesh, K. (2014). Mechanical Properties of Chemically Treated LuffaAegyptiaca Fiber Reinforced Epoxy Matrix Composites. *International Journal of Scientific Research and Management, 2*(10), 1515-1524.

Thomason, J. L. (1995). The interface region in glass fibre-reinforced epoxy resin composites: 1. Sample preparation, void content and interfacial strength. *Composites*, *26*(7), 467–175. doi:10.1016/0010-4361(95)96804-F

Verma, C. S., Chariar, V. M., & Purohit, R. (2012). Tensile Strength Analysis of bamboo and Layered Laminate Bamboo composites. *International Journal of Engineering Research and Applications.*, *2*(2), 1253–1264.

Verma, D., Jain, S., Zhang, X., Vaidya, A. A., Gaugler, M., & Smith, D. A. (2016). Green route to modification of wood waste, cellulose and hemicellulose using reactive extrusion. [PubMed]. *Carbohydrate Polymers*, *136*, 1238–1250.

Vignesh, P., & Sengottaiyan, M. (2014). A Study on Mechanical Behaviour of Hybrid Reinforced Composite. *IOSR Journal of Mechanical and Civil Engineering*, *11*(6), 68–70. doi:10.9790/1684-11626870

Chapter 9
Polymer Composite Materials for Microelectronics Packaging Applications:
Composites for Microelectronics Packaging

Noureddine Ramdani
Harbin Engineering University, China & Research and Development Center, Algeria

Mehdi Azibi
Research and Development Center, Algeria

ABSTRACT

This chapter reports the recent advances in the fabrication methods, properties, and microelectronics packaging applications of various inorganic fillers and reinforced-polymer composites. Recently, inorganic particles, including ceramics and carbon-based material reinforced polymeric matrices, have attracted both academic and industrial interest because they exhibit good thermal and mechanical properties. The low dielectric constant and dielectric loss, the low thermal expansion coefficient, and high thermal conductivity make these kinds of composites suitable for microelectronics packaging. The filler ratio, surface modification, and preparation methods of these composites have a marked effect on the final properties of these materials. Herein, the preparation methods, thermal and dielectric properties, shortcomings, and microelectronics applications of polymers/inorganic composites are summarized and discussed along with detailed examples collected from the extensive scientific literature.

DOI: 10.4018/978-1-5225-5216-1.ch009

INTRODUCTION

In recent years, polymeric thermoplastic, thermoset, and rubber materials have been widely explored and used in many ways to satisfy the needs of industries in the automotive, aerospace, and microelectronics sectors (Fieler, 2000). Such significant interest in the materials derives from various factors, including their ease of processing, light-weight, low cost, high productivity, good corrosion-resistance, low electrical characteristics, a moderate thermal expansion coefficient (CTE), and good customizability for these properties via various structural designs. By some contrast, polymers exhibit inferior thermal conductivity of about 0.2 W/m.K. This makes them useful in microelectronics packaging sector also, particularly as printed circuit boards (PCB), thermal interface materials (TIM), insulation for motor, adhesives for light-emitting diodes (LED), as well as materials for solar cells and phase change applications (Moore & Ashi, 2014).

To satisfy the use conditions of high-performance microelectronic packaging systems, *System-in-Package* (SiP) technique had been thought as the most efficient packaging method due to its area grid array and flip-chip attach characteristics (Song, Kim, & Lee, 2003). However, this packaging technology has many shortcomings, including difficulty to design and elaborate a high-density multilayer packaging substrate with high I/O density, high working frequencies, and high good thermal stability. The required characteristics for any high-density packaging substrate are: fine-pitch lines (<30 μm), fine via-hole diameters (<30 μm), high operation frequency (>3 GHz), better thermal stability, excellent flip-chip mounting reliability, and low production costs (Shimoto, Baba, & Matsui, 2005).

The use of polymers generally produced conventional packaging substrates exhibiting several serious challenges for a high-density IC packaging system (Mishra, Raj, & Tishler, 2016). This includes the higher cost of building up the multilayers; the deformation of curvature and unevenness of the core laminates at higher temperature; impedance mismatching that resulted form the through-holes drilled in the core laminates, which significantly limited their uses in the high-speed and high-frequency systems (Pleşa, Noţingher, & Schlögl, 2016).

Also, in those applications, the small surface of a substrate exhibits dense thermal energy, which directly influences signal transmission and can even destroy the materials, if the generated heat does not effectively or quickly dissipate (Huang, Jiang, & Tanaka, 2011). To increase the reduced thermal conductivity of polymeric materials, two approaches are available: the molecular design or the filling with thermally-conductive agents (Bigg, 1995). In response, the preparation of some micro- and nanocomposites based these polymeric matrices reinforced with various conductive inorganic particles and fibers are arguably the best means to overcome that serious problem. Along with their high thermal conductivity, the composite materials need to exhibit both lower CTE values, which are considered as the most important factors that caused failure in the microelectronic equipment, while the decrease in the dielectric constant lead to decreasing the delay of signal propagation. Research shows that metallic reinforcing agents cannot be used due to their high electrical conductivity, while ceramic fillers are still the best candidate for increasing the thermal conductivity of polymers, where generally a compromise is obtained that improves the thermal conductivity of the insulating molded components (He, Fu, & Han, 2007).

Ceramic micro- and nano-fillers polymer matrix-based composites are extensively used to enhance the inferior thermal conductivity of organic plastics designated for microelectronic packaging purposes, as they have the desired thermal conductivity and satisfied the needed electrical parameters (Chen, Ginzburg, & Yang, 2016). Many types of promising ceramic fillers such as silica (SiO_2), alumina (Al_2O_3), boron nitride (BN), polyhedral oligomeric silsesquioxane (POSS), zinc oxide (ZnO), silicon carbide (SiC), silicon nitride (Si_3N_4), aluminum nitride (AlN), and zinc oxide (ZnO) had been extensively investigated for enhancing the thermal conductivity of various thermosets and thermoplastics such as polyester, epoxy, polybenzoxazine, polyimide, bismalimide, cyanate ester, low-density polyethylene composites (LDPE), polystyrene, high-density polyethylene composite (HDPE), polymethyl methacrylate (PMMA), polypropylene and polymethyl methacrylate (PMMA) (Chen, Ginzburg, & Yang, 2016; Li & Wang, 2002; Zhou, Liu, & Wang, 2017). In these studies, researchers also used different ceramic particle sizes, volume fractions, shapes, and also binary and ternary filler systems to better enhance the reduced thermal conductivity values of polymeric matrices. By using conductive inorganic micro-particles, a very high filler ratio is needed for reaching the percolation threshold and thus affording higher thermal conductivity values (> 60 vol.%), by forming homogenous thermally-conductive pathways within the matrix. But, this approach can seriously affect the mechanical performance of such composites, specifically bending and impact properties which can limit their potential application. To mitigate this problem, many of these investigations reported the role of ceramic particles surface-modification on the inherent thermal conductivity of these composite materials and stated that by tailoring the matrix/fillers interface, this can significantly enhance the thermal conductivity of these materials. Silane coupling agents and chemical functionalization through cross-link are two methods have been applied to ameliorate the contact of fillers with polymeric matrix and to improve the state of dispersion into the polymeric matrices, which resulted in better mechanical performances.

Other efficient inorganic particles can be employed to increase the thermal conductivity of polymeric materials are based on electrically conductive fillers including, carbon-based fillers, metallic fillers, graphene-based platelet or sheets, and carbon nanotubes, which exhibit an improved inherent thermal conductivity

(Haggenmueller, Fischer, & Winey, 2007; Han & Fina, 2011). Although, these fillers can modify the insulating characteristics of the polymeric compounds, thus increasing the electrical conductivity and dielectric properties, they cannot be used alone to develop thermal-conductive composites. However, graphene oxide (GO) was successfully used to produce thermally-conductive polymer composites due to its insulating feature. The use of carbon nanotubes (CNTs) and carbon black to reinforce polymers can provide thermally-conductive nanocomposites exhibiting good mechanical properties. However, due to the higher surface area of these nanoparticles, the scattering of heat bearing phonons can limit the development of these nanomaterials. Thus, some new techniques have been used for producing highly thermally-conductive systems by incorporating binary fillers in the composition of CNTs, other conductive ceramic particles or covalently attached CNTs to an electrically insulating but thermally conductive filler as the thermal resistance at the nanoparticles junction can be markedly decreased.

The prediction the thermal conductivity values of materials constitutes a major part of the thermal-conductance published studies (He, Fu, & Han, 2007). Most reports are related to the thermal conductivity of polymer composites, using various thermal conductivity empirical and semi-empirical models including *Effective Medium Theory* (EMT), *Nielsen* model and *Agari* model, and *Maxwell* to predict the thermal conductivity. In literature, there is huge number of publications which used these models to estimate the thermal conductivity of several thermally-conductive polymer micro- and nanocomposites filled with various types of fillers, and most of them affirmed their applicability for some models like Neilson model. Researcher reported that polymeric composite materials are usually not investigated separately despite the filler volume ratio is not similar when their thermally conductivity values are studied by the mentioned theoretical models. For various filler volume fractions, inorganic thermally-conductive reinforcing agents existed within any composite can be considered as both polymeric matrix or reinforcing phase, according to this fraction. At the equal matrix/filler ratio, the fillers tend to contact each other which are known as the percolation threshold phenomenon (Agari & Uno, 1986), where both phases can be treated like a matrix. The aim of modeling thermal conductivity is to justify the involved mechanisms and compare the different results.

In this issue, the present study summarized the role of adding conductive inorganic fillers for improving the thermal and dielectric behavior of various polymeric matrices and also reported the applicability of some heat conduction models to reproduce the experimental thermal conductivity, dielectric, and CTE data for these composite materials.

CONDUCTIVE INORGANIC FILLERS

The most used thermally-conductive to improve the inferior thermal conductivity of any polymer are generally inorganic fillers, including ceramics, carbon-based fillers, metallic powders, or engineering fibers fillers. These reinforcing agents have higher thermal conductivity; reduced dielectric constant, and inferior CTE value and higher thermal-resistance. For instance, carbon nanotubes (CNT) and diamond exhibit a thermal conductivity value of couple of thousands W/m.K, whereas that of unfilled polymers does not exceed the value of 0.2 W/m.K. Thus, the reinforcing with such nano-fillers could be intended to markedly improve the low thermal conductivity of polymeric matrices. However, the high dielectric constant of carbon-based and metallic particles fillers make them not attractive for microelectronic packaging applications. Ceramic fillers actually still represent the solution for heat dissipation problems as they exhibited higher thermal conductivity, decreased dielectric constant and loss as well as better thermal stability. In addition, to produce thermally conductive polymer composites generally a higher filler ratio that exceeds 35 vol.%, is required for effectively attained a suitable thermal conductivity value of polymers. But, by incorporation a high inorganic filler ratio, this will markedly deteriorate the mechanical properties of polymer and increasing their densities as well as their processing such as extrusion possibility or injection molding will be more difficult. Consequently, developing polymeric composite materials having thermal conductivity values that exceed 4 W/m.K is remain a challenging till now (Chen, Ginzburg, & Yang, 2016). Another kind of anisotropic conductive filler systems, likes engineering fibers and layers, can more significantly improve the thermal conductivity of polymers along a parallel direction in comparison to the perpendicular sense. Therefore, the use of a binary or ternary filler systems or surface-treatment of filler could be efficient ways to ameliorate the isotropic thermal conductivity of polymeric matrices and ensure a homogenous dissipation of heat generated by the integrated small devices. It is also worth noticing that a huge difference in the reported data of thermal conductivity for the various inorganic particles is related to many factors, including the purity of filler, its dominant crystalline phase, its particle dimension and the used measurement method to evaluate the thermal conductivity property (Yung & Liem, 2007).

Carbon-Based Thermally-Conductive Fillers

Carbon-based fillers, including graphite (G), vapour growth carbon fiber (VGCF) Graphene oxide (GO), carbon nanotubes (CNT), carbon fiber (CF) and carbon black (CB) are examined as the most suitable fillers for polymeric due to their higher thermal conductivity values and low densities. Theses fillers are diverse and can be found in different forms and morphologies as shown in Figure 1. The Best conductive carbon-based filler is graphite which exhibits a thermal conductivity as high as 5300 W/m.K, has a low cost and has good dispersion into polymers (Causin, Marega, & Marigo, 2006; Tu & Ye, 2009). Graphene oxide (GO) nano-sheets however are abundantly oxidized graphene (containing functional groups such as -NH_2, -OH, epoxy, diols, C=O, and -COOH), which could significantly change the physical intermolecular interactions and increase the compatibility with the pristine polymeric matrices. In addition, expanded graphite and reduced graphene oxide with thickened layers of 20–100 nm, have equally been reinforced polymers for improving their thermal conductivity values (Ganguli, Roy, & Anderson, 2008), hence the exfoliation degree, the aspect ratios, the state of filler dispersion into a polymeric matrix are key factors for controlling thermal conductivity enhancement. Even that the CNTs exhibit better mechanical properties than graphene, graphene is stilled preferably used nano-filler than CNT in some specific applications. Another advantage of graphene nano-filler is its low cost compared with that of SWNT (>$10^5/Kg) or VGCF ($90–110/Kg). The applications of exfoliated graphite flakes are divers especially in electromagnetic shielding applications, thermally-conductivity encapsulates, gas permeability-resistance membranes, and low flammability applications.

Carbon fibers are carbon-based materials that are used to enhance the thermal conductivity of polymers (Tibbetts, Lake, & Strong, 2007). Due to their annular geometries, they exhibit high thermal conductivity along the fiber axis totally differ from that recorded in the transverse axis (calculated to reach 2000 W/m.K in the perpendicular sense compared to 10–110 W/m.K in the transverse one (Chen & Ting, 2002), which does directly affect the thermal conductivity of its aligned polymer composites (Mohammed & Uttandaraman, 2009). On the other hand, carbon black filler is composed from several aggregates of graphite having a particle size in the range of 10–500 nm with a surface area in the range of 25 and 150 m^2/g (Pierson, 1993). This filler is known as good contributor for enhancing the electrical conductivity of polymers instead of thermal conductivity (Wong, Lo, & Shin, 2001), but several studies have been achieve to increase the later one.

Thermally-Conductive Metallic-Based Fillers

Metallic fillers can be used to fill the various polymers resulted in an enhancement of either thermal conductivity or electrical conductivity for their produced compounds. In addition, the density of these kinds of composite were significantly increased by adding the heavier metal particles to the polymers, therefore their applications is limited in the case of lightweight desertion such as in microelectronic packaging. Till now, several kinds of metallic fillers had been applied to improve the inferior thermal conductivity; this includes aluminum powders, silver particles, copper, and nickel powders. Metal oxide semiconductors like ZnO are mainly used as filler for

Figure 1. Eight allotropes of carbon: a) diamond, b) graphite, c) lonsdaleite, d) C60 buckminsterfullerene, e) C540, Fullerite f) C70, g) amorphous carbon, and h) single-walled carbon nanotube
Reprinted from https://en.wikipedia.org/wiki/Allotropes_of_carbon

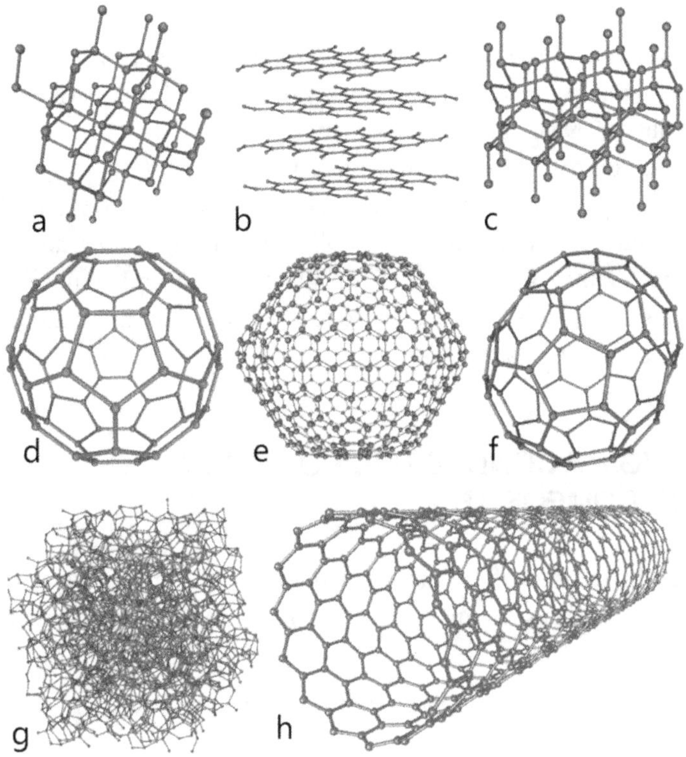

rubbery materials. Due to the inherent high thermal conductivity and non-linearity of metallic powders, their resulted composite materials are used in controlling electric stress for high-voltage devices. Polymer/metallic composites are the subject of increasing interest as important materials which may efficiently attenuate the electromagnetic field. Since, they have many advantages, including low specific weight, high corrosion resistance, plasticity and simple, and ease of processing. They also produce remarkable improvements in the mechanical performance of polymers like microhardness, tensile modulus, and bending properties.

Thermally-Conductive Ceramic-Based Fillers

The high thermal conductivity of ceramic fillers is depended mainly on their inherent composition as depicted in Figure 2. Ceramic particles reinforced polymer matrices based micro-composites and nanocomposites had been widely applied in micro-electronic packaging purposes. Due to their advantageous low electrical conductivity compared to metallic fillers, numerous kinds of ceramic-based fillers such as boron nitride (BN), beryllium oxide (BeO), aluminum nitride (AlN), silicon carbide (SiC), and silicon nitride (Si_3N_4) are attracted much focus as thermally-conductive micro- and nano-reinforcing agents because of the increased high thermal conductivity and electrical resistivity of such fillers (Nu et al., 2008; Ishida & Rimdusit, 1998). Thermal conductivity values of their resulted composites materials can be controlled by the packing-density of filler (Ohashi, Kawakami, & Yokogawa, 2005), particle dimension and (Yu, Hing, & Hu, 2002) and (Mu, Feng, & Diao, 2007), surface-modifications (Gu, Zhang, & Dang, 2009), and preparation methods (Zhou, Zhang, & Yang, 2007). In literature, several semi-empirical and theoretical equation models have been used to predict the thermal conductivity data of polymer/ceramic composites.

PROCESSING OF CONDUCTIVE POLYMER/ INORGANIC COMPOSITES

In accordance with the kind of thermally-conductive agent and the type of polymeric matrix, the selected processing method of thermally-conductive polymer/inorganic

Figure 2. The inherent thermal conductivity of some ceramic filler

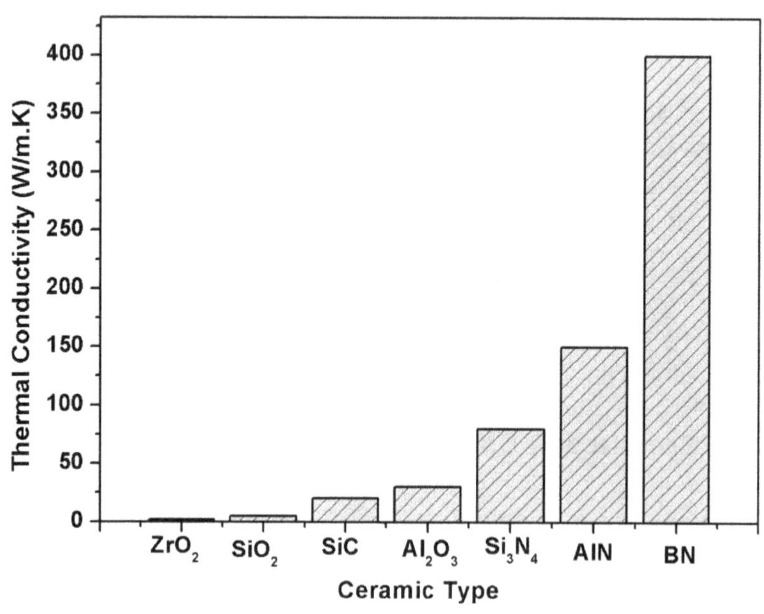

composites can be chosen. In fact three main methods are used to prepare these kinds of materials, which are melt-processing technique, sol-gel technique, solution blending process, and sol-gel technique, (Wang, Xia, & Zhang, 2000). Solution blending is the most used method for developing polymeric composites, hence the conductive filler is dispersed in a suitable solvent by strong mechanical string or high-power ultrasonication, then the polymeric matrix will incorporated and finally the solvent removed using distillation or sample evaporation. In this process, the compatibility of the solvent with the selected polymer and the type of conductive filler can significantly affect the good dispersion state of filler (Tee, Mariatti, & Azizan, 2007). In many cases, the fillers outer surface is modified by silane coupling agents for increasing the adhesion of these fillers to incompatible polymers, especially when thermosetting polymers are used, which will directly influence its bulky properties. Additionally, chemical functionalization can similarly increase solubility and improve the interaction of conductive inorganic filler such as graphene, carbon nanotubes with the various polymers.

Melt powder processing technique is a practical method especially for reinforcing thermoplastic matrices. This method based on elevated temperature and higher shears force for effectively dispersing the conductive reinforcing agents into polymers. By using high temperature this will soften polymers which allow better improvements in the dispersion state of inorganic filler. This process has some drawbacks since at higher filler ratios; a low degree of dispersion is obtained when conductive nano-filler are used to fill polymers (beyond the percolation threshold) because of the significant increase in viscosity of these systems (Wang & Yi, 2004). Also, during mixing by strong shear forces, the nanoparticles sheets can easily buckled, rolled or even shortened. However, in point of security view, this method avoids using toxic solvent.

In-situ polymerization is another famous used technique use especially for producing ceramic-thermosetting polymer based micro- and nanocomposites. In this method, ceramic filler or its derivative is directly dispersed into the liquefied thermosetting monomer to ensure a full wetting process. In this process there is a great possibility for obtaining strong covalent linkage between the modified-inorganic nanoparticle layers and polymeric matrices by different chemical processes and chemical cross-link in the case of thermoset. The dominant shortcoming of this method is the rapid increase in viscosity of composite system under a progressive polymerization process which hinders manipulation and decreases load fraction of fillers.

MODELING AND MEASUREMENT OF THERMAL AND DIELECTRIC FACTORS

The thermal conductivity of polymeric materials is measured by several techniques (Rides et al., 2009) which are categorized as classical steady-state or non-steady-state methods. The first ones evaluate the change in the temperature across the composite samples according to the useful heating flux, either as an absolute value or by comparison to a reference specimen placed in parallel or in series to the tested sample. But, the shortcomings of such techniques are the time spending and need somehow bulky samples to achieve the test. The second ones are non-steady-state methods that use hot wire and hot plate sample tests, flash laser methods, and temperature wave technique (Nunes dos Santos, 2007). However, the most used method in this type of measurement is the laser-flash thermal diffusivity method which is faster and can be applied for only a small specimen dimension (Nunes dos Santos, 2007). In this method, the surface of the tested specimen is irradiated using a very short laser pulse while the change in temperature will be measured on its opposite side, alloying facile measure of the thermal diffusivity of this specimen, after suitable mathematical modification. The value of thermal conductivity (λ) can be determined as shown in Eq. (1):

$$\lambda = Cp * \alpha * \rho \tag{1}$$

Since C_p, α, and ρ are the heat capacity, thermal diffusivity, and density, respectively.

Thermal and Dielectric Parameters Modeling

Models of Thermal Conductivity (λ)

Several key factors can control the thermal conductivity prediction of any polymer-based composite such as inherent thermal conductivity of each phase, the filler ratio, the filler dimension and structure, and the degree of filler dispersion into the polymeric matrix. The practical thermal conductivity of thermally-conductive polymer materials could be predicted using several empirical and semi-empirical equations as extensively published in literature. One principal equation model for calculating the effective thermal conductivity of such compounds is rule of mixtures (Cheng & Vachon, 1969). Another well-known model is the *Maxwell*, which consider a random dispersion of nano-filler spheres into a polymeric phase for predicting a thermal conductivity for a two phase-system composites. The derivation of the *Maxwell* equation for a solid spherical particles loaded into a continuous polymeric suspension is related to the *Bruggeman* model (Progelhof, Throne, & Ruetsch, 1976). On the other hand, Lewis and Nielsen introduce the effect of the filler shapes and their directions or their packing models. Therefore, for a binary-phase composite system, the best alternative could be that the materials are ordered in either parallel or series with respect to heat flow (Agari, Ueda, & Omura, 1997). The *Effective Medium Theory (EMT)* and *Nielsen* model generally reproduced accurately the experimental thermal conductivity data for the majority of polymer composites at relatively low filler volume concentrations. However, *Agari* model can well-predicted the experimental data in the full range, but a large error percentage is obtained (He, Fu, & Han, 2007).

The most investigated model equations in literature are established as a function with the selected filler volume ratio being V_f, which is stated as:

$$V_f = \frac{W_f / \rho_f}{W_f / \rho_f + \left(1 - W_f\right) / \rho_m} \tag{1}$$

where, W_f. is the weight ratio of filler, while ρ_f, ρ_m are the filler and matrix density, respectively.

Some famous equation models generally found in literature are announced as follows:

The *parallel* conduction model is stated

$$\frac{1}{\lambda_c} = \frac{1 - V_f}{\lambda_m} + \frac{V_f}{\lambda_f} \tag{2}$$

The *Maxwell* model is written as:

$$\lambda_c = \lambda_m \frac{\left[2\lambda_m + \lambda_f + 2V_f\left(\lambda_f - \lambda_m\right)\right]}{\left[2\lambda_m + \lambda_f - 2V_f\left(\lambda_f - \lambda_m\right)\right]} \tag{3}$$

uggeman model is given as:

$$1 - V_f = \frac{\left(\lambda_f - \lambda_c\right)\left(\lambda_m / \lambda_c\right)^{\frac{1}{3}}}{\lambda_f - \lambda_m} \tag{4}$$

The *Nielsen-Lewis* model is presented as:

$$\frac{\lambda_C}{\lambda_m} = \frac{1 + ABV_f}{1 - B\psi V_f} \tag{5}$$

where:

$$B = \frac{\left(\lambda_f / \lambda_m - 1\right)}{\left(\lambda_f / \lambda_m + A\right)}; \psi \approx 1 + \left[\left(1 - V_{max}\right) / V_{max}^2\right] V_f$$

where, A is related to the geometry of fillers, which is taken as 2 for spherical shape, while V_{max} is taken as 0.637, for randomly packed spherical particles, respectively (Progelhof, Throne, & Ruetsch, 1976).

ell model is described as:

$$\lambda_c = \lambda_m \frac{\left[V_f^{2/3} + \frac{\lambda_m}{\lambda_f}\left(1 - V_f^{2/3}\right)\right]}{\left[V_f^{\frac{2}{3}} - V_f + \frac{\lambda_m}{\lambda_f}\left(1 + V_f - V_f^{2/3}\right)\right]} \tag{6}$$

The *Agari* model is the most used semi-empirical equation model to reproduce the experimental thermal conductivity data of two components thermally-conductive polymer micro- and nanocomposite and is expressed as:

$$\log(»_c) = C_f . V_f . \log(»_f) + \left(1 - V_f\right) . \log\left(C_p . »_p\right) \tag{7}$$

Since, C_p is a parameter related to the nature of the matrix structure, such as crystallinity and physical state, while C_f is a parameter used for quantifying the ease of fillers conductive-chains formation. The values of these factors might be situated in the range of 0 to 1, and Agari is reported that as the C_f values is closed to 1, the thermally-conductive pathways are more easily formed within the polymeric matrix (Agari, 1986).

Dielectric Constant (ε) Models

Several theoretical models have been issued to reproduce the dielectric constant experimental data of polymer-based materials such as the *effective field method (EFM)* that has several benefits in the calculation of this physic value. In this prediction, it is considered that any applied electric force acts on each inclusion or inclusions in the polymer composites in the same manner. Using this method principle, *Kananu* recently established their mode, as found in Eq. 8 (Kananu, 2003). Also, the logarithmic mixing rule (Ling, Gu, & Liang, 2010) has been successfully applied to reproduce the experimental dielectric constant data for numerous types of polymer composites.

$$\varepsilon_c = \varepsilon_m + \frac{2V_f \varepsilon_m \varepsilon_f}{2\varepsilon_m + \left(1 - V_f\right)\varepsilon_f} \tag{8}$$

where, $\varepsilon_c, \varepsilon_m, and\, \varepsilon_f,$ are the inherent dielectric constant of the composite, the matrix, and the filler, respectively, whereas V_f is the inorganic fillers volume ratio.

It is appeared form Eq.9 that the logarithmic mixing rule also could be used to reproduce the experimental dielectric constant data of thermally-conductive polymer composites (Ling, Gu, & Liang, 2010).

$$\log\left(\varepsilon_c\right) = V_m \log\left(\varepsilon_m\right) + V_f \log\left(\varepsilon_f\right) \tag{9}$$

Modeling of CTE

The CTE models of polymeric materials are diverse and can be subdivided into theoretical and semi-empirical, which are known to well-predict the *CTE* experimental data. The majority of these equation models are in part based on the elastic theory and partly outstanding on some mechanics approach (Vo, Todd, & Shi, 2001). In the literature, the most widely used models to reproduce the CTE experimental data of matrix/inorganic composites are *ROM* and the *Turner* models. The first one assumes the absence interaction exists between the polymer matrix and the reinforcing inorganic agent as shown in Eq.10 (Li & Chung 1994). The second introduces one (Vo, Todd, & Shi, 2001) introduce in its equation formula the mechanical interactions within the composite's constituents as stated in Eq.11.

$$\alpha_c = \alpha_m \left(1 - V_f\right) + V_f \alpha_f . \tag{10}$$

where:

α_c, α_m, and α_f, respectively.

The mechanical interactions between the fillers and the matrix are taken in consideration in the model of turner, which is announced as follows:

$$\alpha_c = \left(\alpha_m V_m Y_m + \alpha_f V_f Y_f\right) / \left(V_m Y_m + V_f Y_f\right) \tag{11}$$

The bulky modulus of the filler and matrix are designated $as\, Y_m$ and Y_f, respectively.

THERMALLY-CONDUCTIVE POLYMER/ INORGANIC COMPOSITES

In the last few decades, polymer/ceramic composite materials have seen extensive development for use in the electronic packaging industry. This is because of their superior thermal, dielectric, and mechanical properties, together with their low cost, and eases of processing. Silica-nanoparticle based-ceramic fillers have a relatively enhanced thermal conductivity value of nearly 1.5 W/m.K, a low dielectric constant of about 2.41, and a *Young's* modulus constant higher than 15 GPa. These nano-fillers have been widely used to prepare thermally-conductive polymeric materials. For instance, the value of thermal conductivity for cyanate ester/silane-modified SiO_2 composites has been greatly increased, while their dielectric constants were reduced with the addition of surface-treated fillers using silane molecules due to the beneficial effect of filler/matrix contact (Ling, 2010). It was also reported that for epoxy/cyanate ester blends based encapsulates, SiO_2 nano-filler significantly increased the thermal conductivity, *Young*'s modulus, while slightly increasing the dielectric constant but the encapsulant thermal expansion (CTE) was markedly decreased (Wooster, Abrol, & Hey, 2004). The silane-treated $Ni_{0.5}Ti_{0.5}NbO_4$ filled cyanate ester matrix was prepared by the traditional casting method with a good degree of dispersion. These composites exhibit excellent dielectric properties and lower water-uptake suitable for microelectronic packaging application. On the other hand, the dielectric constant was better predicted using the *EMT* model with a morphological factor of 0.1407 (Zhang, Ye, & Gao, 2016). Newly prepared cyanate ester/BN nanocomposites exhibited outstanding properties such as good dimensional stability, improved dynamic-mechanical properties, and higher thermal conductivity, together with a tailored dielectric constant, and loss, while preserving excellent thermal stability for high-temperature microelectronic applications (Wu & Kessler, 2015). The hybrid filler systems used were increased the thermal conductivity of the prepared composite more significantly, that could be resulted from the enhanced connectivity offered by the new structure of filler. At a maximum filler ratio of about 58.4 vol.%, the highest values of thermal conductivities in the two composite systems were attained 3.4 W/m.K and 2.8 W/m.K, respectively, while the volume content of large particles to lower particle size reached 7:3. These properties were apparent as the system was reinforced at its maximum packing density and its minimum surface area at a fixed volume ratio (Choi & Kim, 2013). The influence of AlN particle dimension on the polybenzoxazine matrix thermal conductivity, dielectric constant, and CTE of has been studied (Ramdani, Derradji, & Wang, 2016). Choi & Kim found that smaller particle size resulted in a large enhancement in thermal conductivity, which is justified by the formation of numerous conductive-pathways within the polybenzoxazine resin as drawn in Figure 3 (Choi & Kim, 2013). Polybenzoxazine/ Si_3N_4 nanocomposites also exhibited higher thermal conductivity that exceed 5.4 at 70 wt.% of nano-filler content (Ramdani, Derradji, & Feng, 2015).

It has been shown that the thermal conductivity of epoxy/BN composites can be increased by almost 216% using silane surface-modifying fillers and multi-dimension particle size hybrids (two kinds of h-BN particle sizes and one of c-BN) before the fabrication. The measurement and investigation of the curing behavior of these anhydride cured epoxy continuous resin composites, filled with the two particle dimension BN as thermally-conductive filler, were discussed. The results demonstrated the critical of surface modification and multi-size mixing class for inorganic filler filled-epoxy matrix composites (Yung, 2007). The highest value of thermal conductivity for the composite of AlN/epoxy tested by the hot-disk technique reached about 3.4 W/m.K, which is approximately 15 times greater than the thermal conductivity of pure matrix. The *Agari* and *Uno* model gave accurate predictions for the measured thermal conductivity data (Lee, Lee, & Shanefield, 2008). In another study, h-BN platelets were modified by the KH-550 silane coupling agent prior to incorporation in a polytetrafluoroethylene matrix using a cold pressing and sintering method. The thermal conductivity of the prepared composite with a 30 vol.% filler loading reached 0.72 W/m.K, representing a 2.7 times compared to that of the unmodified polymer. The enhanced interfacial adhesion and reduced surface hydrophilicity of the h-BN platelets significantly reduced the interfacial polarization, which lowered both the dielectric constant and dielectric loss tangent (Pan, Kou, & Jia, 2017). An h-BN/polymethyl-vinyl siloxane rubber (h-BN/VMQ) composite which had a low dielectric but was thermally-conductive was produced by kneading

Figure 3. The concept of using particles with different sizes resulted in more efficient packing of fillers. Reproduced with permission from Elsevier Ltd (Choi & Kim, 2013)

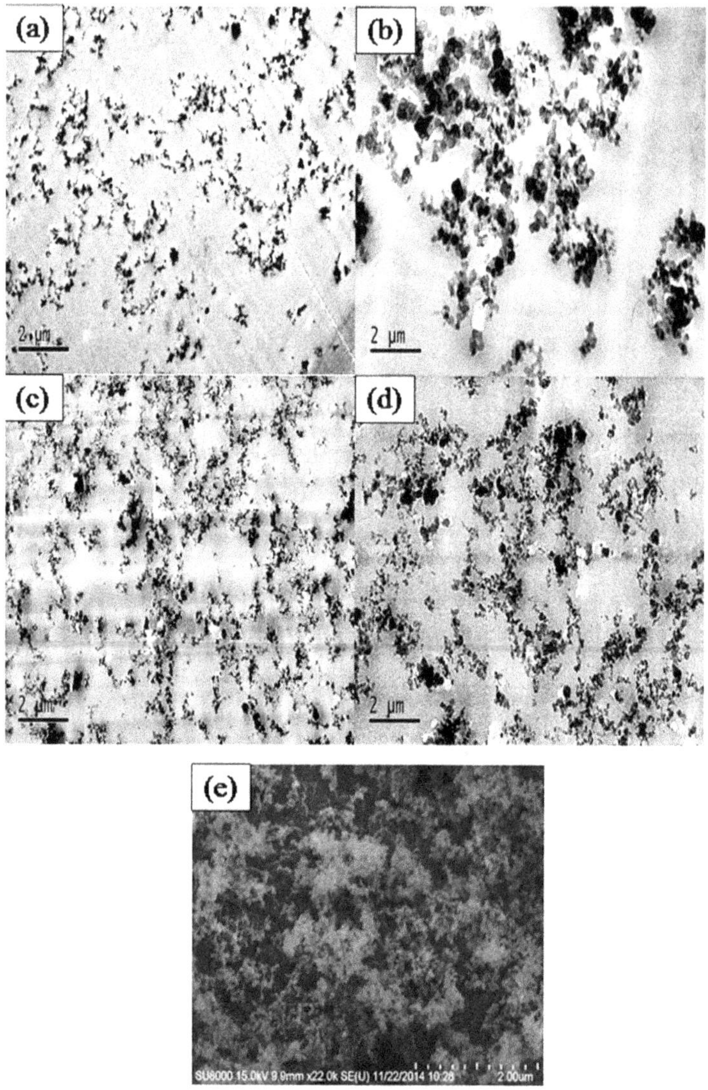

followed by a hot compression process. With a filler content of 40 vol.% of h-BN fillers, the recorded thermal conductivity attained about 1 W/m.K while dielectric constant and dielectric loss tangent attained 3.5 and 5.4×10^{-3}, respectively. These results indicated that these composites could be suitable for wrapping in a constant integration process and for miniaturisation of electronic equipment (Gu, Meng, & Tang, 2017). To enhance the energy effectiveness for solar energy systems, a polyethylene glycol/Ti_4O_7/SiO_2 form-stable composite having the ability of light-heat conversion with enhanced thermal conductivity was produced through the *in*-situ doping with Ti_4O_7 particles using sol-gel method in presence of acid and alkali catalysts (Tang, Wei, & Zhao, 2017). Jang et al., found that the improvements in thermal conductivity values can be tailored through the modification the surface features of BN particles to promote dispersion in the epoxy resin. Conductivities, exceeding 3.5 W/m.K were generated by lowering hostility to the thermal exposure within the BN/epoxy interface (Jang, Shin, & Yang, 2017). Kim et al. stated in BN/polyphenylene sulfide (PPS) composites, the interfacial affinity was increased by secondary interactions between functional polar groups, which increased the through-plane thermal conductivity by almost 2.1 W/m.K at a 70 wt.% BN filler loading. This was due to the spherical particles that formed three-dimensional thermal diffusion paths (Kim, Ryu, & Kim, 2017). A thermally-conductive boron nitride/polyamide micro-composite was prepared with a 30 wt.% loading of BN micro-filler using an *in*-situ polymerization, thermal imidization process followed by hot pressing (Gu, Lv, & Wu, 2017). It was reported that the addition of surface-treated BN nanosheets can markedly increase the thermal conductivity of styrene-butadiene rubber materials (Wu, Yang, Kuang, 2017). With a high percentage of filler, Wong *et al.* reported that the *Agari* model gave better predictions of the thermal conductivity data for the SiO_2/epoxy composites than the *Maxwell* model (Wong, Lo, & Shin, 2001). The thermal conductivity of epoxy composites was controlled by incorporating various fillers of different size and shape. Particularly, the highest thermal conductivity among the epoxy composites exceeds 6.5 W/m.K for an 80 wt.% content of Al_2O_3, which is high for epoxy/Al_2O_3 composites. The experimental results were well-predicted by the *Agari* model (Yeo, Islam, & You, 2017). Polypropylene/aluminum oxide (Al_2O_3) composites were produced by using two preparation techniques: mechanical grinding and melt mixing. The filler was incorporated in the interface of the polymer and the reinforcing phase; the segregated structure helped improving the composite's thermal conductivity (Li, Li, & Xie, 2017). Droval and his co-workers studied the impact of BN, Al_2O_3, AlN, Al_2O_3, and talc fillers on the thermal conductivity of polymers (Droval, Feller, & Salagnac, 2006). In addition, various models such as the *Cheng & Vachon*, *Lewis & Nielson*, *Agari* and *Uno* models were applied to estimate experimental thermal conductivity data at various particle ratios. The authors reported that all the models were good predictors of thermal conductivity. In a further study, at 24 wt.% boron nitride nanotubes (BNNT) filled polystyrene (PS), poly(vinyl butyral) (PVB), polymethylemethacrylate (PMMA), and poly(ethylene vinyl alcohol) (PEVA) based composites exhibited thermal conductivity values of 3.6, 1.8, 3.2 and 2.5 W/m.K, respectively (Zhi, Bando, & Terao, 2009). Moreover, thermally-conductive polypropylene-based nanocomposites have also been widely investigated (Vakili, Ebadi-Dehaghani, & Haghshenas-Fard, 2011; Ebadi-Dehaghani,

Reiszadeh, & Chavoshi, 2014). In one study, three nanocomposites incorporating till 15 wt.% of $CaCO_3$ and ZnO nanofillers were produced by an extrusion process. By adding nanoparticles this would gradually improve both the crystallinity and thermal conductivity of the resultant materials and a linear relationship increase between these two characteristics was observed. A very high thermal conductivity of 5.1 W/m.K was recorded for a new type of epoxy composite filled with 65 vol.% graphene oxide (sintered at 1000°C)/T-ZnO hybrid fillers (Jiang, Sun, & Zhang, 2017). Furthermore, the thermal conductivity of β-Si_3N_4 nanowire filled-epoxy matrix nanocomposites has been evaluated and compared to the results when commercially available irregular β-Si_3N_4 powders added to the mix at fixed loading fractions (Kusunose, 2013). The in-plane thermal conductivities of these materials were 2.5 and 2.6 W/m.K, respectively. Moreover, the effective thermal conductivity of polybenzoxazine/β-Si_3N_4 can reach its maximum value at 5.2 W/m.K as reported by Ramdani et al., (Ramdani, Derradji, & Feng, 2015). In these nanocomposites, the silane-treated β-Si_3N_4 tends to form more conductive pathways into the polybenzoxazine matrix, as shown in Figure 4. New flexible polydimethylsiloxane matrix-based composites reinforced with boron nitride foam and nanosheets have been prepared and showed a synergistic thermal stability, thermal conductivity, electromagnetic shielding, low moisture uptake, and low dielectric constant (Fang, Bai, & Wong, 2017).

Polymer/Carbon-Based Filler Composites

Carbon-based nanofibers have also been found to increase the thermal conductivity of polymeric materials (Sui et al., 2008; Elgafy & Lafdi, 2005). However, the most extensively investigated nanoparticles for improving thermal conductivity are SWCNT, CB, MWCNT, and graphene based-fillers. These materials have attracted a great deal interest in the literature (Marconnet, Yamamoto, & Panzer, 2011). They have an increased thermal conductivity and good aspect proportion and are therefore easily organized into conductive networks within polymeric matrices at very low filler ratios. Several carbon-based fillers had been applied to ameliorate the low thermal conductivity of polymeric matrices. For examples, with the addition of 7 vol.% expanded graphite nanoparticles filled high density polyethylene (HDPE) exhibited an increased thermal conductivity of about 1.6 W/m.K compared to only 0.78 W/m.K for the micro-composites ones (Ye, Shentu, & Weng, 2006). Thermal conductivity measurements showed that graphene (Gn), of only a few layers ($n \sim 4$) and a 2 nm thickness, functions as very efficient filler for epoxy composites. When embedded in an epoxy matrix, G4 GNPs resulted in a thermal conductivity improvement exceeding 3000% (loading of 25 vol.%), and thermal conductivities can reach 6.5 nearly W/m.K surpassing the properties level achievable using standard fillers which demand a ratio of ~70 vol.% to reach these values. The outstanding thermal

Figure 4. TEM micrographs of the surface of the polybenzoxazine/Si_3N nanocomposites at various Si_3N_4 content: (a) 30 wt.% without silane, (b) 30 wt.% silane-treated, (c) 40 wt.%, and (d) 50 wt.%. (e) SEM micrograph of polybenzoxazine/Si_3N nanocomposite filled with 70 wt.% of nano-filler. Reprinted with permission from Elsevier (Ramdani et al., 2015)

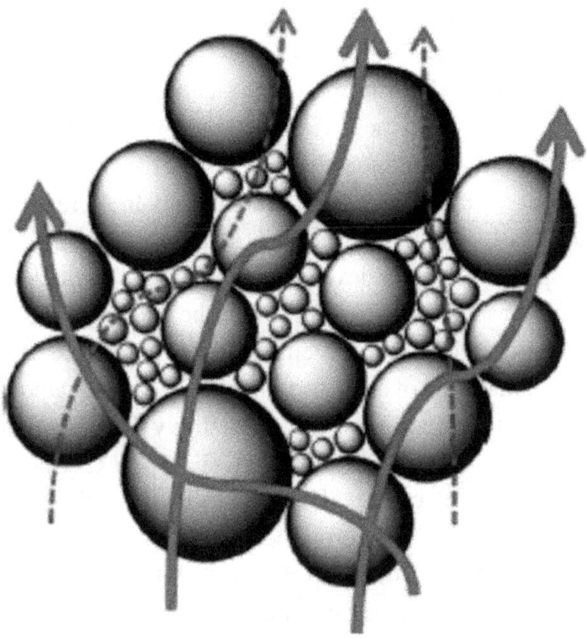

properties of this material can be attributed to the combination of the high aspect ratio, two-dimensional geometry, stiffness, and low thermal interface resistance of GNPs (Yu, Bekyarova, & Zhao, 2007). The combination of multi-graphene platelets and MWCNTs, resulting I improved tensile and impact properties and an increased thermal conductivity of the reinforced-epoxy resin (Yang, Lin, & Huang, 2011). The chemically-modified graphite filled an epoxy nanocomposite exhibited higher thermal conductivity that reached 28-fold over the virgin epoxy resin by incorporating modified graphite at a ratio of 20 wt.% ratio (Ganguli, Roy, & Anderson, 2008). A nanoscale binary filler system of SWCNTs and graphite nano-fillers gave a better improvement in the epoxy composite's thermal conductivity value, which can be justified by the construction of an efficient percolating system that relatively lowered the thermal resistance in the filler/matrix interface (Yu, Ramesh, & Sun, 2008). The effect of the incorporation of benzenetricarboxylic acid functionalized MWCNTs on the formation of thermally-conductive pathways and the thermal conductivity of epoxy resin has also been studied. The thermal conductivity of these composites was markedly enhanced from about 0.1 to 1 W/m.K through the filling with only a small ratios of treated-MWCNTs ranging from 1-5 vol.%, respectively, (Yang, Ma, & Teng, 2010). A homogeneous dispersion state of graphene flakes into epoxy matrix was achieved using un-oxidized graphene with non-covalent acid treatment. The resulting filled epoxy resins had improved thermal conductivities (~1.5 W/m.K) and good tensile modulus values (~1 GPa) (Song, Park, & Kim, 2013). Dodecylamine-treated graphene flakes and silane-modified MWCNTs were applied to develop a new kind of cyanate ester (CE) thermally conductive nanocomposites. With the incorporation of only 5 wt.% of these fillers to the pure cyanate ester resin, their resulted composites thermal conductivities reached 3.2 and 2.5 times higher than that of pure thermoset. Using hybrid fillers resulted in an increase of 185% (Zhao, Xu, & Qin, mailto:zhaocblw@163.com2014). Additional kinds of cyanate ester/nano-diamond composite has been developed with a high thermal conductivity and excellent mechanical properties, which promote their uses in various applications such as in microelectronic devices and engineering materials (Sen & Kahraman, 2014). Furthermore, a new type of epoxy composite filled with vapour grown carbon fiber (VGCF) had been developed exhibiting amazingly thermal conductivity of 695 W/m.K, low density, and a negative CTE value (−1.5 ppm/°C (Chen & Ting, 2002). New nanocomposites have also been designed to lowered the dielectric constant value of new polybenzoxazine resin through the addition of various amounts of GO as illustrated in Figure 5 (Sasi kumar, Ariraman, & Alagar, 2015). In this work, the concept of polarization was exploited, enabling the use of prepared films as dielectrics in microelectronics applications (kumar, Ariraman, & Alagar, 2015). The thermal conductivity of polystyrene (PS) was remarkably increased by incorporation of a colloidal graphite filler, and the maximum thermal conductivity of the prepared composites reached 1.95 W/m.K with a 34 vol.% filler loading (Tu & Ye, 2009). It was reported that the thermal conductivity of silicon rubber resin could be further enhanced by adding graphite nano-fillers via a solution intercalation technique rather than by using a melt mixing process (Mu & Feng, 2007).

Graphene sheets have also been to increase the thermal conductivity of polybenzoxazine thermoset and the resultant materials were found to be highly suitable for using as bipolar plates for fuel cell purposes (Plengudomkit, Okhawilai, & Rimdusit, 2016). In addition, expanded perlite/n-eicosane composites have been prepared for heat energy stock applications. The filling with CNTs did highly increase the thermal conductivity of these composites (Karaipekli, Biçer, & Sarıc, 2017). Furthermore, it has been discovered that the value of thermal conductivity for a polycarbonate (PC) nanocomposite reinforced by 9.9 wt.% of EG and 0.1 wt.% MWCNT fillers was markedly raised by almost 49% higher than that of a PC composite filled with 10 wt.% EG alone (Kim, Kim, & Yang, 2017). Morphologically tailored MWCNT/polyamide-6/poly(p-phenylene sulfide) hybrids, having high

Figure 5. Preparation of graphene oxide-polybenzoxazine film. Reprinted from (Sasi kumar et al., Ariraman, & Alagar, 2015) with permission from RSC publishers

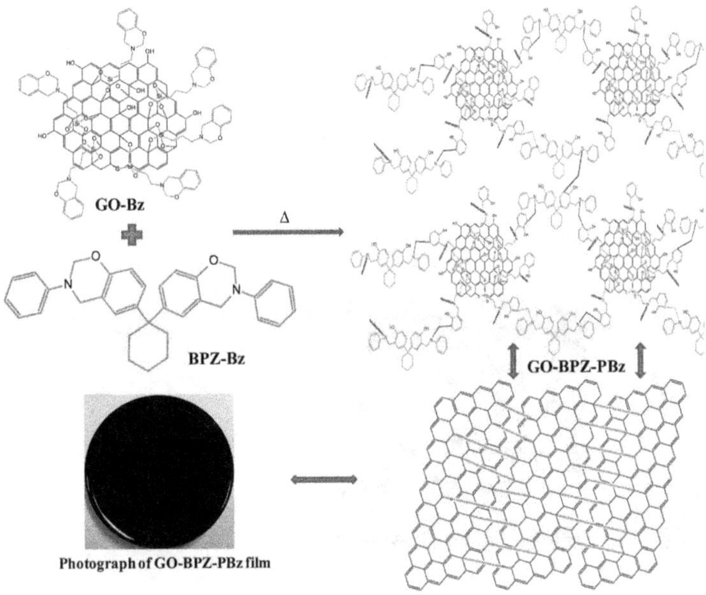

thermal conductivity, better electrical insulation and elastic modulus, have also been successfully produced (Morishit, Katagiri, & Matsunag, 2017). Furthermore, polyamide 6/expanded graphene composites have been developed by an *in*-situ intercalation polymerization approach, at an expanded graphene loading of only 12 wt.%, these materials exhibited a high thermal conductivity of 2.5 W/m.K, corresponding to an increase of 678% from to that of the pure polymer (Meng, Huang, & Guo, 2017). The addition of graphene into a poly(vinylidene fluoride) resin resulted in significant improvements in thermal conductivity and tensile strength. More specifically, with a graphene ratio of 20 wt.%, the thermal conductivity of the resulted membrane was raised by almost 213% from that of the unfiled thermoset, this was further increased to 226% when an electric field was applied during processing (Guo, Li, & Li, 2017). In this work, a perpendicular electrical field was applied to the composite membrane surface during casting, which improved the alignment of the graphene within the composite as shown in Figure 6.

Polymer/Metals Composites

The effective thermal conductivity of aluminum powder reinforced HDPE thermoplastic has been numerically studied by various theoretical and empirical approaches as a function of the filler fractions (Kumlutas, Tavman, & Coban,

Figure 6. The application of electric field on the graphene/PVDF membranes. Reprinted from (Guo, Li, & Li, 2017) with permission from Elsevier

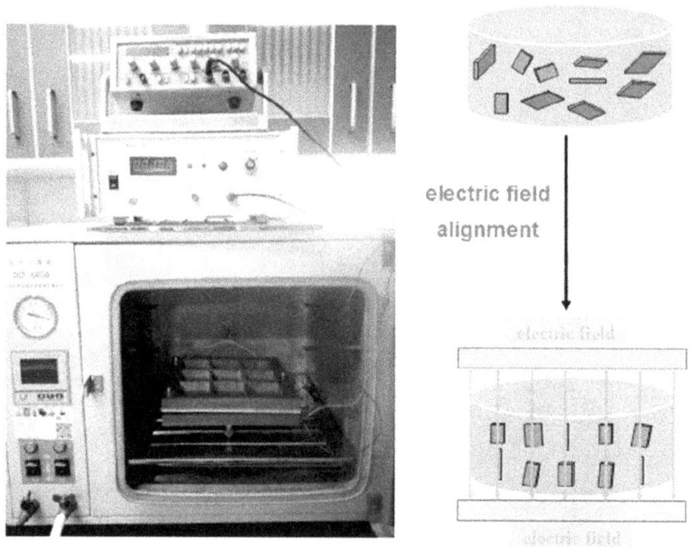

2003). The obtained thermal conductivity of an aluminium particle-reinforced polymer composite has also been predicted numerically using micro-structural photos. Since each pixel is identified by means of a finite difference equation and is accompanied by suitable image-treatment, the thermal conductivity of the prepared hybrids can be numerically evaluated. At low filler loading, a high correlation between the experimental and numerical data has been achieved. However, for filler concentrations that exceed 10 wt.%, the calculated thermal conductivity was increased exponentially and all the used models to date failed to reproduce thermal conductivity values in this domain. By contrast, the finite element numerical simulation provides satisfactory data in whole particle concentration ranges. The electrical and thermal conductivity, specific heat, thermal diffusivity, and effusively of polypropylene thermoplastic reinforced with two different sized copper fillers were studied (Kumlutaş et al., 2003). Good heat dissipation ability was detected for the composites reinforced by small particles and the *Agari* model enabled accurate predication of the thermal conductivity data for these composites over the full range of particle ratios. Metallic particles filled polymeric matrices include polyethylene (Kumlutaş et al., 2003; Tavman, 1996), polypropylene (Boudenne, Ibos, & Fois, 2005), polyamide (Tekce, Kumlutas, & Tavman, 2007), polyvinylchloride and epoxy resins (Mamunya, Davydenko, & Pissis, 2002), and such composites reveal good thermal conductivity that directly depends on the inherent thermal conductivity of each metallic fillers, the particle shape and dimensions of the filler, the filler volume ratio and the dispersion of the filler into these matrices. The thermal conductivity of some metal powder modified-polymer feedstocks for powder injection molding was also investigated (Kowalski, Duszczyk, & Katgerman, 1999). It has been also reported that by filling a polyacrylate polymer with only 0.9 vol.% of thermally-conductive single-crystalline copper nanowires (Cu_{NWs}) the resulting nanocomposites can exhibit thermal conductivity as high as 2.46 W/m.K (Wang, Cheng, & Wang, 2014). Using *Scanning Electron Microscopy* (SEM), Mu and colleagues found that a percolation threshold of 31.4 vol.% can be attained when ZnO filler is used to form a composite with a silicone rubber matrix. Whereas, the use of hybrid fillers resulted in more densely packed structures compared with those of single fillers (Mu, Feng, & Diao, 2007). A high-temperature short annealing time of micro-sized silver flakes found that they played a critical role in enhancing the electrical conductivity of their related electrically-conductive adhesives showing a very low resistivity value of 1.28×10^{-4} Ω·cm at 70 wt.% filler content (Li, Li, & Cheng, 2017).

Hybrids Filler Reinforced Polymer Composites

The thermal conductivity of reinforced-polymeric composites filled with both MWCNT nanoparticles and spherical Al_2O_3 micro-particles was numerically and experimentally improved. The proposed thermal model predicted by the *Finite Elements Method (FEM)* demonstrated agreement between the experimental and predicted thermal conductivity data (Sanada, Tada, & Shindo, 2009). By filling an epoxy resin with a 6 wt.% MWCNT or nearly 72 wt.% micro-SiC, the thermal conductivity of the resulted hybrids attained its maximum value of 2.9, which is about 21 times higher than that of the pure thermoset. To beyond ameliorate the thermal conductivity, the combination of the two fillers (5 wt.% MWCNTs + 55 wt.% micro-SiC) was used to obtained the best thermal conductivity improvement, which was attained 24 times that of the matrix (Zhou, Qi, & An, 2007; Zhou, Wang, & Liu, 2017). SiO_2-modified MWCNTs were elaborated using a sol–gel technique and then used to reinforce an epoxy resin. The thermal conductivity of the epoxy/MWCNT@SiO_2 nanocomposites increased by almost 51% and 67% at low filler levels of 0.5 wt.% and 1 wt.%, respectively. Simultaneously, SiO_2 shell retained the high electrical resistivity of the prepared nanomaterials (Cui, Du, & Zhao, 2011). A new MgO@MWNT-filled epoxy nanocomposite retains electrical insulation despite of increasing filler content. However, the thermal conductivity of epoxy was increased with an increase in MgO@MWNT content. When the MgO@MWNT content reached 2 wt.%, thermal conductivity was raised by 89% in comparison to unfilled epoxy, higher than that of untreated MWNT nanocomposites a fixed filler content (Du, Tang, & Huang, 2013). The thermal conductivity data of pure epoxy resin filled with SiO_2 and CNT increased with an increase of nanoparticle loading. These results were then compared with the three models which are Lewis-Nielsen or Halpin-Tsai analytical models, reflecting that the thermal conductivity is directly dependent on the state of filler dispersion state of the nano-fillers. Moreover, the influence of temperature on the thermal properties of the prepared nanocomposites is discussed in detail (Tessem, Zhao, & Moll, 2017). The inclusion of a hybrid filler of 30 vol.% modified BN and 1 vol.% functionalized MWCNTs into epoxy resin provided a significant enhancement of thermal conductivity attained almost 743% as shown in Figure 7 (Teng, Ma, & Chiou, 2011).

CONCLUSION

For several years ago, metals had been widely used to dissipate heat generated in electric engines and equipment, electric machines, exchangers of heat for power generation, and automotive. However, due to their heavy feature and high electrical conductivity, these materials are gradually replaced by thermally-conductive insulating polymer composites due to the tiny integration and the miniaturized electronic systems, the dissipation of generated heat become a serious problem for using and designing such devices. These newly developed composites are produced with low cost and their properties satisfied the requirement of microelectronic industries, including

Figure 7. Thermal conductivity of epoxy composites with various types of hybrid filler. Reprinted from (Teng, Ma, & Chiou, 2011) with permission from Elsevier

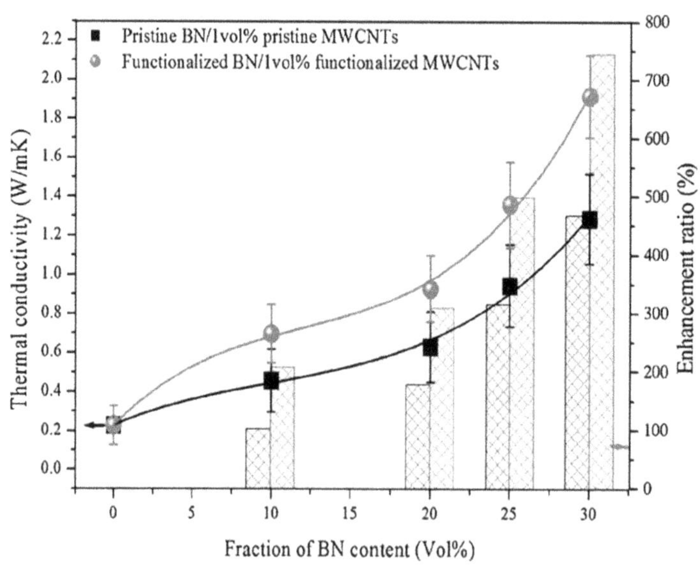

a low thermal expansion coefficient, increased thermal conductivity, reduced dielectric constant and loss, as well as light weight. Till now, numerous kinds of thermally-conductive elastomers, thermoplastics or thermosetting polymers based micro- and nanocomposites have been developed by adding the various kinds of inorganic conductive fillers, which includes metallic powders, ceramic particles, and carbon based fillers using mainly melt processing, solution blending, and sol-gel polymerization techniques. In addition, the effect of particles size, particles surface-treatment, and filler ratio on the thermal conductivity of reinforced-polymer systems are extensively investigated and predicted by different theoretical and semi-empirical equation models. Moreover, studies on these polymer/ceramic composites revealed that generally a compromise between the thermal and mechanical properties must be established hence using higher filler ratio will alter the mechanical properties while it significantly improved the thermal properties. On the other hand, using nano-sized particles will rapidly result in the percolation threshold for carbon based filler, which influence the dispersion of nanoparticles into polymers. The improvement mechanism of thermal conductivity in the produced polymeric composites is related to the construction of uniform thermally-conductive chains within the polymeric matrix, which play the role of heat sink can effectively dissipate the generated heat.

ACKNOWLEDGMENT

The authors are very grateful Mr. Oukil Khaled Ibn El-walid and Mr. Abid Nacereddine for his grateful help. In addition, we are grateful to all the publishers, including Elsevier, Wiley, and Springer for permission to reuse in our review illustrations issued from their publications.

REFERENCES

Agari, Y., Ueda, A., Omura, Y., & Nagai, S. (1997). Thermal diffusivity and conductivity of PMMA–PC blends. *Polymer*, *38*(4), 801–807. doi:10.1016/S0032-3861(96)00577-0

Agari, Y., & Uno, T. (1986). Estimation on Thermal Conductivities of Filled Polymers. *Journal of Applied Polymer Science*, *32*(7), 5705–5712. doi:10.1002/app.1986.070320702

Bigg, D. M. (1995). Thermal conductivity of heterophase polymer compositions. *Advances in Polymer Science*, *119*, 1–30. doi:10.1007/BFb0021279

Boudenne, A., Ibos, L., Fois, M., Majeste, J. C., & Géhin, E. (2005). Electrical and thermal behavior of polypropylene filled with copper particles. *Composites. Part A, Applied Science and Manufacturing*, *36*(11), 1545–1554. doi:10.1016/j.compositesa.2005.02.005

Causin, V., Marega, C., Marigo, A., Ferrara, G., & Ferraro, A. (2006). Morphological and structural characterization of polypropylene/conductive graphite nanocomposites. *European Polymer Journal*, *42*(12), 3153–3161. doi:10.1016/j.eurpolymj.2006.08.017

Chen, H., Ginzburg, V. V., Yang, J., Yang, Y., Liu, W., Huang, Y., ... Chen, B. (2016). Thermal conductivity of polymer-based composites: Fundamentals and applications. *Progress in Polymer Science*, *59*, 41–85. doi:10.1016/j.progpolymsci.2016.03.001

Chen, Y. M., & Ting, J. M. (2002). Ultra-high thermal conductivity polymer composites. *Carbon*, *40*(3), 359–362. doi:10.1016/S0008-6223(01)00112-9

Cheng, S. C., & Vachon, R. I. (1969). The Prediction of the Thermal Conductivity of Two and Three Phase Solid Heterogeneous Mixtures. *International Journal of Heat and Mass Transfer*, *12*(3), 249–264. doi:10.1016/0017-9310(69)90009-X

Choi, S., & Kim, J. (2013). Thermal conductivity of epoxy composites with a binary-particle system of aluminum oxide and aluminum nitride fillers. *Composites. Part B, Engineering*, *51*, 140–147. doi:10.1016/j.compositesb.2013.03.002

Cui, W., Du, F., Zhao, J., Zhang, W., Yang, Y., Xie, X., & Mai, Y. W. (2011). Improving thermal conductivity while retaining high electrical resistivity of epoxy composites by incorporating silica-coated multi-walled carbon nanotubes. *Carbon*, *49*(2), 495–500. doi:10.1016/j.carbon.2010.09.047

Droval, G., Feller, J. F., Salagnac, P., & Glouannec, P. (2006). Thermal conductivity enhancement of electrically insulating syndiotactic poly(styrene) matrix for diphasic conductive polymer composites. *Polymers for Advanced Technologies*, *17*(9-10), 732–745. doi:10.1002/pat.777

Du, F. P., Tang, H., & Huang, D. Y. (2013). Thermal Conductivity of Epoxy Resin Reinforced with Magnesium Oxide Coated Multiwalled Carbon Nanotube. *International Journal of Polymer Science*, *2013*, 1–5. doi:10.1155/2013/541823

Ebadi-Dehaghani, H., Reiszadeh, M., Chavoshi, A., Nazempour, M., & Vakili, M. H. (2014). A study on the effect of zinc oxide (ZnO) and calcium carbonate ($CaCO_3$) nanoparticles on the thermal conductivity (TC) of polypropylene. *Journal of Macromolecular Science, Part B: Physics*, *53*(1), 1364–1376. doi:10.1080/00222348.2013.810032

Elgafy, A., & Lafdi, K. (2005). Effect of carbon nanofiber additives on thermal behavior of phase change materials. *Carbon*, *43*(15), 3067–3074. doi:10.1016/j.carbon.2005.06.042

Fang, H., Bai, S. L., & Wong, C. P. (2017). Thermal, mechanical and dielectric properties of flexible BN foam and BN nanosheets reinforced polymer composites for electronic packaging application. *Composites. Part A, Applied Science and Manufacturing*, *100*(0), 71–80. doi:10.1016/j.compositesa.2017.04.018

Fieler, D. P. (2000). Hybrid polymer-inorganic nanocomposites. *Russian Chemical Reviews*, *69*(1), 53–80. doi:10.1070/RC2000v069n01ABEH000506

Ganguli, S., Roy, A. K., & Anderson, D. P. (2008). Improved thermal conductivity for chemically functionalized exfoliated graphite/epoxy composites. *Carbon*, *46*(5), 806–817. doi:10.1016/j.carbon.2008.02.008

Gu, J., Lv, Z., Wu, Y., Guo, Y., Tian, L., Qiu, H., ... Zhang, Q. (2017). Dielectric thermally conductive boron nitride/polyimide composites with outstanding thermal stabilities via in-situ polymerization-electrospinning-hot press method. *Composites. Part A, Applied Science and Manufacturing*, *94*, 209–216. doi:10.1016/j.compositesa.2016.12.014

Gu, J., Meng, X., Tang, Y., Li, Y., Zhuang, Q., & Kong, J. (2017). Hexagonal boron nitride/polymethyl-vinyl siloxane rubber dielectric thermally conductive composites with ideal thermal stabilities. *Composites. Part A, Applied Science and Manufacturing*, *92*, 27–32. doi:10.1016/j.compositesa.2016.11.002

Gu, J., Zhang, Q., Dang, J., Zhang, J., & Yang, Z. (2009). Thermal conductivity and mechanical properties of aluminum nitride filled linear low-density polyethylene composites. *Polymer Engineering and Science*, *49*(5), 1030–1034. doi:10.1002/pen.21336

Guo, H., Li, X., Li, B., Wang, J., & Wang, S. (2017). Thermal conductivity of graphene/poly(vinylidene fluoride) nanocomposite membrane. *Materials & Design*, *114*, 355–363. doi:10.1016/j.matdes.2016.11.010

Haggenmueller, R., Fischer, J. E., & Winey, K. I. (2007). Single Wall Carbon Nanotube/Polyethylene Nanocomposites: Thermal and Electrical Conductivity. *Macromolecules*, *40*(7), 2417–2421. doi:10.1021/ma0615046

Han, Z., & Fina, A. (2011). Thermal conductivity of carbon nanotubes and their polymer Nanocomposites: A review. *Progress in Polymer Science*, *36*(7), 914–944. doi:10.1016/j.progpolymsci.2010.11.004

He, H., Fu, R., Han, Y., Shen, Y., & Song, X. (2007). Thermal conductivity of ceramic particle filled polymer composites and theoretical predictions. *Journal of Materials Science*, *42*(16), 6749–6754. doi:10.1007/s10853-006-1480-y

Huang, X. Y., Jiang, P. K., & Tanaka, T. (2011). A review of dielectric polymer composites with high thermal conductivity. *IEEE Electrical Insulation Magazine*, *27*(4), 8–16. doi:10.1109/MEI.2011.5954064

Ishida, H., & Rimdusit, S. (1998). Very high thermal conductivity obtained by boron nitride filled polybenzoxazine. *Thermochimica Acta*, *320*(1-2), 177–186. doi:10.1016/S0040-6031(98)00463-8

Jang, I., Shin, K. H., Yang, I., Kim, H., Kim, J., Kim, W.-H., ... Kim, J.-P. (2017). Enhancement of thermal conductivity of BN/epoxy composite through surface modification with silane coupling agents. *Colloids and Surfaces. A, Physicochemical and Engineering Aspects*, *518*, 64–72. doi:10.1016/j.colsurfa.2017.01.011

Jiang, Y., Sun, R., Zhang, H. B., Min, P., Yang, D., & Yu, Z. Z. (2017). Graphene-coated ZnO tetrapod whiskers for thermally and electrically conductive epoxy composites. *Composites. Part A, Applied Science and Manufacturing*, *94*, 104–112. doi:10.1016/j.compositesa.2016.12.009

Kananu, S. K. (2003). Dielectric properties of matrix composite materials with high volume concentrations of inclusions (effective field approach). *International Journal of Engineering Science*, *41*(12), 1287–1312. doi:10.1016/S0020-7225(03)00042-9

Karaipekli, A., Biçer, A., Sarıc, A., & Tyagie, V. V. (2017). Thermal characteristics of expanded perlite/paraffin composite phase change material with enhanced thermal conductivity using carbon nanotubes. *Energy Conversion and Management*, *134*(15), 373–381. doi:10.1016/j.enconman.2016.12.053

Kim, H. S., Kim, J. H., Yang, C. M., & Kim, S. Y. (2017). Synergistic enhancement of thermal conductivity in composites filled with expanded graphite and multi-walled carbon nanotube fillers via melt-compounding based on polymerizable low-viscosity oligomer matrix. *Journal of Alloys and Compounds*, *690*, 274–280. doi:10.1016/j.jallcom.2016.08.141

Kim, K., Ryu, S., & Kim, J. (2017). Melt-processable aggregated boron nitride particle via polysilazane coating for thermal conductive composite. *Ceramics International, 43*(2-1), 2441–2447.

Kowalski, L., Duszczyk, J., & Katgerman, L. (1999). Thermal conductivity of metal powder polymer feedstock for powder injection moulding. *Journal of Materials Science*, *34*(1), 1–5. doi:10.1023/A:1004424401427 PMID:15347988

Kumar, R. S., Ariraman, M., & Alagar, M. (2015). Studies on dielectric properties of GO reinforced bisphenol-Z polybenzoxazine hybrids. *RSC Advances*, *5*(30), 23787–23797. doi:10.1039/C5RA00472A

Kumlutas, D., Tavman, I. H., & Coban, M. T. (2003). Thermal conductivity of particle filled polyethylene composite materials. *Composites Science and Technology*, *63*(1), 113–117. doi:10.1016/S0266-3538(02)00194-X

Lee, E. S., Lee, S. M., Shanefield, D. J., & Cannon, W. R. (2008). Enhanced thermal conductivity of polymer matrix composite via high solids loading of aluminum nitride in epoxy resin. *Journal of the American Ceramic Society*, *91*(4), 1169–1174. doi:10.1111/j.1551-2916.2008.02247.x

Li, B., Li, R., & Xie, Y. (2017). Properties and effect of preparation method of thermally conductive polypropylene/aluminum oxide composite. *Journal of Materials Science*, *52*(5), 2524–2533. doi:10.1007/s10853-016-0546-8

Li, C., Li, Q., Cheng, L., Li, T., Lu, H., Tang, L., & Zhang, K., E, S., Jun Zhang, J., Li, Z., & Yao, Y. (2017). Conductivity enhancement of polymer composites using high-temperature short-time treated silver fillers. *Composites. Part A, Applied Science and Manufacturing*, *100*(0), 64–70.

Li, K.S., & Wang, Q. (2002). Advances in thermal conductive polymeric materials. *Journal of Functional Materials of China, 33*(1), 201-232.

Li, L., & Chung, D. D. L. (1994). Thermally conducting polymer–matrix composites containing both AlN particles and SiC whiskers. *Journal of Electronic Materials*, *23*(6), 557–564. doi:10.1007/BF02670659

Ling, W., Gu, A., Liang, G., & Yuan, L. (2010). New composites with thermal conductivity and low dielectric constant for microelectronic packaging. *Polymer Composites*, *31*(2), 307–313.

Mamunya, Y. P., Davydenko, V. V., Pissis, P., & Lebedev, E. V. (2002). Electrical and thermal conductivity of polymers filled with metal powders. *European Polymer Journal*, *38*(9), 1887–1897. doi:10.1016/S0014-3057(02)00064-2

Marconnet, A. M., Yamamoto, N., Panzer, M. A., Wardle, B. L., & Goodson, K. E. (2011). Thermal Conduction in Aligned Carbon Nanotube–Polymer Nanocomposites with High Packing Density. *ACS Nano*, *5*(6), 4818–4825. doi:10.1021/nn200847u PMID:21598962

Meng, F., Huang, F., Guo, Y., Chen, J., Chen, X., Hui, D., ... Zhou, Z. (2017). In situ intercalation polymerization approach to polyamide-6/graphite nanoflakes for enhanced thermal conductivity. *Composites. Part B, Engineering*, *117*, 165–173. doi:10.1016/j.compositesb.2017.02.043

Mishra, D., Raj, P. M., Tishler, J., Sun, T., Shipton, E., & Tummala, R. (2016). Multilayered Ferromagnetic Polymer Composite Structures for High-Density Power Inductors. *IEEE Transactions on Magnetics*, *52*(11), 1–5. doi:10.1109/TMAG.2016.2589925

Mohammed, H. A., & Uttandaraman, S. (2009). A review of vapor grown carbon nanofiber/polymer conductive composites. *Carbon*, *47*(1), 2–22. doi:10.1016/j.carbon.2008.09.039

Moore, A. L., & Shi, L. (2014). Emerging challenges and materials for thermal management of electronics. *Materials Today*, *17*(4), 163–174. doi:10.1016/j.mattod.2014.04.003

Morishit, T., Katagiri, Y., Matsunag, T., Muraok, Y., & Fukumori, K. (2017). Design and fabrication of morphologically controlled carbon nanotube/polyamide-6-based composites as electrically insulating materials having enhanced thermal conductivity and elastic modulus. *Composites Science and Technology*, *142*, 41–49. doi:10.1016/j.compscitech.2017.01.022

Mu, Q., & Feng, S. (2007). Thermal conductivity of graphite/silicone rubber prepared by solution intercalation. *Thermochimica Acta*, *462*(1-2), 70–75. doi:10.1016/j.tca.2007.06.006

Mu, Q., Feng, S., & Diao, G. (2007). Thermal conductivity of silicone rubber filled with ZnO. *Polymer Composites*, *28*(2), 125–130. doi:10.1002/pc.20276

Nunes dos Santos, W. (2007). Thermal properties of polymers by non-steady-state techniques. *Polymer Testing*, *26*(4), 556–566. doi:10.1016/j.polymertesting.2007.02.005

Ohashi, M., Kawakami, S., Yokogawa, Y., & Lai, G.-C. (2005). Spherical aluminum nitride fillers for heat conducting plastic packages. *Journal of the American Ceramic Society*, *88*(9), 2615–2618. doi:10.1111/j.1551-2916.2005.00456.x

Pan, C., Kou, K., Jia, Q., Zhang, Y., Wu, G., & Ji, T. (2017). Improved thermal conductivity and dielectric properties of h-BN/PTFE composites via surface treatment by silane coupling agent. *Composites. Part B, Engineering*, *111*, 83–90. doi:10.1016/j.compositesb.2016.11.050

Plengudomkit, R., Okhawilai, M., & Rimdusit, S. (2016). Highly filled graphene-benzoxazine composites as bipolar plates in fuel cell applications. *Polymer Composites, 37*(6), 1715–1727. doi:10.1002/pc.23344

Pleşa, I., Noţingher, P. V., Schlögl, S., Sumereder, C., & Muhr, M. (2016). Properties of Polymer Composites Used in High-Voltage Applications. *Polymers, 8*(5), 173–236. doi:10.3390/polym8050173

Progelhof, R. C., Throne, J. L., & Ruetsch, R. R. (1976). Methods of Predicting the Thermal Conductivity of Composite Systems. *Polymer Engineering and Science, 16*(9), 615–625. doi:10.1002/pen.760160905

Ramdani, N., Derradji, M., Feng, T., Tong, Z., Wang, J., Mokhnache, E., & Liu, W. (2015). Preparation and characterization of thermally-conductive silane-treated silicon nitride filled polybenzoxazine nanocomposites. *Materials Letters, 155*, 34–37. doi:10.1016/j.matlet.2015.04.097

Ramdani, N., Derradji, M., Wang, J., Mokhnache, E., Liu, W., Liu, Y., & Dong, W. (2016). Experimental and modeling of thermal and dielectric properties of aluminum nitride-reinforce polybenzoxazine hybrids. *Journal of Thermal Analysis and Calorimetry, 126*(2), 561–570. doi:10.1007/s10973-016-5492-9

Sanada, K., Tada, Y., & Shindo, Y. (2009). Thermal conductivity of polymer composites with close-packed structure of nano and micro fillers. *Composites. Part A, Applied Science and Manufacturing, 40*(6–7), 724–730. doi:10.1016/j.compositesa.2009.02.024

Sasi Kumar, R., Ariraman, M., & Alagar, M. (2015). Studies on dielectric properties of GO reinforced bisphenol-Z polybenzoxazine hybrids. *RSC Advances, 5*, 23787-23797.

Sen, F., & Kahraman, M. V. (2004). Thermal conductivity and properties of cyanate Ester/nanodiamond composites. *Polymers for Advanced Technologies, 25*(9), 1020–1026. doi:10.1002/pat.3346

Shimoto, T., Baba, K., Matsui, K., Tsukano, J., Maeda, T., & Oyachi, K. (2005). Ultra-thin high-density LSI packaging substrate for advanced CSPs and SiPs. *Microelectronics and Reliability, 45*(3-4), 567–574. doi:10.1016/j.microrel.2004.08.006

Song, S. H., Park, K. H., Kim, B. H., Choi, Y. W., Jun, G. H., Lee, D. J., ... Jeon, S. (2013). Enhanced thermal conductivity of epoxy-graphene composites by using non-oxidized graphene flakes with non-covalent functionalization. *Advanced Materials, 25*(5), 732–735. doi:10.1002/adma.201202736 PMID:23161437

Song, Y. H., Kim, S. G., Lee, S. B., Rhee, K. J., & Kim, T. S. (2003). A study of considering the reliability issues on ASIC/memory integration by SIP (system-in-package) technology. *Microelectronics and Reliability*, *43*(9–11), 1405–1410. doi:10.1016/S0026-2714(03)00251-8

Tang, B., Wei, H., Zhao, D., & Zhang, S. (2017). Light-heat conversion and thermal conductivity enhancement of PEG/SiO_2 composite PCM by in situ Ti_4O_7 doping. *Solar Energy Materials and Solar Cells*, *161*, 183–189. doi:10.1016/j.solmat.2016.12.003

Tavman, I. H. (1996). Thermal and mechanical properties of aluminum powder-filled high density polyethylene composites. *Journal of Applied Polymer Science*, *62*(12), 2161–2167. doi:10.1002/(SICI)1097-4628(19961219)62:12<2161::AID-APP19>3.0.CO;2-8

Tee, D. I., Mariatti, M., Azizan, A., See, C. H., & Chong, K. F. (2007). Effect of Silane based coupling agent on the properties of silver nanoparticles filled epoxy composites. *Composites Science and Technology*, *67*(11-12), 2584–2591. doi:10.1016/j.compscitech.2006.12.007

Tekce, H. S., Kumlutas, D., & Tavman, I. H. (2007). Effect of particle shape on thermal conductivity of copper reinforced polymer composites. *Journal of Reinforced Plastics and Composites*, *26*(1), 113–121. doi:10.1177/0731684407072522

Teng, C. C., Ma, C. C. M., Chiou, K. C., Lee, T. M., & Shih, Y. F. (2011). Synergetic effect of hybrid boron nitride and multi-walled carbon nanotubes on the thermal conductivity of epoxy composites. *Materials Chemistry and Physics*, *126*(3), 722–728. doi:10.1016/j.matchemphys.2010.12.053

Tessem, A., Zhao, D., Moll, J., Xu, S., Yang, R., & Li, C. (2017). Effect of filler loading, geometry, dispersion and temperature on thermal conductivity of polymer Nanocomposites. *Polymer Testing*, *57*, 101–106. doi:10.1016/j.polymertesting.2016.11.015

Tibbetts, G. G., Lake, M. L., Strong, K. L., & Rice, B. P. (2007). A review of the fabrication and properties of vapor-grown carbon nanofiber/polymer composites. *Composites Science and Technology*, *67*(7-8), 1709–1718. doi:10.1016/j.compscitech.2006.06.015

Tu, H., & Ye, L. (2009). Thermal conductive PS/graphite composites. *Polymers for Advanced Technologies*, *20*(1), 21–27. doi:10.1002/pat.1236

Vakili, M. H., Ebadi-Dehaghani, H., & Haghshenas-Fard, M. (2011). Crystallization and Thermal Conductivity of $CaCO_3$ Nanoparticle Filled Polypropylene. *Journal of Macromolecular Science, Part B: Physics*, *50*(8), 1637–1645. doi:10.1080/00222348.2010.543033

Vo, H. T., Todd, M., Shi, F. G., Shapiro, A. A., & Edwards, M. (2001). Towards model based engineering of underfill materials: CTE modeling. *Microelectronics Journal*, *32*(4), 331–338. doi:10.1016/S0026-2692(00)00152-X

Wang, J. J., & Yi, X. S. (2004). Effects of interfacial thermal barrier resistance and particle shape and size on the thermal conductivity of AlN/PI composites. *Composites Science and Technology*, *64*(10-11), 1623–1628. doi:10.1016/j.compscitech.2003.11.007

Wang, Q., Xia, H. S., & Zhang, C. H. (2000). Preparation of polymer/inorganic nanoparticles composites through ultrasonic inadiation. *Journal of Applied Polymer Science*, *80*(7-10), 1478–1488.

Wang, S. L., Cheng, Y., Wang, R. R., Sun, J., & Gao, L. (2014). Highly thermal conductive copper nanowire composites with ultralow loading: Toward applications as thermal interface materials. *ACS Applied Materials & Interfaces*, *6*(9), 6481–6486. doi:10.1021/am500009p PMID:24716483

Wong, Y. W., Lo, K. L., & Shin, F. G. (2001). Electrical and thermal properties of composite of liquid crystalline polymer filled with carbon black. *Journal of Applied Polymer Science*, *82*(6), 1549–1555. doi:10.1002/app.1993

Wooster, T. J., Abrol, S., Hey, J. M., & MacFarlane, D. R. (2004). Thermal, mechanical, and conductivity properties of cyanate ester composites. *Composites. Part A, Applied Science and Manufacturing*, *35*(1), 75–82. doi:10.1016/j.compositesa.2003.09.002

Wu, H., & Kessler, M. R. (2015). Multifunctional cyanate ester nanocomposites reinforced by hexagonal boron nitride after noncovalent biomimetic functionalization. *ACS Applied Materials & Interfaces*, *7*(10), 5915–5926. doi:10.1021/acsami.5b00147 PMID:25726956

Wu, X., Yang, Z., Kuang, W., Tang, Z., & Guo, B. (2017). Coating polyrhodanine onto boron nitride nanosheets for thermally conductive elastomer composites. *Composites. Part A, Applied Science and Manufacturing*, *94*, 77–85. doi:10.1016/j.compositesa.2016.12.015

Yang, S., Lin, W., Huang, Y., Tien, H.-W., Wang, J.-Y., Ma, C.-C. M., ... Wang, Y.-S. (2011). Synergetic effects of graphene platelets and carbon nanotubes on the mechanical and thermal properties of epoxy composites. *Carbon*, *49*(3), 793–803. doi:10.1016/j.carbon.2010.10.014

Yang, S. Y., Ma, C. C. M., Teng, C. C., Huang, Y. W., Liao, S. H., Huang, Y.-L., ... Chiou, K.-C. (2010). Effect of functionalized carbon nanotubes on the thermal conductivity of epoxy composites. *Carbon*, *48*(3), 592–603. doi:10.1016/j.carbon.2009.08.047

Ye, C. M., Shentu, B. Q., & Weng, Z. X. (2006). Thermal conductivity of high density polyethylene filled with graphite. *Journal of Applied Polymer Science*, *101*(6), 3806–3810. doi:10.1002/app.24044

Yeo, H., Islam, A. M., You, N. H., Ahn, S., Goh, M., Hahn, J. R., & Jang, S. G. (2017). Characteristic correlation between liquid crystalline epoxy and alumina filler on thermal conducting properties. *Composites Science and Technology*, *141*, 99–105. doi:10.1016/j.compscitech.2017.01.016

Yu, A., Bekyarova, E., Zhao, B., & Haddon, R. C. (2007). Graphite nanoplatelet–epoxy composite thermal interface materials. *Journal of Physics and Chemistry Part C*, *111*(21), 7565–7569. doi:10.1021/jp071761s

Yu, A., Ramesh, P., Sun, X., Bekyarova, E., Itkis, M. E., & Haddon, R. C. (2008). Enhanced thermal conductivity in a hybrid graphite nanoplatelet-carbon nanotube filler for epoxy composites. *Advanced Materials*, *20*(24), 4740–4744. doi:10.1002/adma.200800401

Yu, S., Hing, P., & Hu, X. (2002). Thermal conductivity of polystyrene–aluminum nitride composite. *Composites. Part A, Applied Science and Manufacturing*, *33*(2), 289–292. doi:10.1016/S1359-835X(01)00107-5

Yung, K. C., & Liem, H. (2007). Enhanced thermal conductivity of boron nitride epoxy-matrix composite through multi-modal particle size mixing. *Applied Polymer Science*, *106*(6), 3587–3591. doi:10.1002/app.27027

Zhang, B., Ye, F., Gao, Y., Liu, S. C., Liu, Q., & Ding, D. H. (2016). Fabrication and dielectric properties of BADCy/$Ni_{0.5}Ti_{0.5}NbO_4$ composites for ultra-low-loss printed circuit board application. *Ceramics International*, *42*(1), 234–241. doi:10.1016/j.ceramint.2015.08.099

Zhao, C., Xu, S., Qin, Y., Chen, H., Zhao, W., Sun, F., & Zhu, X. (2014). Thermally conductive cyanate ester nanocomposites filled with graphene nanosheets and multiwalled carbon nanotubes. *Polymers for Advanced Technologies*, *25*(12), 1546–1552. doi:10.1002/pat.3399

Zhi, C., Bando, Y., Terao, T., Tang, C., Kuwahara, H., & Golberg, D. (2009). Towards thermoconductive, electrically insulating polymeric composites with boron nitride nanotubes as fillers. *Advanced Functional Materials*, *19*(12), 1857–1862. doi:10.1002/adfm.200801435

Zhou, H., Zhang, S., & Yang, M. (2007). The effect of heat-transfer passages on the effective thermal conductivity of high filler loading composite materials. *Composites Science and Technology*, *67*(6), 1035–1040. doi:10.1016/j.compscitech.2006.06.004

Zhou, T., Wang, X., Liu, X., & Xiong, D. (2017). Improved thermal conductivity of epoxy composites using a hybrid multi-walled carbon nanotube/micro-SiC filler. *Carbon*, *48*(4), 1171–1176. doi:10.1016/j.carbon.2009.11.040

Zhou, W., Qi, S., An, Q., Zhao, H., & Liu, N. (2007). Thermal conductivity of boron nitride reinforced polyethylene composites. *Materials Research Bulletin*, *42*(10), 1863–1873. doi:10.1016/j.materresbull.2006.11.047

Zhou, Y., Liu, F., & Wang, H. (2017). Novel organic–inorganic composites with high thermal conductivity for electronic packaging applications: A key issue review. *Polymer Composites*, *38*(4), 803–813. doi:10.1002/pc.23641

KEY TERMS AND DEFINITIONS

Dielectric: An electrical insulator that can be polarized by an applied electric field.

Thermal Coefficient of Expansion (CTE): The tendency of matter to change in shape, area, and volume in response to a change in temperature.

Thermal Conductivity: A measure of the ability of a body to conduct heat.

Chapter 10
Advanced Composites for Civil Engineering Infrastructures

Xia Cui
Dalian University of Technology, China

Shuzhu Zeng
Dalian University of Technology, China

Zhen Li
Dalian University of Technology, China

Qiaofeng Zheng
Dalian University of Technology, China

Xun Yu
New York Institute of Technology, USA

Baoguo Han
Dalian University of Technology, China

ABSTRACT

The development of advanced composites not only enhances strength, ductility, durability of materials, and endows materials with the multifunctional property, but also reduces the construction cost and promotes civil engineering infrastructure to make sustainable development. In this chapter, several representative advanced composites with abundant research achievements and wide applications are systematically introduced with regard to cementitious composites, fiber-reinforced polymer composites, novel thermally functional composites, and 3D printing composites in terms of their definitions, properties, research progress, and applications in civil engineering infrastructures.

DOI: 10.4018/978-1-5225-5216-1.ch010

INTRODUCTION

Civil engineering infrastructures in a narrow sense involve buildings, bridges, roads, railways, tunnels, airports and other engineering facilities for living, transportation and production. Not only are civil engineering infrastructures for human beings the essential facilities of production and life, but also reflect the development of society and civilization. With the improvement of people's living standard and the progress of engineering technology, the demands for safer, smarter, more comfortable, more beautiful and more durable civil engineering infrastructures are growing dramatically. After decades of services, the existing civil engineering infrastructures are confronted with serious problem of material deterioration and functional deficiency. Furthermore, the sustainable development of civil engineering infrastructures has become an urgent issue in the future. Obviously, conventional building materials are difficult to meet the above challenges. Firstly, the defects of conventional materials are likely to result in a risk of structural failure and durability deterioration, such as the brittle failure of concrete and the corrosion of steel. Secondly, most traditional building materials are non-renewable and energy-intensive, thus they put huge pressure on environment with too much pollutant discharge and block the sustainable development of civil engineering infrastructures. Thirdly, modern cities tend to be more convenient, more humanized and smarter, which urges for new building materials possessing not only excellent mechanical properties but also multifunctional properties in various aspects like electricity, heat, sound, light and magnetism (Han et. al., 2015; Han & Zhang, 2017). Therefore, new advanced composites, such as high-performance concrete, multifunctional and smart cementitious composites, green and environmental civil engineering composites, are desired to meet the above demands in civil engineering infrastructures.

In this chapter, several representative advanced composites with abundant research achievements and wide applications are systematical introduced with regard to cementitious composites, fiber reinforced polymer composites, novel thermally functional composites and 3D printing composites, in terms of their definition, properties, research progress and application in civil engineering infrastructures.

CEMENTITIOUS COMPOSITES

During development history of civil engineering composites, cementitious composites become the main and irreplaceable construction materials because of their strong adaptability and low cost. However, poor ductility and single function of cementitious composites bring the challenges and opportunities for advanced cementitious composites. Several advanced cementitious composites with high-performance, multi-functional, green and smart are introduced as follows.

Self-Compacting Concrete

Self-compacting concrete (SCC) can be used for shortening the construction period of major engineering, assuring compaction in the confined zones of structures where vibrating compacting is difficult and eliminating effectively vibration noise for concrete products plants.

Definition and Properties

SCC is a highly-liquidity concrete mixture that can consolidate and densely fill into every corner of a formwork under its deadweight without any vibration. Moreover, SCC has the advantages of high flowability while it is non-segregative and non-bleeding compared with normally vibrated concrete (Domone, 2007; Han, Zhang & Ou, 2017).

Current Progress

The prototype of SCC was completed by Okamura for the first time in 1986 and was employed in settling the durability problems due to the decrease of skilled workers available in Japan's construction industry. Until now, the research and development of SCC is mainly focused on three aspects: self-compacting property of fresh SCC, test methods of self-compacting property, and properties of hardened SCC.

The property of fresh SCC, i.e. self-compacting, is very important to adjust mix proportion and to assess the self-compacting degree of SCC. It is affected by many factors, such as water to cement ratio, types and amount of aggregates, types and dosage of chemical admixtures, types and amount of mineral mixtures, fibers, and nanomaterials (Felekoğlu & Sarıkahya, 2008; Felekoğlu, Türkel, & Baradan, 2007; Ferrara, Park, & Shah, 2007; Kou & Poon, 2009; Sahmaran & Yaman, 2007; Sari, Prat, & Labastire, 1999). The influence of these factors on the performance of SCC is interaction and mutual restriction. Moreover, these factors are dominated by mix design in preparing SCC. Compared with conventional concrete, the amount of coarse aggregates reduces, but the amount of cement increases in process of fabricating SCC.

Workability of SCC indicates the self-compacting property of SCC. It includes self-filling ability, passing ability and high stability that was characterized by the corresponding test methods, as listed in Table 1.

Table 1. Test methods for self-compacting property of SCC

Properties	Methods
Self-filling ability	1. Slump flow Abrams cone 2. T_{50cm} slump flow 3. V-funnel 4. Orimet 5. U-box
Passing ability	1. J-ring 2. L-box
High stability	GTM screen stability

In the meanwhile of concerning the property of fresh SCC, much research efforts also have been devoted to exploring the properties of hardened SCC (Chan, Chen, & Liu, 2003; Domone, 2006; Mazaheripour et al., 2011; Parra, Valcuende, & Gómez, 2011; Rozière et al., 2007; Zhu & Bartos, 2003). Compared with conventional concrete, bond to steel, sulphate resistance and permeation properties of SCC enhance, however, modules of elasticity, shrinkage and creep of SCC are worse. In addition, compressive and tensile strength have no obvious difference with conventional concrete (Goodier, 2003).

Application

SCC can be used for ready-mixed concrete (cast-in-place) and prefabricated products (precast members) as detailed as follows: (1) cast-in-place application, such as construction of bridge, tank, high building, and nuclear power plant. A respective engineering example of SCC is the two anchorages of Akashi-Kaikyo (Straits) Bridge. SCC was mixed on site and pumped through a piping system to the specified point in this project. The anchorage construction period was shortened by 20%, from 2.5 to 2 years. (2) precast application, such as fabrication of super span prestressed beam, ecological revetment member, and prefabricated square columns. SCC has made it possible to precast quality concrete structures. The use of SCC in the precast market of the UK, Europe, and Asia-Pacific region is continuing to increase with the development of module buildings, especially in Asia-Pacific region.

Engineered Cementitious Composites

As we all know, conventional concrete has very poor ductility. Although the development of fiber reinforced concrete (FRC) has improved toughness, strain softening will occur under tensile loading. The appearance of engineered cementitious composites (ECC) enhances the toughness of FRC and meets the demand for strain hardening of FRC.

Definition and Properties

ECC is ultra ductile fiber reinforced cementitious composites systematically designed on the basis of micromechanics and engineered to achieve high ductility under tensile and shear loading (Li, 2002; Li, 2008).

ECC has a great advantage of super high toughness. During the tensile loading process, ECC presents a multiple cracking pattern and a strain-hardening behavior (Li & Leung, 1992). The ultimate tensile strain of ECC is usually in the range of 3% to 7%, although the fiber volume fraction is less than 2%. The crack width is generally smaller than 60μm even in the saturated multi crack stage near failure. In fact, the ultimate strain of ECC is composed of many fine micro cracks, not a single wide crack. Moreover, the fine micro crack width is only determined by inherent property of ECC. The great durability of ECC is because small and fine crack width prevent effectively from external harm (Li, 2003; Li, 2007; Şahmaran & Li, 2009; Zhou et al., 2012).

Current Progress

ECC theory began in 1992 (Li & Leng, 1992). It was the earliest attempt to solve the problem of strengthening and toughening of cementitious composites with polyethylene (PE) fiber. In 1997, Li and Kanda began to use polyvinyl alcohol (PVA) for ECC and prepared PVA fiber reinforced cementitious composites. The tensile strength and tensile elastic modulus of PVA fiber is higher than that of ordinary concrete. Moreover, molecular of PVA fiber contains OH groups that can form high bond strength with cement slurry. The cost of PVA fiber is 1/8 of the same volume of PE fibers, and is significantly lower than steel fiber. At present, the main research is focused on PVA-ECC.

After more than ten years of development, PVA-ECC has made great progress in theoretical and experimental research. Its research and development uses a unique integrated structure and material design method (ISMD). ISMD linked the microscopic and macroscopic systems in the material structure, and joined the design and construction (Li, 2002). Thus, material properties serve as a natural interface between structural engineering and materials engineering. Besides, the researchers studied and developed a series of ECC products through various optimizations and improvements, such as self-healing ECC, green ECC, and so on. Kan et al. (2012) found that ECC with multiple micro-cracks has self-healing behavior. Furthermore, fly ash incorporated into ECC contributes to create more fine cracks with small width, which can enhance self-healing property. Recently, a large number of fly ash (Şahmaran & Li, 2009), slag (Kim et al., 2007) rubber (Huang et al., 2013) and other industrial byproducts have been applied to the preparation of ECC to replace some part of ECC raw materials. The new ECC not only enhance mechanical and durable performance, but also reduce environmental pollution and save energy. Because of high cost of ECC production, its application in engineering is constrained. However, if the geological polymers of the mineral matrix take the place of cement, production cost of ECC will come down (Ohno & Li, 2014).

Application

ECC exhibits great toughness and good strain hardening properties, whether in bending or in tension. In addition, under extreme environmental conditions, ECC has excellent durability and fatigue resistance. Therefore, it has a broad application prospect in seismic structure, bridge road construction, structure reinforcement and reinforcement.

ECC is applied in the connection plate of bridge deck without expansion joint. ECC take the place of original materials of the adjacent bridge panel and cover the joint between the adjacent bridge panels to form the continuous bridge deck without joint (Lepech & Li, 2009). ECC can meet the requirement of structure and durability of bridge deck. Shrinkage deformation and creep of adjacent bridge panel due to wet and dry cycles, temperature and stress action can be adjust by ECC connecting plate, so as to reduce and avoid cracking (Kim & Fischer, 2009).

ECC can be used as a protective layer to improve the durability of concrete. The average crack width of ECC is about 60μm. It can meet the durability requirements of the building. On the other hand, the energy duo to crack of component can be absorbed or weakened by ECC, preventing further development and intercommunication of cracks, and then blocking the harmful impurity entering into concrete interior.

ECC is also applied in earthquake resistant structural member. The strain-hardening and multiple-cracking properties of ECC beams can be used to resist shear, prevent bond splitting and restrain concrete cracking after yielding of members (Li, 2004). Furthermore, fine microcrack is easily to be repaired after damage of ECC components. Reciprocating cycle load test results shows that ECC beam can play the role of constraints on the longitudinal rebar, preventing bending damage prematurely (Kamada & Li, 2000). In summary, ECC can be effectively applied in earthquake resistant structural member.

Reactive Powder Concrete

Reactive powder concrete (RPC) is an extraordinary development in concrete field. Application of RPC in civil engineering structure can reduce effectively the weight of the structure, prolong structural service life and reduce the amount of rebar.

Definition and Principle

RPC is a dense system through ultra-fine particles and fiber reinforced cementitious composites. It features excellent mechanical properties, durability, and volume stability because of high dosage of cement and silica fume, large amounts of fibers, and low dosage of water. It is named because of high content of ingredients with high reactivity and fineness (Richard & Cheyrezy, 1995). The main ingredients of RPC are cement, quartz sand, silica fume, fly ash, superplasticizer and water. The coarse aggregate usually used for concrete is removed. In addition, steel fibers are incorporated in preparation of RPC.

Compared with common concrete, the reinforcement of RPC owe to reactivity and fineness of components and pore, microstructure. Silica fume as a component is critical to the compactness of RPC, because particle size of silica fume is two orders of magnitude lower than that of cement particles. On the other hand, silica fume can from the intensive pozzolanic reaction. Calcium hydroxide and silica fume react to produce ultrafine gels. These gel particles heap and fill the pore, so the microstructure of RPC is refined. This can reduce the ion diffusion rate as well as obtain good corrosion resistance, durability and high strength (Cheyrezy, Maret, & Frouin, 1995).

The steel fibers also play an important part to reinforce the mechanical properties of RPC. In addition, steel fibers give RPC deformability like metals. The fracture energy, which is a measurement of ductility, can reach 40,000 J/m^2 for RPC200, as compared to 100 to 150 J/m^2 for ordinary concretes (Dugat, Roux, & Bernier, 1996). Fracture energy depends on the volume of fibers added to the concrete.

In addition, the pressure-casting and heat treatment are applied during the curing process. The heat treatment can accelerate cement hydration and activate the pozzolanic reaction of the silica fume. Pressurization of 60MPa can remove entrapped air bubbles and expel excess water during setting, thus increasing the composites' compactness by about 5% to 6%.

Current Progress

RPC was developed by Bouygues of French in 1993. High performance concrete and reactive powder concrete international seminar was held in Canada at 1998, in which the principle, performance and application of RPC were widely discussed. The experts agreed that RPC has wide application prospect as a new type of concrete. After the past decade, a widely variety of researches about RPC have been obtained. The compressive strength of RPC is between 200 and 800MPa, and its flexural strength can reach 140MPa. The comparison about mechanical properties between RPC and ordinary concretes are shown in Table 2 (Dugat, Roux, & Bernier, 1996). Han et al. (2016) studied the super-fine stainless wires reinforced RPC. Results showed that flexural strength and fracture energy increase by 103% and 442%, respectively. In addition, the electrical conductivity and self-sensing properties of RPC can be improved by the incorporation of short-cut and super-fine stainless wire (Dong et al., 2016). Chan and Chu (2004) though that silica fume can effectively improve the interfacial properties of fiber, especially in fiber pullout energy, and also concluded that the optimal content of silica fume range from 20% to 30%, in terms of the bond performance. In some researches, the nano fillers are tried to replace steel fiber in RPC. Han et al. (2017) use Nano SiO_2-coated TiO_2 (NSCT) to reinforce RPC. The flexural and compressive strengths of NSCT reinforced RPC at age of 28d achieve increases of 87%/6.69MPa and 12.26%/12.2MPa with respect to RPC without NSCT, respectively. Yazici et al. (2009; 2008) added fly ash or ground granulated blast furnace slag into RPC in order to decrease the content of cement and silica fume. This behavior has no obvious loss for mechanical properties.

Application

In 1997, the world's first large-scale structure to be built with RPC was completed with the construction of the pedestrian bridge in the city of Sherbrooke, Quebec, Canada (Blais & Couture, 1999). The deck and top and bottom chords were made

Table 2. Comparsion between mechanical properties of RPC and those of ordinary concrete

Types	Ordinary Concrete	RPC
Compressive strength (MPa)	20-50	200-800
Flexural strength (MPa)	4-8	15-140
Fracture energy (J/m^2)	130	1000-40000
Ultimate tensile strain (10^{-6})	100-150	2000-8000

of RPC with a compressive strength of 200MPa. For the diagonal web members, the RPC was confined in stainless steel tubes and could withstand 350MPa in compression. The 3 m deep truss spans 60 m across the Magog River (in dowtown Sherbrooke) in a circular arch with a radius of 326 m, carrying both pedestrian and bicycle traffic. Lee et al. (Lee, Wang, & Chiu, 2007) had been using RPC as the repair material. Results indicated that RPC displayed an excellent bond durability to existing concrete.

Recycled Aggregate Concrete

Recycled concrete technology can realize the reproduction of waste concrete, and forming new concrete products. The technology can not only save energy and reduce consumption, but also promoting sustainable development of concrete. As a result, it is one of the main ways to develop green concrete.

Definition

Recycled aggregate concrete (RAC), or recycled concrete for short, refers to the recycled aggregate (RCA) is incorporated into matrix and replace some or all of the natural aggregate in order to form new concrete. The RCA is prepared by the waste concrete that is treated through crushing, washing and grading and then re-mixed according to a certain proportion of mixed formation and gradation.

Current Progress

Research on the reuse of waste concrete can be traced back to 1940s. Development of RAC optimizes the concrete structure and realizes the sustainable development of concrete composites. It is possible to promote the concrete industry towards low-carbon industrialization. Therefore, RAC is paid much attention from around the world. Up to now, more and more researches on RAC have been investigated and nearly covered every aspect of mechanical properties, microstructure and structural performance (Hansen, 1992; Mcneil & Kang, 2013).

The amount of RCA has remarkable influence on the mechanical properties of concrete. In general, the compressive, tensile, flexural and shear strengths of RAC are lower than those of common concrete ((Kou, Chi, & Chan, 2007; Rahal, 2007; Zhuang & Yang, 2006). Moreover, strength of RAC decreases with the increase of the RCA content. However, introduction of water reducing agent and high concentrations of cement is beneficial to enhance strength RAC. Previous study showed that RCA proportion had almost no effect on the bond strength between RAC and rebar (Corinaldesi & Moriconi, 2009). The modulus of elasticity for RAC generally reduces as RCA proportion increase (Kou, Chi, & Chan, 2007). However, the peak strain is larger than that of common concrete (Xiao, Li, & Zhang, 2005), the reason for which is that decrease of elasticity modulus leads to a large deformation in RAC. In addition, RAC has adverse effect on durability of concrete. Both the carbonation resistance (Sagoe-Crentsil, Brown, & Taylor, 2001) and the chloride penetration (Jian et al., 2009; Sim & Park, 2011) of RAC reduce with the increasing RCA proportion. The drying shrinkage of RAC increases as the RCA proportion and water/cement ratio increase, but it decreases when an amount of fly ash and water reducing agent are added into the RAC (Guo et al., 2011). There is research showed that creep of RAC was higher than that of common concrete. Moreover, the creep of RAC increases as RCA proportion increases. However, the addition of slag and high-quality RCA is helpful to decrease the creep of RAC (Zou, Wang, & Hu, 2009).

The microstructure of RAC was much complicated as compared to the common concrete (Marí, Etxeberria, & Vázquez, 2006; Tam, Gao, & Tam, 2005). RAC possessed two the interfacial transition zones (ITZs), one is between the RCA and new mortar matrix, and the other is between the RCA and the old mortar attached (old ITZ). The old ITZ consist of many voids and cracks, which consume water of composites, thus leading to less water for hydration at the ITZ zones of RAC. Therefore, the micro-hardness of the interface between the RCA and the new mortar matrix was the lowest, which embodied in lower macro mechanical properties (Du eta al., 2010).

The structural behavior of RAC members is usually slightly weaker than that of members made with natural aggregates. With the increasing RCA proportion, the seismic performance has no obvious decline, which indicates that it is feasible to apply RAC frame or joints to practical engineering. As a result, RAC can be used as a structural material from the view points of the loading capacity as well as seismic behavior through proper construction and durability design (Corinaldesi, Letelier, & Moriconi, 2011).

Application

As we know, RAC is a low carbon concrete that can save resources and reduce carbon emissions. In addition, RAC can basically meet the performance requirements of ordinary concrete through reasonable design, thus it is feasible to apply RAC in civil engineering infrastructure.

In Japan, the United States, China and many other countries, RAC have been applied in the engineering practice, and the relevant standards and specifications are formulated. RAC had been widely used in highway construction in the United States. In France, the broken concrete and crushed bricks were used to produce brick concrete blocks. Moreover, the concrete blocks conform to the relevant standards. In

Japan, RAC was early applied in rigid pavements and building structures. Recently, RAC was developed as new pervious materials for pavement.

Pervious Concrete

Definition and Principle

Pervious concrete, also known as porous concrete, has the characteristics of air permeability, water permeability and light weight. Pervious concrete used in urban roads is of great significance to regulate microclimate of urban, protect groundwater, maintain ecological balance and relief heat island effect of urban.

Permeability is the most important performance of pervious concrete. Coarse aggregates as the skeleton of concrete will form the plenty of pores and gaps, and these voids often ranges from 2 mm to 8 mm in the case of concrete without fine aggregate. Therefore, water can penetrate through the pervious concrete quickly (Chandrappa & Biligiri, 2016; Dang et al., 2014; Neithalath, Bentz, & Sumanasooriya, 2010).

Current Progress

The concept of pervious concrete has been put forward for nearly 60 years. Since 1980s, pervious concrete has been widely used in urban road. In fact, the research of pervious concrete is mainly focused on the effect of various factors on properties of it. For pervious concrete, permeability coefficient and strength are main index to evaluate pervious concrete. Therefore, in the mix proportion design of pervious concrete, the appropriate parameters of mix proportion should be chosen to balance the requirements of its strength and permeability. The type and gradation of aggregate are the key factors that affect the void ratio, permeability coefficient and compressive strength of pervious concrete (Ćosić, et al., 2015). With the increasing size of aggregate, void content and permeability coefficient of pervious concrete increased, while the compressive strength decreased. Therefore, the porosity is usually designed as 15-25% so as to balance the contradiction between permeability and strength of pervious concrete on the basis of relevance between porosity and permeability, porosity and strength (Kevern, et al., 2006). In generally, the aggregates of common concrete is about 60-75 vol. %, while that in pervious concrete is about 50-65 vol. % (Chandrappa & Biligiri, 2016). The coarse aggregates with size of 2.36-9.5 mm were used to balance the strength properties and water permeability in several studies (Lian & Zhuge, 2010). In addition, the aggregate/cement ratio ranged from 4:1 to 6:1. The water/cement ratio as a very important variable is in range of 0.27-0.33 (Nagarajan, 2008). The proper dosage of superplasticizer is beneficial to improve workability, and increase compressive strength of pervious concrete (Yang & Jiang,

2003). Except for using high-strength cement and superplasticizer, supplementary materials, organic intensifiers and other special chemical reinforcing agents are also tried to develop pervious concrete (Chandrappa & Biligiri, 2016; Chandrappa & Biligiri, 2016; Neithalath, Bentz, & Sumanasooriya, 2010). The appropriate amount of polymer emulsion can improve the workability and reduce the void content of the fresh concrete. In addition, the mixing process of cement wrapped stone has little influence on the permeability coefficient of porous pervious concrete, but it can increase the compressive strength, but can decrease the void fraction (Jiang, Sun, & Wang, 2005).

Application

The pervious concrete can be widely applied in parking lots, squares, pedestrian walkways and other areas with light traffic. A new type of pervious concrete, in which crushed granite was used as coarse aggregate, was developed to divert quickly rainwater during storms. The water through this pervious concrete to the ground level can reach 36000 millimeters an hour (as shown in Figure 1 (Chris, W., 2015)). Compared with the impervious pavement, pervious pavement has many ecological advantages, which are listed as follows. (1) Pervious concrete is used to construct the sustainable drainage systems, which can collect, store, treat, redistribute and recycle water. The recycled water is used in the development of ecosystems and landscapes. (2) Pervious concrete can improve the ecological balance of urban road. Because pervious concrete is applied in pavement, the soil underneath can keep wet. In addition, introduction of pervious concrete pavements also decreases greatly the phenomenon of hot island in cities. (3) The pervious concrete provides a comfort and safety urban road environment. The noise of vehicles can be absorbed due to porosity of pervious concrete. Moreover, the pervious concrete pavement can reduce dangers of plash and glisten in rainy days. By applying the properties of the pervious concrete above, cities can be built into sponge eco-cities. It is beneficial for preventing flood, replenishing water resources, promoting biodiversity development, and improving air quality (Ghafoori & Dutta, 1995; Scholz & Grabowiecki, 2007; Yang & Jiang, 2003).

Smart Cementitious Composites

Smart cementitious composite is essential for intelligent building system, which possesses not only basic structural function, but also one or more types of intelligent or functional behaviors. Smart cementitious composite is tailored to meet some specific requirements such as safety, longevity, and function of infrastructures, as well as to reduce life cycle costs, resource consumption, and environment pollution.

Figure 1. Test of pervious concrete

Definition and Properties

Smart cementitious composite is a kind of multifunctional material which consists of intelligent components on basis of the ordinary concrete. It has properties different from those of conventional concrete, such as self-sensing, self-adaptive and self-healing. Smart cementitious composites can monitor the internal damage of concrete effectively, and repair oneself according to the detection results. As a result, smart cementitious composites can remarkably improve the safety and durability of concrete structures.

Current Progress

Smart cementitious composite possesses adaptive functions to external stimuli, such as loads or changes of environmental condition, with inherent intelligence. In the late 1980s, the concept of smart cementitious composite was first proposed (Chen & Chung, 1993; Han et al., 2015; Kamila, 2013). Since then, Yanagida developed "self-test cementitious composites" by incorporating glass and carbon fibers into cementitious composites in 1992. In 1989, Chung added short-cutting carbon fiber (CF) into the cementitious composites and reported the relationship between the electrical resistivity and inner structure of cementitious composites, which marks the beginning of smart CF cementitious composites (Chen & Chung, 1993). The "smartness" of cementitious composites is achieved by not only material composition, but also the introduction of other functional components. For example, electric resistance strain gauges are usually embedded in the cementitious composites, attached to the cementitious composites surfaces or installed on reinforcement bars within the cementitious composites structure for strain monitoring. The possibility of smart cementitious composites is based on biomimetic design, and multifunctional, multiscale, and multicomponent compositions (Gandhi & Thompson, 1992; Han et al., 2015).

Smart cementitious composites are a synthesis of many functions, such as self-sensing, self-healing, self-heating, self-cleaning, self-adjusting, electromagnetic (EM) shielding/absorbing, energy harvesting, light-transmitting, aircraft arresting and so on. However, it is still very difficult to prepare ideal smart cementitious composites with the current technological level. At present, self-sensing cementitious composites, self-healing cementitious composites and self-repairing cementitious composites are the primary stage of intelligent cementitious composites, which have laid a very solid foundation for the study of smart cementitious composites. These cementitious composites can be seen as a simplified form of smart cementitious composites. In the future, smart cementitious composites are expected to assemble two or more smart cementitious composites.

Application

Smart cementitious composites have great potential for practical applications. Up to now, they have been applied individually or as an ensemble in many fields of construction industry. Smart cementitious composites are capable to sense the inside conditions and environmental changes including stress (or loading), strain (or deformation), damage, crack, temperature and humidity by means of incorporating functional fillers or sensing components. The physical parameters of smart cementitious composites will change with the subjected external forces, deformations or environmental actions, hence presenting sensing capability. Therefore, it can be drawn that the smart cementitious composites can be well applied in the field of structural health monitoring (SMH), border/military security and traffic detection. Take traffic detection as an example. The pavements or bridge components assembled with smart cementitious composites will be able to detect important traffic data such as traffic flow speed, traffic concentration and vehicular speed, even realize weighing vehicles in motion. In addition to self sensing, the smart cementitious composites can also be used in industrial anti-static structure, the cathodic protection in the steel reinforced cementitious composites structure, deicing of highway pavement and runway, electric heating structure of residential and farm.

FIBER REINFORCED POLYMER COMPOSITES

Fiber reinforced polymer (or fiber reinforced plastic) (FRP) composites are promising structural materials for civil and military infrastructures because of their superior corrosion resistance, excellent thermomechanical property, and high strength-to-weight ratio. FRP composites combines fibers and resin together, in which fibers form a network and provide the load-bearing component of the composites while the surrounding resin transfer loads to the fiber network and maintain fiber orientation (Uddin, 2013). Not only can FRP composites improve structural performance and reduce dead-weight, but also meet the demands of green construction material and low-carbon society. In the past three decades, FRP composites have increasingly been applied in infrastructures, such as bridge decks, structural shapes, internal reinforcements and externally bonded reinforcements (Bakis, 2002).

Classification and Properties

Fibers and matrixes are fundamental to fabric FRP composites. Fiber provides strength and stiffness for FRP composites because of its high modulus and high tensile strength. Generally, FRP composites are fabricated by impregnating reinforced fibers or fiber fabrics with resins. There exists carbon fiber, steel fiber, glass fiber and aramid fiber FRP composites according to types of fibers. Table 3 shows the major fibers applied in FRP composites, carbon fiber and glass fiber more popular in civil engineering (Uddin, 2013; GüneÅ, 2004). The surrounding matrix supports the fibers in a desired orientation and location, transferring stress between them and protecting them from external damages and environmental corrosions. Matrix is divided into two general categories: thermoplastics and thermosets, shown in Table 4 (Hull, 1996). The most typical matrixes are thermosetting resins like polyester, epoxy, vinyl ester, and phenolics resins. In this form, both fibers and matrix retain their intrinsic properties, producing a combination of behaviors that cannot be gotten by either of them alone. FRP composites are typically fabricated into various forms of fabrics, rebars, cables, bridge decks and pultruded shapes for practical applications.

Compared to traditional construction materials, FRP composites with optimum selection of fiber and matrix combine their advantages and display excellent mechanical properties and durability. Generally, FRP composites possess much higher tensile strengths and moduli than steel, but fail in a brittle way. FRP composites show various degrees of anisotropy depending on the orientation of fibers, so their mechanical properties are also anisotropic. The tensile strength and modulus of a unidirectional fiber reinforce laminate are maximum when measured in the fiber axis direction. The shear strength, compressive strength and ductility of a structural member can be improved by wrapped or wound FRP reinforcement and confinement

Table 3. Mechanical properties of fibers

Fiber	Type	Young's Modulus (GPa)	Tensile Strength (GPa)	Failure Strain (%)
Carbon	PAN	230-540	3.53-7.06	0.7-2.0
	Pitch	42-820	0.85-2.20	0.2-2.1
Glass	E	72	3.45	1.8-3.2
	C	69	3.30	-
	S	87	3.50	4.0
Boron	-	400	4.00	1.0
Kevlar	29	70.5	2.92	3.6
	49	112.4	3.00	2.4

Table 4. Mechanical properties of matrixes

Type	Matrix	Young's Modulus (GPa)	Tensile Strength (GPa)	Failure Strain (%)
Thermosets	Epoxy resin	3.0-6.0	0.035-0.1	1-6
	Polyesters	2.0-4.5	0.04-0.09	2
Thermoplastics	Nylon	1.4-2.8	0.06-0.07	40-80
	Polypropylene	1.0-1.4	0.02-0.04	300
	PEEK	3.6	0.17	50

(Uddin, 2013). FRP composites with stiff fibers like boron fiber and carbon fiber show excellent fatigue resistance. FRP composites also display high impact resistance and high damping characteristic. FRP composites are inherently corrosion-resistant than common building materials such as steel and wood, showing substantial cost benefits when used in aggressive environments. But high temperature and UV radiation can reduce the lifespan of FRP composites and fire is terrible for them because most of the organic resin matrixes are easy to melt and be on fire (GüneÅ, 2004; Micelli, Mazzotta, Leone, & Aiello, 2015).

Current Progress

FRP composites were first developed in the 1930s, but originally, they were too expensive to be applied in civil engineering. From 1930s to 1970s, the applications of them were limited to aerospace and defense industries. With the promotion of manufacturing technology and the reducing of manufacturing cost, FRP composites became universally affordable for construction projects and drew more attentions of engineers in the late 1980s (Bakis, 2002). FRP composite reinforcements used for concrete were developed in the 1960s in the US(Dolan, 1999), and later in the 1970s in Europe (Taerwe & Matthys, 1999) and Japan (Fukuyama, 1999).Since 1980s, FRP composites became external reinforcements for reinforced concrete bridges and columns (Bakis, 2002). In recent years, the number of construction projects employing FRP composites are increasing dramatically and numerous researches on them are hot.

Applications

Since 1980s, FRP composites in infrastructure applications has been increasingly popular and used for pultruded bars and shapes, bridge decks, internal reinforcements and external reinforcements. Due to their excellent mechanical properties and durability, FRP composites are perfect materials for structural shapes and have been made into various structural shapes for the applications of buildings and bridges. FRP composite is stronger, stiffer, lighter, higher material efficiency and more resistant to deicing salt than steel and concrete, so it is an ideal material for large-span bridge decks. FRP composites for internal reinforcements are primarily used for improving corrosion resistance and electromagnetic transparency in structures. They are divided into non-prestressed and prestressed ones per stress, and supplied with three shapes: one-dimensional bars and strands, two- and three- dimensional rigids (Nanni, 1993). One-dimensional FRP bars and strands are used for FRP reinforcements instead of steel ones. The FRP rigids are used for flexural reinforcement in slabs and FRP cages are used for both shear and axial reinforcement in beams. So far, there are a large number of non-prestressed and prestressed structures like bridges, tunnels, highways and skyscrapers employing FRP reinforcements, bridges comprising the majority. FRP composites for external reinforcements are mainly used in structural strengthening and maintaining existing structures, such as bridges, foundations, masonry buildings, historic monuments (Bakis, 2002). In practical applications, structures are strengthened by FRP composite strands/sheets and plates, which are strong, flexible, easy to cut and install.

NOVEL THERMALLY FUNCTIONAL COMPOSITES

With the development of living level, the requirement for environmental amenity is becoming higher and higher. However, the construction energy consumption (air-conditioning energy consumption included) raising makes energy exhausted more quickly and environment polluted more seriously. Buildings liberated 8.3 GT carbon dioxides, which is more than thirty percent of the carbon dioxides emissions in many countries, in 2005 (Baetens, Jelle, & Gustavsen, 2011). How to balance the environmental amenity and energy consumption has become the eternal theme in architectural designs and building energy-saving aspects. To replace traditional insulation materials by thermally functional materials is one kind of the most effective means to reduce greenhouse gas emissions. This chapter will provide an introduction to two kinds of novel thermally functional composites based on different principles, with attentions to its properties, classification, current progress, and applications.

Phase Change Composites

Phase change composite is one of the novel thermally functional composites which can be made by adding the phase change materials (PCMs) into their matrixes. The PCM is a material that can change state with temperature and provide latent heat. During the phase change process, it can exchange heat with the environment. Hence, it can achieve the purposes of reducing the fluctuation of ambient temperature and improving the energy efficiency (Kuznik, Virgone, & Noel, 2008; Regin, Solanki, & Saini, 2008).

Properties and Classification

Phase change composites can be divided into solid-solid phase change composites, liquid-gas phase change composites, liquid-solid phase change composites and gas-solid phase change composites due to the process of phase change. Taking the solid-liquid phase change composite as an example, it changes from solid to liquid along with absorbs a lot of heat during melting process when the temperature beyond the limit temperature (the melting temperature). Conversely, the material will change from liquid to solid when the temperature is lower than the limit temperature. The stored or released energy during phase transition process is called the latent heat. Although the temperature of material itself is almost unchanged after the phase change is completed, the latent heat of phase change is quite large.

According to the component of the materials, phase change composites can be divided into organic, inorganic and mixed phase change composites. Each type of phase change composites has its typical phase transition temperature and phase transition enthalpy. The classification of phase change composites and their characteristics are summarized in Table 5.

Current Progress

Phase change composites has been taken into account for thermal storage in buildings ever since 1980 (Castellón et al., 2007). The United States Department of Energy

Table 5. Classification of the phase change composites

Types	Advantages	Disadvantages
Organic phase change composites	(1) Wide phase transition temperature(20-70°C); (2) Low vapor pressure during melting process; (3) No phase separation; (4) Good thermal properties.	(1) Low thermal capacity; (2) Small density; (3) Large volume change during the process of phase transition. (4) High price.
Inorganic phase change composites	(1) High latent heat storage capacity(250-400MJ/m^3); (2) Good thermal conductivity; (3) Low price.	(1) Most of them are corrosive; (2) Prone to occur phase separation.
Mixed phase change composites	(1) The constraints of the melting point and price of the single PCM can be ignored, so it could have a wider range of options; (2) The phase transition temperature can be adjusted by the composition and the component content of the composition; (3) The advantages of some components can be preserved.	Need to ensure the durability and thermal conductivity of the mixed PCMs.

Solar Energy initiated a study on the phase change composites which was used in building materials in 1982. In 1999, University of Dayton Research Institute successfully developed a new type of liquid-solid eutectic phase change composite with the eutectic temperature of 23.3°C(Zhong, 2006). The material could be added into lightweight concrete prefabricated panels to make the room maintaining a comfort temperature. With the development of PCM, the cost is constantly reduced, and the compatibility with civil engineering materials will be improved.

Application

These phase change composites can be used for reducing the environmental temperature fluctuations and keeping a comfortable temperature indoors. Hirayama et al. added phase change material into concrete and found that the thermal mass and the specific heat capacity of concrete has been increased. Researchers stated that the size of the air conditioning systems can be reduced when the building fabricated with this kind of concrete composite (Hirayama, Jolly, & Batty, 1997). In the daytime, the phase change material in the composite absorbs heat through phase change to make the indoor temperature decrease. At night, the phase change material in the composites would release the heat absorbed during the daytime through phase change to improve the indoor temperature. In this way, the phase change composites can make the room maintain a comfort temperature. Cabeza et al. (2007) made the phase change composite which is a blend of conventional concrete, with microcapsules filled in it. They built two small houses by using the composites and the conventional concrete, respectively. They observed that the highest temperature in the house built of the conventional concrete reaches 39°Cwhile that of the phase change composites is only 36°C when the outdoor ambient temperature is 32°C.

Aerogel Composites

Aerogels are dried gels whose skeletal structures are made of mutual cross-linked nanoparticles or polymer molecule. They were first discovered by Kistler in the early 1930s (1931) and their porosity can reach up to 98%. They have the excellent performance of low refractive index and thermal conductivity, high optical transmittance (about 93%), low dielectric constant (<1.01), low density (0.003g/ cm^3), waterproof, fireproof. As the research moves along, aerogel has been regarded as one kind of thermal insulation materials with great promise in building fields.

Classification and Properties

There are various kinds of aerogels prepared after nearly one century of development. According to its chemical composition, it can be divided into inorganic aerogel, organic aerogel and carbon aerogel. To prepare an aerogel needs sol-gel process and drying process. The former process is mainly to obtain alcogel with a network structure and the latter is to take out the solvent in the network to get an aerogel material. Gel drying technology mainly includes supercritical drying, subcritical drying, conduction drying, freeze-drying, distillation drying and ambient pressure drying.

The travel path of heat in aerogels can be prolonged and the efficiency of heat can be reduced due to the unique nanoscale skeleton and pore structure. Therefore, the aerogel' s thermal conductivity is only 0.2% of that of common material. At present, the thermal conductivity of aerogel is the lowest in the world. The comparison between aerogel and several common insulation materials is shown in Table 6.

Current Progress

Hayase et al. (2012) have prepared a flexible silica aerogel using methyl trimethoxysilane (MTMS) as the precursor and cetyltrimethyl ammonium chloride (CTAC) as the surfactant. The aerogel has high optical transmittance up to 91% by controlling the content of the surfactant and the size of pore structure. However, the aerogel is easy to be shattered due to its slender nano-porous structure, which limits its use. Therefore, improving the mechanics properties of aerogel has been the emphasis of current researches.

Table 6. Comparison between aerogel and several common insulation materials.

Material	Temperature Range (°C)	Thermal Conductivity (W·/(m·k))	Flame Class
Aerogel	-200-1400	0.0136-0.038	A
Ceramsite mortar	≤800	≤0.100	A
Rock wool	0-600	0.040	A
Polystyrene board	≤100	0.031-0.040	B1-B2
Polyurethane PU	0-120	0.017-0.020	B1-B2
Phenolic Foam Board	-190-200	0.033	B1-B2

In 2001, Tan et al. fabricated cellulose nanocrystal aerogel by using cellulose derivative as main raw material. The cellulose nanocrystal aerogel not only has the advantage of high porosity but also has the characteristic of good toughness and workability. These characteristics make it have potential application prospects in the special coating, heat insulation, sound insulation and membrane materials fields. In addition, it can be used as dual or multiple composite materials template to achieve the adjustment of aerogels' performance. Chen et al. (2011) adopted the freeze-drying method and prepared the cellulose nanocrystal aerogel. However, the cellulose nanocrystal aerogel is liable to absorb water because it is an hydrophilicity biomacromolecule material. The porous structure of the cellulose nanocrystal aerogel with poor mechanical properties is easy to collapse.

Hayase et al. (2013) adopted the two-step synthesis consisting of a simple sol-gel process to obtain vinyl-modified super hydrophobic monolithic gel. The perfluoroalkyl groups have been grafted onto the pore surface of the gel to decrease their surface energy. This kind of aerogel has super hydrophobic property just like Figure 2.

Cai et al. (2012) carried out a series of experiments to prepared cellulose-silica nanocomposite aerogels and found this kind of aerogels can be made by in-situ formation. The aerogel not only has the transparent property, but also has the flexible property. A demonstration of the transparent and flexibility properties of the aerogel composites can be seen in Figure 3.

Application

Aerogel composites are applied in civil engineering applications in the form of aerogel plates, aerogel insulation mortars, aerogel insulation coatings, aerogel collectors and aerogel glasses. Kim et al. (2003a) mixed the PVB particles with shredded SiO_2 aerogel by certain ratio and prepared an translucent aerogel. They used the material in a curtain wall of Yale University. Common cementitious composite can be added

Figure 2. A demonstration of the super hydrophobic property of aerogel composites

Figure 3. A demonstration of the transparent and flexibility properties of the aerogel composites

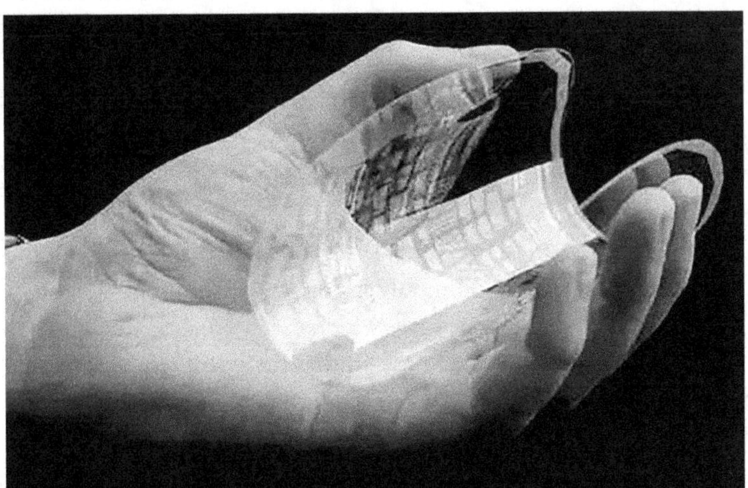

to the aerogel to prepare aerogel insulation mortar and the new mortar can be used for wall external insulation material and wall insulation layer filling material (Ni et al., 2010). Aerogel insulation coating has been applied to zero carbon house in the Shanghai World Expo and the laboratory building of Vanke successfully (Koebel, Rigacci, & Achard, 2012). Applying aerogel to the storage tank of water heater, pipe and collector can doubled the heat-collecting efficiency and decrease the heat loss to 30% (Dowson et al., 2012). By combining aerogel and glass, Kim et al. (2003b) manufactured the aerogel film by the method of wet gel films. The transmittance of window adopting the aerogel glazing was over 90%. Furthermore, when the aerogel thickness was 100μm, the thermal conductivity was merely 0.2 W/(m·K).

THREE DIMENSIONAL PRINGTING COMPOSITES

The building and construction industry is the foundation of economy. However, in recent years because of the shortage of labor resources, automation and environmental concerns, it becomes more and more difficult to meet the targeted demand (e.g. reduced construction time, less pollution and wastages). Three-dimensional printing (3DP) technology provides an innovative construction method that can optimize construction time, cost, design flexibility, error reduction, and environmental protection.

3DP Technology

3DP technology is based on the design of the 3D model using the 3DP equipment to manufacture three-dimensional products layer by layer. The progressive stacking forming technology is also called as material additive construction (Werkheiser et al., 2015). 3DP is a kind of rapid prototyping technology, involved digital modeling technology, information technology, many cutting-edge technologies and materials science/chemistry. It is known as "the core technology of the third industrial revolution".

To make the concrete self-shaped, it is firstly needed to design a building path for the robotic machine. The most commonly used method for creating a building path is to utilize a 3D-to-2D slicing software, which can help slice the 3D-shape object into flat thin layers of constant thickness so that the extruded concrete ink can be layered one by one(Gosselin et al., 2016). Digital signals of the structure are then delivered to printing machine to complete the operation. The main principle during printing is that the nozzle circumnavigates according to the predetermined paths with concrete ink extruded out of it one layer at a time. In this way, a completed 3D structure component is printed.

Compared with the traditional manufacturing technology, 3DP needs not solid mold, creates no waste materials and demands no complicated forging process during the manufacturing process. Therefore, 3DP can achieve the optimization on structure, saving material and energy in the production. 3DP technology is not only suitable for the development of new products, fast single piece and small batch parts manufacturing, complex shape parts mold design and manufacture, but also suitable for difficult machining materials manufacturing, inspection, assembly examination and design of rapid reverse engineering (Feng et al., 2015). Therefore, 3DP industry has attained more and more attention. It will become a promising sunrise industry.

Current Progress of 3DP Composites

3DP composite is an important material foundation for the development of 3DP technology. At present, 3DP composites mainly include engineering plastics, rubber materials, metal materials, photosensitive resins and ceramic materials. For civil engineering, 3DP concrete is widely applied. Compton and Lewis (2014) used oriented fiber-filled epoxy composites as the printing composites and found that they had almost same absolute stiffness with both wood and commercial 3D printed structures. Results exhibits that tensile bars printed using this epoxy "inks" have longitudinal Young's modulus values very close to that of wood cell walls, which are 10-20 times higher than the most of commercial, 3D printed polymers, and 2 times higher than the best commercial polymer composites. Malaeb et al. (2016) have tried to formulate geopolymer-based materials for criteria of commercially available powder-based 3D printers. Results indicates that geopolymer-based material has sufficient deposit ability and wettability, while the printed cubic structures are anisotropic in dimensional accuracy and the compressive strength of composites have raised from 0.9 MPa to 16.5 MPa of maximum after post-processing. Moreover, plaster cementitious composite, whose major ingredients are plaster, humectant and water, vinyl polymer and carbohydrate, was also adopted as concrete ink of 3D printers (Le et al., 2012).

Application of 3DP Composites

The application of 3DP technology has been reported worldwide. At present, three main concrete printing technologies, D-shape, contour crafting (CC) and concrete printing, are in public domain (Zhang & Khoshnevis, 2013). The core principle of D-shape is powder deposition process, which alternatively binds the powder using a chemical binder. CC is based on concrete ink extruded against a trowel that results in smoother surface finish to the printed components. Similar to CC, concrete printing has also an extrusion-based manufacturing procedure which fabricates complex component geometries by layers without heavy labor work. When it comes to practical application, both D-shape and concrete printing are suitable for gantry based off-site printing while CC is more adapted to on site, in-situ applications (Lim et al., 2012).

Thanks to the researcher's unremitting efforts, not only large and complex structural components but also full-scale constructions are now possible to be printed. Behrokh Khoshnevis, who developed CC, said he envisioned a 3D concrete-extruding printer which will be capable to print a 2,500-square-foot building within 20 hours (Figure 4 (Jeff, L., 2015)). The expectation has now come true.

In 2015, Eindhoven University of Technology in the Netherlands made a breakthrough. Researchers there developed a new type 3D printer, which was able to provide 11m × 5m × 4m building space with high printing accuracy of self-shaping concrete. The device was built by ROHACO, a Dutch company, and has been currently put into practice (as shown in Figure 5 (ROHACO, 2015)). In the same year, a five-story apartment block was printed using glass fiber reinforced concrete by a company in Shanghai, China. It is said that the printed apartment houses were fully compliant with the relative national standards.

CONCLUSION

In this chapter, several representative advanced composites are systematical introduced with abundant research achievements and wide applications in civil engineering infrastructures. Development of advanced composites has been achieved

Figure 4. 3D concrete-extruding printer of CC

Figure 5. Giant 3D printer and 3DP concrete by ROHACO

by the introduction of new materials, improvements in manufacturing processes and developments of new analytical and testing methods. The advancements of composites are reflected as follows.

1 Brittleness of conventional cementitious composites has been greatly improved. The cementitious composites with high toughness and strength (such as ECC, RPC) are developed in order to prolong the service life of structure and reduce the deadweight of structure (such as RPC, FRP,). In addition, the application of the advanced cementitious composites with high flowability (such as SCC) promotes the development of the prefabricated structure.

2. The energy-saving and emission-reduction targets are achieved, promoting the sustainable development of civil engineering (such as RAC, PCM, aerogel composites). Relying on multifunctional properties, the green and smart city is expected to be realized in the future (such as PCM, permeable concrete, smart cementitious composites).

3. The superior corrosion resistance and high strength-to-weight ratio of the advanced composites (such as FRP) is more propitious to reduce the deadweight of structure and enhance durability of structure, as compared to traditional reinforced concrete.

4. An innovative construction method is developed by relying on 3DP technology. 3DP composites can automatically generate structures, which optimize construction time, cost, design flexibility, error reduction, and environmental protection.

In conclusion, in the future, development of composites for civil engineering infrastructures will tend to high performance, multifunction and smart, along with green, energy conservation and environmental protection.

ACKNOWLEDGMENT

This research was supported by the National Science Foundation of China [51578110] and [51428801].

REFERENCES

Au, C., & Buyukozturk, O. (2005). Effect of fiber orientation and ply mix on fiber reinforced polymer-confined concrete. *Journal of Composites for Construction*, *9*(5), 397–407. doi:10.1061/(ASCE)1090-0268(2005)9:5(397)

Baetens, R., Jelle, B. P., & Gustavsen, A. (2011). Aerogel insulation for building applications: A state-of-the-art review. *Energy and Building*, *43*(4), 761–769. doi:10.1016/j.enbuild.2010.12.012

Bakis, C., Bank, L. C., Brown, V. L., Cosenza, E., Davalos, J. F., Lesko, J. J., ... Triantafillou, T. C. (2002). Fiber-reinforced polymer composites for construction-state-of-the-art review. *Journal of Composites for Construction*, *6*(2), 73–87. doi:10.1061/(ASCE)1090-0268(2002)6:2(73)

Blais, P. Y., & Couture, M. (1999). Precast, prestressed pedestrian bridge-world's first reactive powder concrete structure. *PCI Journal*, *44*(5), 60–71. doi:10.15554/pcij.09011999.60.71

Cabeza, L. F., Castellón, C., Nogués, M., Medrano, M., Leppers, R., & Zubillaga, O. (2007). Use of microencapsulated PCM in concrete walls for energy savings. *Energy and Building*, *39*(2), 113–119. doi:10.1016/j.enbuild.2006.03.030

Cai, J., Liu, S., Feng, J., Kimura, S., Wada, M., Kuga, S., & Zhang, L. (2012). Cellulose-silica nanocomposite aerogels by in situ formation of silica in cellulose gel. *Angewandte Chemie*, *124*(9), 2118–2121. doi:10.1002/ange.201105730 PMID:22275132

Castellón, C., Medrano, M., Roca, J., Nogués, M., Castell, A., & Cabeza, L. F. (2007). *Use of microencapsulated phase change materials in building applications*. Retrieved July 30, 2017, from http://celsius.co.kr/phase_change_materials/download/energy/Use_of_Microencapsulated_PCM_in_Building_Applications.pdf

Chan, Y. W., Chen, Y. S., & Liu, Y. S. (2003). Development of bond strength of reinforcement steel in self-consolidating concrete. *ACI Structural Journal*, *100*(4), 490–498.

Chan, Y. W., & Chu, S. H. (2004). Effect of silica fume on steel fiber bond characteristics in reactive powder concrete. *Cement and Concrete Research*, *34*(7), 1167–1172. doi:10.1016/j.cemconres.2003.12.023

Chandrappa, A. K., & Biligiri, K. P. (2016). Pervious concrete as a sustainable pavement material-research findings and future prospects: A state-of-the-art review. *Construction & Building Materials*, *111*, 262–274. doi:10.1016/j.conbuildmat.2016.02.054

Chen, P. W., & Chung, D. D. L. (1993). Carbon-fiber-reinforced concrete smart structures capable of nondestructive flaw detection. In *Proceedings of North American Conference on Smart Structures and Materials* (*vol. 1*, pp. 22). Academic Press. doi:10.1117/12.148502

Chen, W., Yu, H., Li, Q., Liu, Y., & Li, J. (2011). Ultralight and highly flexible aerogels with long cellulose I nanofibers. *Soft Matter*, *7*(21), 10360–10368. doi:10.1039/c1sm06179h

Cheyrezy, M., Maret, V., & Frouin, L. (1995). Microstructural analysis of RPC (reactive powder concrete). *Cement and Concrete Research*, *25*(7), 1491–1500. doi:10.1016/0008-8846(95)00143-Z

Chris, W. (2015). *This 'thirsty' concrete absorbs 880 gallons of water a minute-here's how it works*. Retrieved July 30, 2017, from http://www.techinsider.io/how-magical-concrete-absorbs-water-2015-9/

Compton, B. G., & Lewis, J. A. (2014). 3D-printing of lightweight cellular composites. *Advanced Materials*, *26*(34), 5930–5935. doi:10.1002/adma.201401804 PMID:24942232

Corinaldesi, V., Letelier, V., & Moriconi, G. (2011). Behaviour of beam-column joints made of recycled-aggregate concrete under cyclic loading. *Construction & Building Materials*, *25*(4), 1877–1882. doi:10.1016/j.conbuildmat.2010.11.072

Corinaldesi, V., & Moriconi, G. (2009). Influence of mineral additions on the performance of 100% recycled aggregate concrete. *Construction & Building Materials*, *23*(8), 2869–2876. doi:10.1016/j.conbuildmat.2009.02.004

Ćosić, K., Korat, L., Ducman, V., & Netinger, I. (2015). Influence of aggregate type and size on properties of pervious concrete. *Construction & Building Materials*, *78*, 69–76. doi:10.1016/j.conbuildmat.2014.12.073

Dang, H. N., Sebaibi, N., Boutouil, M., Leleyter, L., & Baraud, F. (2014). A modified method for the design of pervious concrete mix. *Construction & Building Materials*, *73*(73), 271–282.

Dolan, C. W. (1999). FRP prestressing in the USA. *Concrete International*, *21*(10), 21–24.

Domone, P. L. (2006). Self-compacting concrete: An analysis of 11 years of case studies. *Cement and Concrete Composites*, *28*(2), 197–208. doi:10.1016/j.cemconcomp.2005.10.003

Domone, P. L. (2007). A review of the hardened mechanical properties of self-compacting concrete. *Cement and Concrete Composites*, *29*(1), 1–12. doi:10.1016/j.cemconcomp.2006.07.010

Dong, S., Han, B., Ou, J., Li, Z., Han, L., & Yu, X. (2016). Electrically conductive behaviors and mechanisms of short-cut super-fine stainless wire reinforced reactive powder concrete. *Cement and Concrete Composites*, *72*, 48–65. doi:10.1016/j.cemconcomp.2016.05.022

Dowson, M., Pegg, I., Harrison, D., & Dehouche, Z. (2012). Predicted and in situ performance of a solar air collector incorporating a translucent granular aerogel cover. *Energy and Building*, *49*(2), 173–187. doi:10.1016/j.enbuild.2012.02.007

Du, T., Wang, W. H., Lin, H. L., Liu, Z. X., & Liu, J. (2010). Experimental study on interfacial strength of the high performance recycled aggregate concrete. In *Proceedings of the Workshop on Biennial International Conference on Engineering* (pp. 2821-2828). Academic Press. doi:10.1061/41096(366)266

Dugat, J., Roux, N., & Bernier, G. (1996). Mechanical properties of reactive powder concretes. *Materials and Structures*, *29*(4), 233–240. doi:10.1007/BF02485945

Felekoğlu, B., & Sarıkahya, H. (2008). Effect of chemical structure of polycarboxylate-based superplasticizers on workability retention of self-compacting concrete. *Construction & Building Materials*, *22*(9), 1972–1980. doi:10.1016/j.conbuildmat.2007.07.005

Felekoğlu, B., Türkel, S., & Baradan, B. (2007). Effect of water/cement ratio on the fresh and hardened properties of self-compacting concrete. *Building and Environment*, *42*(4), 1795–1802. doi:10.1016/j.buildenv.2006.01.012

Feng, P., Meng, X., Chen, J. F., & Ye, L. (2015). Mechanical properties of structures 3D printed with cementitious powders. *Construction & Building Materials*, *93*, 486–497. doi:10.1016/j.conbuildmat.2015.05.132

Ferrara, L., Park, Y. D., & Shah, S. P. (2007). A method for mix-design of fiber-reinforced self-compacting concrete. *Cement and Concrete Research*, *37*(6), 957–971. doi:10.1016/j.cemconres.2007.03.014

Fukuyama, H. (1999). FRP composites in Japan. *Concrete International*, *21*(10), 29–32.

Gandhi, M. V., & Thompson, B. D. (1992). *Smart materials and structures*. Springer Science and Business Media.

Ghafoori, N., & Dutta, S. (1995). Development of no-fines concrete pavement applications. *Journal of Transportation Engineering*, *121*(3), 283–288. doi:10.1061/(ASCE)0733-947X(1995)121:3(283)

Goodier, C. I. (2003). Development of self-compacting concrete. *Structures & Buildings*, *156*(4), 405–414.

Gosselin, C., Duballet, R., Roux, P., Gaudillière, N., Dirrenberger, J., & Morel, P. (2016). Large-scale 3D printing of ultra-high performance concrete-a new processing route for architects and builders. *Materials & Design*, *100*, 102–109. doi:10.1016/j.matdes.2016.03.097

Güne, Å., & Uz, O. Ä. (2004). *A fracture-based approach to understanding debonding in FRP bonded structural members*. Academic Press.

Guo, Y. C., Wang, X., Sun, K. W., & Cheng, L. F. (2011). Experiment study on drying shrinkage deformation of recycled aggregate concrete. In *Proceedings of Electric Technology and Civil Engineering 2011 International Conference* (pp. 1219-1222). Academic Press. doi:10.1109/ICETCE.2011.5776036

Han, B., Ding, S., & Yu, X. (2015). Intrinsic self-sensing concrete and structures: A review. *Measurement*, *59*, 110–128. doi:10.1016/j.measurement.2014.09.048

Han, B., Dong, S., Zhang, C., Wang, Y., Yu, X., & Ou, J. (2016). Microstructure related mechanical behaviors of short-cut super-fine stainless wire reinforced reactive powder concrete. *Materials & Design*, *96*, 16–26. doi:10.1016/j.matdes.2016.02.004

Han, B., Li, Z., Zhang, L., Zeng, S. Z., Yu, X., Han, B., & Ou, J. (2017). Reactive powder concrete reinforced with nano SiO_2-coated TiO_2. *Construction & Building Materials*, *148*, 104–112. doi:10.1016/j.conbuildmat.2017.05.065

Han, B., Sun, S., Ding, S., Zhang, L., Yu, X., & Ou, J. (2015). Review of nanocarbon-engineered multifunctional cementitious composites. *Composites. Part A, Applied Science and Manufacturing*, *70*(70), 69–81. doi:10.1016/j.compositesa.2014.12.002

Han, B., Wang, Y., Dong, S., Zhang, L., Ding, S., Yu, X., & Ou, J. (2015). Smart concretes and structures: A review. *Journal of Intelligent Material Systems and Structures*, *26*(11), 1303–1345. doi:10.1177/1045389X15586452

Han, B., Zhang, L., & Ou, J. (2017). *Smart and multifunctional concrete toward sustainable infrastructures*. Springer. doi:10.1007/978-981-10-4349-9

Hansen, T. C. (1992). *Recycling of demolished concrete and masonry*. CRC Press.

Hayase, G., Kanamori, K., Hasegawa, G., Maeno, A., Kaji, H., & Nakanishi, K. (2013). A superamphiphobic macroporous silicone monolith with marshmallow-like flexibility. *Angewandte Chemie*, *52*(41), 10788–10791. doi:10.1002/anie.201304169 PMID:24009025

Hayase, G., Kanamori, K., & Nakanishi, K. (2012). Structure and properties of polymethylsilsesquioxane aerogels synthesized with surfactant n-hexadecyltrimethylammonium chloride. *Microporous and Mesoporous Materials*, *158*(4), 247–252. doi:10.1016/j.micromeso.2012.03.049

Hirayama, Y., Jolly, S., & Batty, W. J. (1997). Investigation of thermal energy storage within building thermal mass in northern Japan through dynamic building and building services simulation. In *Proceedings of Seventh International Conference on Thermal Energy Storage* (pp. 355-360). Sapporo, Japan: Academic Press.

Huang, X., Ranade, R., Ni, W., & Li, V. C. (2013). On the use of recycled tire rubber to develop low e-modulus ECC for durable concrete repairs. *Construction & Building Materials*, *46*(8), 134–141. doi:10.1016/j.conbuildmat.2013.04.027

Hull, D. (1996). *An introduction to composite materials*. Cambridge University press. doi:10.1017/CBO9781139170130

Jeff, L. (2015). *3D Printing Concrete: A 2,500-Square-Foot House in 20 Hours and an Eye on a Moon Shot*. Retrieved July 30, 2017, from https://redshift.autodesk.com/3d-printing-concrete/

Jiang, Z. W., Sun, Z. P., & Wang, P. M. (2005). Effects of some factors on properties of porous pervious concrete. *Journal of Building Materials*, *8*(5), 513–519.

Kamada, T., & Li, V. C. (2000). The effects of surface preparation on the fracture behavior of ECC/concrete repair system. *Cement and Concrete Composites*, *22*(6), 423–431. doi:10.1016/S0958-9465(00)00042-1

Kamila, S. (2013). Introduction, classification and applications of smart materials: An overview. *American Journal of Applied Sciences*, *10*(8), 876–880. doi:10.3844/ajassp.2013.876.880

Kan, L. L., & Shi, H. S. (2012). Investigation of self-healing behavior of engineered cementitious composites (ECC) materials. *Construction & Building Materials*, *29*(4), 348–356. doi:10.1016/j.conbuildmat.2011.10.051

Kevern, J., Wang, K., Suleiman, M. T., & Schaefer, V. R. (2006). Mix design development for pervious concrete in cold weather climates. *Binder Content*. In *Proceedings of the 2005 Mid-Continent Transportation Research Symposium*. Ames, IA: Academic Press.

Kim, G. S., & Hyun, S. H. (2003a). Effect of mixing on thermal and mechanical properties of aerogel-PVB composites. *Journal of Materials Science*, *38*(9), 1961–1966. doi:10.1023/A:1023560601911

Kim, G. S., & Hyun, S. H. (2003b). Synthesis of window glazing coated with silica aerogel films via ambient drying. *Journal of Non-Crystalline Solids*, *320*(1), 125–132. doi:10.1016/S0022-3093(03)00027-9

Kim, J. K., Kim, J. S., Ha, G. J., & Yun, Y. K. (2007). Tensile and fiber dispersion performance of ECC (engineered cementitious composites) produced with ground granulated blast furnace slag. *Cement and Concrete Research*, *37*(7), 1096–1105. doi:10.1016/j.cemconres.2007.04.006

Kim, Y. Y., & Fischer, G. (2009). Performance of bridge deck link slabs designed with ductile engineered cementitious composite. *ACI Structural Journal*, *101*(6), 792–801.

Kistler, S. S. (1931). Coherent expanded aerogels and jellies. *Nature*, *127*(3211), 741. doi:10.1038/127741a0

Koebel, M., Rigacci, A., & Achard, P. (2012). Aerogel-based thermal super insulation: An overview. *Journal of Sol-Gel Science and Technology*, *63*(3), 315–339. doi:10.1007/s10971-012-2792-9

Kou, S. C., Chi, S. P., & Chan, D. (2007). Influence of fly ash as cement replacement on the properties of recycled aggregate concrete. *Journal of Materials in Civil Engineering*, *19*(9), 709–717. doi:10.1061/(ASCE)0899-1561(2007)19:9(709)

Kou, S. C., & Poon, C. S. (2009). Properties of self-compacting concrete prepared with coarse and fine recycled concrete aggregates. *Cement and Concrete Composites*, *31*(9), 622–627. doi:10.1016/j.cemconcomp.2009.06.005

Kuznik, F., Virgone, J., & Noel, J. (2008). Optimization of a phase change material wallboard for building use. *Applied Thermal Engineering*, *28*(11), 1291–1298. doi:10.1016/j.applthermaleng.2007.10.012

Le, T. T., Austin, S. A., Lim, S., Buswell, R. A., Law, R., Gibb, A. G. F., & Thorpe, T. (2012). Hardened properties of high-performance printing concrete. *Cement and Concrete Research*, *42*(3), 558–566. doi:10.1016/j.cemconres.2011.12.003

Lee, M. G., Wang, Y. C., & Chiu, C. T. (2007). A preliminary study of reactive powder concrete as a new repair material. *Construction & Building Materials*, *21*(1), 182–189. doi:10.1016/j.conbuildmat.2005.06.024

Lepech, M. D., & Li, V. C. (2009). Application of ECC for bridge deck link slabs. *Materials and Structures*, *42*(9), 1185–1195. doi:10.1617/s11527-009-9544-5

Li, V. C. (2002). Reflections on the research and development of engineered cementitious composites (ECC). *Proceedings of the JCI International Workshop on Ductile Fiber Reinforced Cementitious Composites (DDFRCC)-Application and Evaluation.*

Li, V. C. (2003). Hygral behavior of engineered cementitious composites (ECC) / vergleich der hygrischen eigenschaften von ECC mit beton: Restoration of buildings and monuments. *Restoration of Buildings and Monuments*, *9*(5), 513–534.

Li, V. C. (2004). High performance fiber reinforced cementitious composites as durable material for concrete structure repair. *International Journal for Restoration*, *10*, 163–180.

Li, V. C. (2007). Progress and application of engineered cementitious composites. *Journal of the Chinese Ceramic Society*, *35*, 531–536.

Li, V. C. (2008). Engineered cementitious composites (ECC)-material, structural, and durability performance. In Nawy (Ed.), Concrete construction engineering handbook. CRC Press.

Li, V. C., & Leung, C. Y. (1992). Theory of steady state and multiple cracking of random discontinunous fiber reinforced brittle matrix composites. *Journal of Engineering Mechanics, 118*(11), 2246–2264. doi:10.1061/(ASCE)0733-9399(1992)118:11(2246)

Lian, C., & Zhuge, Y. (2010). Optimum mix design of enhanced permeable concrete-an experimental investigation. *Construction & Building Materials, 24*(12), 2664–2671. doi:10.1016/j.conbuildmat.2010.04.057

Lim, S., Buswell, R. A., Le, T. T., Austin, S. A., Gibb, A. G. F., & Thorpe, T. (2012). Developments in construction-scale additive manufacturing processes. *Automation in Construction, 21*(1), 262–268. doi:10.1016/j.autcon.2011.06.010

Malaeb, Z., Hachem, H., Tourbah, A., Maalouf, T., Zarwi, N. E., & Hamzeh, F. (2016). 3D concrete printing: Machine and mix design. *International Journal of Civil Engineering and Technology, 6*(6), 14–22.

Marí, A., Etxeberria, M., & Vázquez, E. (2006). Microstructure analysis of hardened recycled aggregate concrete. *Magazine of Concrete Research, 58*(10), 683–690. doi:10.1680/macr.2006.58.10.683

Mazaheripour, H., Ghanbarpour, S., Mirmoradi, S. H., & Hosseinpour, I. (2011). The effect of polypropylene fibers on the properties of fresh and hardened lightweight self-compacting concrete. *Construction & Building Materials, 25*(1), 351–358. doi:10.1016/j.conbuildmat.2010.06.018

Mcneil, K., & Kang, H. K. (2013). Recycled concrete aggregates: A review. *International Journal of Concrete Structures and Materials, 7*(1), 61–69. doi:10.1007/s40069-013-0032-5

Micelli, F., Mazzotta, R., Leone, M., & Aiello, M. A. (2015). Review study on the durability of FRP-confined concrete. *Journal of Composites for Construction, 19*(3), 4014056. doi:10.1061/(ASCE)CC.1943-5614.0000520

Nagarajan, V., & Subramanian, K. (2008). Studies on the influence of water cement ratio on the properties of pervious concrete made with natural and recycled aggregate. *Journal on Future Engineering and Technology, 3*(3), 62–67.

Nanni, A. (1993). *Fiber-reinforced-plastic for concrete structures: Properties and applications*. Amsterdam: Elsevier Science.

Neithalath, N., Bentz, D. P., & Sumanasooriya, M. S. (2010). Predicting the permeability of pervious concrete. *Concrete International*, *32*, 35–40.

Ni, X., Wang, B., Shen, J., Zhang, Z., Zhou, B., Wu, G., & Wang, H. (2010) Insulator properties of SiO2 nano aerogel in energy-saving structural wallboard. In *Proceedings of 7th China Functional Materials and Applications of Academic Meeting* (pp. 350-352). Changsha, China: Academic Press.

Ohno, M., & Li, V. C. (2014). A feasibility study of strain hardening fiber reinforced fly ash-based geopolymer composites. *Construction & Building Materials*, *57*(5), 163–168. doi:10.1016/j.conbuildmat.2014.02.005

Parra, C., Valcuende, M., & Gómez, F. (2011). Splitting tensile strength and modulus of elasticity of self-compacting concrete. *Construction & Building Materials*, *25*(1), 201–207. doi:10.1016/j.conbuildmat.2010.06.037

Peters, S. T. (1998). *Introduction, composite basics and road map*. Springer. doi:10.1007/978-1-4615-6389-1_1

Rahal, K. (2007). Mechanical properties of concrete with recycled coarse aggregate. *Building and Environment*, *42*(1), 407–415. doi:10.1016/j.buildenv.2005.07.033

Regin, A. F., Solanki, S. C., & Saini, J. S. (2008). Heat transfer characteristics of thermal energy storage system using PCM capsules: A review. *Renewable & Sustainable Energy Reviews*, *12*(9), 2438–2458. doi:10.1016/j.rser.2007.06.009

Richard, P., & Cheyrezy, M. (1995). Composition of reactive powder concretes. *Cement and Concrete Research*, *25*(7), 1501–1511. doi:10.1016/0008-8846(95)00144-2

ROHACO. (2015). *TU eindhoven helps builders pioneer recyclable concrete constructions with giant 3D printer.* Retrieved July 30, 2017, from https://www.designboom.com/technology/tu-eindhoven-concrete-3d-printer-06-28-2016/

Rozière, E., Granger, S., Turcry, P., & Loukili, A. (2007). Influence of paste volume on shrinkage cracking and fracture properties of self-compacting concrete. *Cement and Concrete Composites*, *29*(8), 626–636. doi:10.1016/j.cemconcomp.2007.03.010

Sagoe-Crentsil, K. K., Brown, T., & Taylor, A. H. (2001). Performance of concrete made with commercially produced coarse recycled concrete aggregate. *Cement and Concrete Research*, *31*(5), 707–712. doi:10.1016/S0008-8846(00)00476-2

Şahmaran, M., & Li, V. C. (2009). Durability properties of micro-cracked ECC containing high volumes fly ash. *Cement and Concrete Research*, *39*(11), 1033–1043. doi:10.1016/j.cemconres.2009.07.009

Sahmaran, M., & Yaman, I. O. (2007). Hybrid fiber reinforced self-compacting concrete with a high-volume coarse fly ash. *Construction & Building Materials*, *21*(1), 150–156. doi:10.1016/j.conbuildmat.2005.06.032

Sari, M., Prat, E., & Labastire, J. F. (1999). High strength self-compacting concrete original solutions associating organic and inorganic admixtures. *Cement and Concrete Research*, *29*(6), 813–818. doi:10.1016/S0008-8846(99)00037-X

Scholz, M., & Grabowiecki, P. (2007). Review of permeable pavement systems. *Building and Environment*, *42*(11), 3830–3836. doi:10.1016/j.buildenv.2006.11.016

Sim, J., & Park, C. (2011). Compressive strength and resistance to chloride ion penetration and carbonation of recycled aggregate concrete with varying amount of fly ash and fine recycled aggregate. *Waste Management (New York, N.Y.)*, *31*(11), 2352–2360. doi:10.1016/j.wasman.2011.06.014 PMID:21784626

Spoelstra, M. R., & Monti, G. (1999). FRP-confined concrete model. *Journal of Composites for Construction*, *3*(3), 143–150. doi:10.1061/(ASCE)1090-0268(1999)3:3(143)

Taerwe, L. R., & Matthys, S. (1999). FRP for concrete construction: Activities in Europe. *Concrete International*, *21*(10), 33–36.

Tam, V. W. Y., Gao, X. F., & Tam, C. M. (2005). Microstructural analysis of recycled aggregate concrete produced from two-stage mixing approach. *Cement and Concrete Research*, *35*(6), 1195–1203. doi:10.1016/j.cemconres.2004.10.025

Tan, C., Fung, B. M., Newman, J. K., & Vu, C. (2001). Organic aerogels with very high impact strength. *Advanced Materials*, *13*(9), 644–646. doi:10.1002/1521-4095(200105)13:9<644::AID-ADMA644>3.0.CO;2-#

Uddin, N. (2013). *Developments in Fiber-Reinforced Polymer (FRP) Composites for Civil Engineering*. Elsevier. doi:10.1533/9780857098955

Werkheiser, M. J., Fiske, M., Edmunson, J., & Khoshnevis, B. (2015). On the development of additive construction technologies for application to development of lunar/martian surface structures using in-situ materials. In *Proceedings of the AIAA SPACE 2015 Conference and Exposition*. AIAA. doi:10.2514/6.2015-4451

Xiao, J., Li, J., & Zhang, C. (2005). Mechanical properties of recycled aggregate concrete under uniaxial loading. *Cement and Concrete Research*, *35*(6), 1187–1194. doi:10.1016/j.cemconres.2004.09.020

Yang, J., & Jiang, G. (2003). Experimental study on properties of pervious concrete pavement materials. *Cement and Concrete Research*, *33*(3), 381–386. doi:10.1016/S0008-8846(02)00966-3

Yazıcı, H., Yardımcı, M. Y., Aydın, S., & Karabulut, A. Ş. (2009). Mechanical properties of reactive powder concrete containing mineral admixtures under different curing regimes. *Construction & Building Materials*, *23*(3), 1223–1231. doi:10.1016/j.conbuildmat.2008.08.003

Yazıcı, H., Yiğiter, H., Karabulut, A. Ş., & Baradan, B. (2008). Utilization of fly ash and ground granulated blast furnace slag as an alternative silica source in reactive powder concrete. *Fuel*, *87*(12), 2401–2407. doi:10.1016/j.fuel.2008.03.005

Zhang, J., & Khoshnevis, B. (2013). Optimal machine operation planning for construction by contour crafting. *Automation in Construction*, *29*(1), 50–67. doi:10.1016/j.autcon.2012.08.006

Zhang, J., Li, Q., Du, J., & Li, X. (2009). Experimental study on mineral admixture and recycled aggregates affecting the rapid chloride permeability of high-performance recycled concrete. *Concrete (London)*, *232*, 94–96.

Zhong, K. (2006). Review of the application of phase change energy-storing material on building energy saving. *Guangzhou Architecture*, *2*, 37–40.

Zhou, J., Qian, S., Ye, G., Copuroglu, O., Breugel, K. V., & Li, V. C. (2012). Improved fiber distribution and mechanical properties of engineered cementitious composites by adjusting the mixing sequence. *Cement and Concrete Composites*, *34*(3), 342–348. doi:10.1016/j.cemconcomp.2011.11.019

Zhu, W., & Bartos, P. J. M. (2003). Permeation properties of self-compacting concrete. *Cement and Concrete Research*, *33*(6), 921–926. doi:10.1016/S0008-8846(02)01090-6

Zhuang, X. J., & Yang, L. (2006). Investigation on the tensile behavior of recycled aggregate concrete. *Journal of Building Materials*, *9*(2), 154–158.

Zou, C. Y., Wang, Y., & Hu, Q. (2009). Experimental study and model predictive of recycled aggregate concrete creep. *Journal of Wuhan University of Technology*, *12*, 94–98.

Chapter 11
Laminated Composite Hypar Shells as Roofing Units:
Static and Dynamic Behavior

Sarmila Sahoo
Heritage Institute of Technology, India

ABSTRACT

This chapter presents a broad perspective of the recent research done on laminated composite hypar shells used as roofing units. Different types of analysis including bending, vibration, buckling, impact, and failure are included. The chapter is collated and categorized based on various aspect of research. The first aspect concentrates on typical analyses including problems in which various boundary conditions and laminations are considered. Then it focuses on the structural complexities which include stiffened shells, shells with cutouts, etc. The basic results of theoretical and experimental investigations of stress-strain state, vibration, buckling, and failure are summarized in this chapter incorporating the review of materials published in scientific and technical journals and proceedings in recent times.

INTRODUCTION

One prime achievement of structural engineering profession is the design, construction and analysis of thin walled constructions and buildings in the shape of different shell surfaces. Laminated composite hyper shells (Hyperbolic paraboloids bounded by straight edges) are landmarks and testimonials to this achievement. They are true three dimensional representations of our equilibrium equations and affirmations of our analytical techniques, design standards and construction practices and choice

DOI: 10.4018/978-1-5225-5216-1.ch011

of smart materials. These shells are useful as roofing units in civil engineering industry to cover large column free areas such as in stadiums, airports, shopping malls and parking lots. Being a doubly curved and doubly ruled surface, it satisfies aesthetic and ease of casting requirements of the industry. Moreover, hyper shell allows entry of daylight and natural air which is preferred in food processing and medicine units. Cutout is sometimes necessary in roof structure for the passage of light, to provide accessibility of other parts of the structure, for venting and also sometimes for alteration of resonant frequency. Shell structures that are normally thin walled, when provided with cutout, exhibits improved performances with stiffeners. Interest in composite shell behavior has grown day by day due to increased usage of laminated composite shells in many engineering applications. Laminated composites offer higher strength-to-weight and stiffness-to weight ratios, improved chemical and environmental resistance and ability to tailor properties compared to traditional materials. Additionally, the advances in composite manufacturing methods have also contributed to the increased usage of laminated composite materials in many modern applications.

Shell is an arbitrarily curved structural surface capable of resisting any externally superimposed load by combined in-plane thrusts and bending of the surface. Coupling of in-plane forces and bending moments provides high strength to the shell forms. Essentially a shell form gets its strength through its form rather than mass. The doubly curved, non-developable shell surfaces have added efficiency both in terms of strength and aesthetics. Some of the earliest applications of shell structures in the history of civilization include the first prototype submarine in 1620, the first ever pressure cooker in 1688, the use of balloon as a thin walled pressure vessel, the pneumatic tyre in 1845 etc. Shell roofs of civil engineering interests include the Pantheon of Rome built in AD 125, the Mosque of Santa Sophia, Istanbul built in AD 538 etc. Subsequently, shell forms were used in civil engineering practices as roofs, grain-storage structures, chimneys, water tanks, etc. From the early years of human civilization up to the present time, the course of shell construction, application and research has been developed quite a lot with the focus being shifted from one aspect to another with time. Structures made of thin plates/shells and strengthened by integrating them with a row of ribs appeared as a revolutionary concept.

Wide use of stiffened structural elements in engineering came into existence in the nineteenth century, mainly with the application of steel plates for hulls of ships and with the development of steel bridges and aircraft structures. As the use of stiffened structures gained popularity, the engineers started thinking of using different materials in different applications depending on the case specific requirements. The search in the field of advanced materials resulted in the development of lightweight laminated composites with high stiffness-to-weight and strength-to-weight ratios. The potentially high resistance of these materials to environmental degradation results in lower life-cycle costs and they may be tailored to any desired shape. Due to the wide spectrum of advantages the laminated composites can offer, these materials started being applied in different branches of engineering. The use of composites in engineering structures evolved as a result of a necessity during the Second World War.

Broadly composites are combinations of two or more materials to achieve specific properties. From engineering point of view a composite may be defined as a combination of materials, which differ in composition or form on macroscopic level and remain chemically inactive with respect to each other to form a useful material. Commonly used composites are of two types i.e. particulate and fibrous. A particulate composite consists of particle reinforcements embedded in a continuous phase (matrix) of other material. Both, particle and matrix may be metallic or non-metallic. Fibrous composites consist of a large number of strong, stiff, continuous or chopped fibres embedded in a matrix having a large length to diameter ratio (100 or more) to ensure a reinforcing action. They are made of either inorganic (glass, carbon, boron) or organic (aramid or kevlar) materials. The matrices can also be metallic (aluminium) or non-metallic (polyester, epoxy, phenolic resins and ceramics). Lamina is formed when an array of fibres is given a resin bath and hardened. The matrix transfers the loads to fibres. A number of laminae or plies are bonded together in different orientations to form a laminate. Laminae are homogeneous and anisotropic in macroscopic scale, and laminate is orthotropic. These are used in different structural forms such as beams, plates, shells, stiffened plates, stiffened shells etc. The low cost of assembly, ease of repair, high specific stiffness and strength, excellent damage tolerance and superior fatigue response characteristics increased the use of laminated composites in several important engineering fields.

The exact analysis of stiffened plates/shells based on the theory of elasticity is rarely performed due to its tedious and difficult computational procedures for evaluation of deflection and stresses. Analysis of stiffened shells has been approached from different angles. Out of these, the finite element method is considered to be the most versatile due to its ability to cater to any arbitrary geometry, loading and boundary conditions. The review of the literature shows that the researchers from different corners of the world are engaged in analysis of composite stiffened plates and shells by finite element approach and the present research also uses this technique to analyse the bending and free vibrations of composite stiffened skewed hypar shells which have wide applications as roofing units.

Stiffened shell structures are extensively used in aerospace, civil, marine and other engineering applications as mentioned earlier. In civil engineering construction, cylindrical, elliptic paraboloid, hyperbolic paraboloid, conoidal and hypar shells are commonly used as roofing units. Among these the doubly curved shells are architecturally appealing and frequently favoured for roofing large column-free areas. The anticlastic doubly curved hypar shells having only the twist of curvature are preferred in many situations due to their architectural beauty. The surfaces are doubly ruled and hence easy to cast and fabricate. The shells are often stiffened to achieve greater strength with relatively less amount of material which results in considerable weight reduction.

As major structural components of ships and aircrafts the stiffened plates/shells are subjected to both static and dynamic loading of varying magnitudes. The static loads are caused by the internal and/or external pressure in fuel cells of aircrafts, long pipes and conduit structures, dead and live loads on the ship decks and similar other loads. Dynamic effects are caused by the aerodynamic loadings on aircrafts, hydrodynamic loadings on ship hulls, engine vibrations and such other forces. Shell forms used in civil engineering as roofing units have to withstand various distributions of dead and live loads apart from the dynamic loads from wind, earthquake, snowfall etc. Hence, the knowledge of their static as well as dynamic behaviour is important from the standpoint of analysis and design.

In May 1966, Cornel students put finishing touches on the hyperbolic paraboloid structure on the Cornell Plantations. The Hypar, affectionately known Flying Diaper was built as part of an interdisciplinary course devoted to teaching the fundamentals of surface (shell) structures to structural engineering and architecture students. At the time it was built, the hypar may have been the first of its kind the United States. Fifty years later, the structure still stands at the entrance to the Cornell Plantations. Shells became popular in countries where labour was relatively cheap and materials, particularly steel, was expensive. For locations within temperate zones, where snow loads were non-existent and it was only necessary to carry its own weight, even optimal shapes could be designed. Among all of the possible shapes, the hyperbolic paraboloid, "hypar", is particularly unique in that the structures which initially appear curved in two directions, it can really be constructed out of straight line segments. The "anticlastic" surface support tensile stresses in the hanging direction and compressive stresses in the arch direction.

Extensive research about bending, vibrations, stability etc. are going on both about isotropic and composite shells all over the world. The advent of high speed computers is helping the applications of the numerical techniques like finite difference, finite element, boundary element and others to analyse arbitrary shell geometry subjected to complicated loadings and boundary conditions. Now-a-days, civil engineers use laminated composites to fabricate these shell forms as the high specific stiffness and strength properties of these materials result in less gravity forces and mass-induced forces (seismic force) on the laminated shells compared to their isotropic counterparts. All these taken together reduce the foundation costs to a great extent. Realizing the importance of laminated composite shells in the industry, several aspects of shell behaviour such as bending, buckling, vibration, impact etc. are being reported by different researchers.

Bending Behaviour of Laminated Composite Hypar Shell

Study of the bending behaviour of doubly curved composite shells started long ago (Narita & Leissa, 1984; Reddy & Chandrashekhara, 1985; Schwarte, 1994) where the practical boundary conditions like corner point support etc. have been considered along with other aspects. A skewed hypar shell has only a twist of curvature and thus structurally resembles a twisted plate. Consequently, characteristics of cantilever twisted plate have been studied (Kielb, Leissa & Macbein, 1985; Seshu & Ramamurti, 1989; Qatu & Leissa, 1991). Extensive bending analysis of doubly curved composite conoidal and paraboloidal revolution shells were also carried out (Dey, Bandyopadhyay & Sinha, 1992a, 1992b, 1994) taking simply supported and clamped boundary conditions. Combinations of different boundary conditions were also considered (Qatu & Leissa, 1993; Sivasubramonian, Kulkarni, Rao & Krishnan, 1997) to study free-vibration characteristics of doubly curved panels excluding skewed hypar shells. Skewed hypar shells with different boundary conditions were considered (Chakravorty, Bandyopadhyay & Sinha, 1998) for identical boundary conditions in all four edges. Bending analysis of skewed hypar composite shells with different combinations of boundary conditions which may include different edge situations at the four sides is scanty. Bending behaviour of hypar shell (Figure 1) has been studied for different lamination and boundary conditions (Sahoo & Chakravorty, 2004).

Authors considered different types of boundary constraints (shown in Figure 2) and mentioned that an increase in the number of support constraints always reduces the deflection but may increase other shell actions like inplane forces and bending moments etc. They also mentioned that free edges introduce marked flexibility in shells and whether the other edges are simply supported or clamped matter less significantly. Besides among shells with two edges clamped and other two simply supported, the ones with adjacent edges clamped show lesser deflection for all cross ply and symmetric angle ply laminates. For these boundaries antisymmetric angle ply emerges as the best choice from static deflection consideration. Again shells involving two free edges, symmetric cross ply laminate exhibits least deflection and the ones with two adjacent edges free always exhibit greater static deflections for all combinations of laminations. A change in the arrangement of restraints in the individual edges, keeping the total number of support fixities same, has a marked influence on maximum deflection if free edges are involved. It is also observed that maximum deflection always occurs at a point on the free edge in case of shells involving free boundaries. Finally it is seen that the relative performances of single layered orthotropic and anisotropic shells can form the basis of predicting that of multilayered cross and angle ply shells except FCFC shells.

Figure 1. Typical hypar shell with laminations

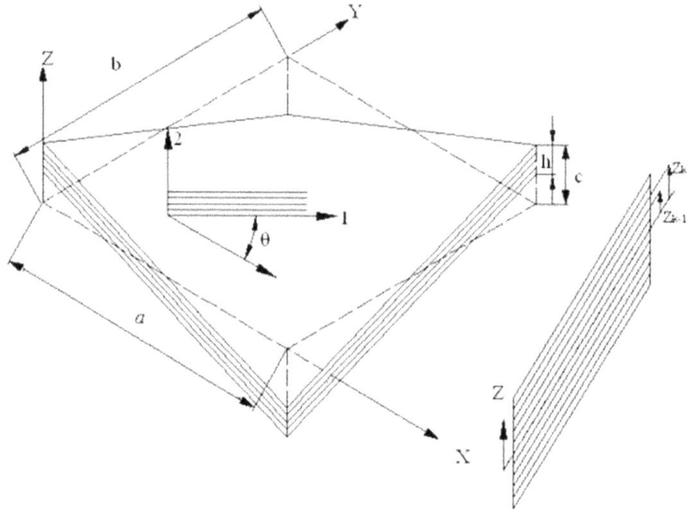

Figure 2. Arrangement of boundary conditions

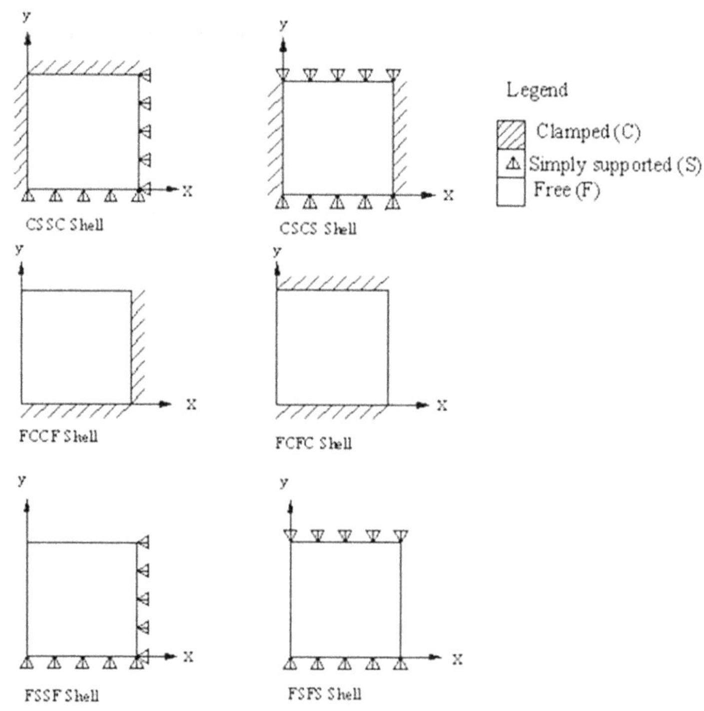

Bending of Stiffened Shell

The advent of laminated composites in different branches of engineering applications has provided a new impetus to the researchers to explore the different aspects of composite structural elements including different forms of shells. Shell structures that are normally thin walled exhibit improved performances with stiffeners, particularly when the shell surface is provided with cutouts or subjected to concentrated loads. Basic knowledge of the bending behaviour of stiffened composite skewed hypar shells is a prerequisite for using these forms confidently. In an excellent review article, Sinha & Mukhopadhyay (1995) discussed the developments in the field of static and dynamic analysis of stiffened shells upto 1995. Chakravorty, Sinha & Bandyopadhyay (1998) reported results on free and forced vibrations of laminated composite hypar shells where in order to evaluate the dynamic magnification factors some results of static deflection, force and moment resultants of point supported unstiffened hypar shells under distributed loading were furnished. Sahoo & Chakravorty (2004) carried out detailed analysis of static deflection, force and moment resultants of composite unstiffened hypar shells with different practical boundary conditions. Results of bending of both isotropic and composite shell panels are scanty in the existing body of literature. Only a few papers (Kohnke & Schnobrich, 1972, Rao, 1978, Sinha, Sheikh & Mukhopadhyay, 1992; Prusty & Satsangi, 2001) deal with static analysis of stiffened shell panels but information regarding bending characteristics of stiffened composite hypar shell is limited in the literature. Sahoo & Chakravorty (2008) studied the deflections, force and moment resultants of stiffened composite skewed hypar shells with different boundary condition and subjected to central concentrated load with different parametric variations. A typical hypar shell panel with biaxial stiffeners eccentric at shell bottom is shown in Figure 3. Authors observed that the maximum deflection occurring under the load always decreases with increase in boundary constraints. CSSC and CSCS boundaries have equal number of restraints but the performance of the former is either better or comparable with later in terms of static deflection. Provision of a single stiffener along one of the plan directions can adequately stiffen a hypar shell to such an extent that a stiffened SSSS shell may exhibit a deflection lower than a bare CCCC shell. They also observed that the performances of angle ply shells are better than the cross ply ones when static deflections are studied but interestingly when all the shell actions are considered together it is found that in general cross ply shells should be preferred. Again eccentrically stiffened shells perform better than concentrically stiffened ones in terms of all shell actions and should be preferred unless restricted by any functional requirements. Increase in stiffener depth always have positive effect in arresting the differential shell actions and hence other criteria like aesthetic, headroom shall be considered to fix up the stiffener depth. They also observed that the complex

Figure 3. A typical hypar shell panel with biaxial stiffeners eccentric at shell bottom

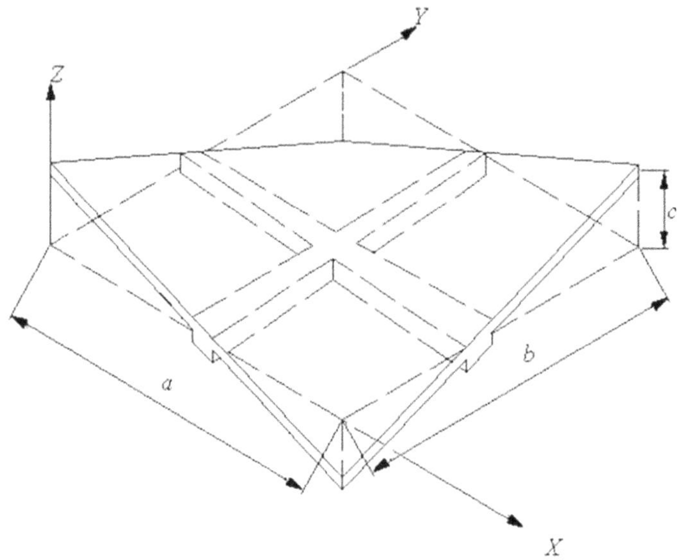

interaction between boundary conditions and stacking sequences with curvature make it difficult to predict variation of shell actions with curvature.

But the deflection and governing moment improve with increase in curvature. They furnished the critical values of different shell actions and locations for various boundary conditions, stacking sequences and stiffening schemes will help design engineers using composite stiffened skewed hypar shells. They developed the relative performance matrix and analysed the same (considering two shell options combining laminations, boundary conditions and stiffening schemes) which will help an engineer to make a conclusive decision between the options by assigning appropriate weightage factor to different shell actions. They mentioned that if a designer has the limitation of providing only one central stiffener he may preferably run the stiffeners along *X* direction because central *x* stiffeners perform either equally or better than central *y* stiffeners for the class of shells they considered.

Bending of Stiffened Shell With Cutout

Review of available literature on the bending behavior of laminated composite hypar shells with a cutout reveals that very little study has been reported so far on this aspect. Sanders Jr. (1990) and Ghosh & Bandyopadhyay (1994) have considered the bending of isotropic shells with a cutout. The static behavior of a cylindrical composite panel in presence of cutouts has been reported using a geometrically nonlinear theory (Dennis & Palazotto, 1990) while vibration of cylindrical panel with square cutout has been studied based on finite element method (Sivasubramanian, Kulkarni, Rao & Krishnan, 1997). The vibration of isotropic shallow spherical shell with circular cutout has also been analyzed (Hwang & Foster Jr., 1992). Madenci & Barut (1994) studied buckling of composite panels in presence of cutouts. Non-linear post-buckling analysis of composite cylindrical panels with central circular cutouts but for unstiffened ones was studied by Noor, Starnes & Peters (1996) to consider the effect of edge shortening as well as uniform temperature change. Later Sai Ram & Babu (2001a, b) investigated the bending behavior of axisymmetric composite spherical shell both punctured and un-punctured using the finite element method based on a higher order theory.

Surface of a skewed hypar shell with cutout is shown in Figure 4. Bending analysis of stiffened hypar shell with cutout (Chaudhury, Mitra and Sahoo, 2016) reveals that an increase in support restraints always reduces the deflection and static stress resultants near the boundary. Authors also observed that among shells with two edges clamped and other two simply supported, the ones with adjacent edges clamped show lesser deflection for all the antisymmetric lamination. Among shells involving two free edges, ones with two adjacent edges free always exhibit greater static deflection for all antisymmetric laminations considered. Free edges introduce marked flexibility in shells and whether the other edges are simply supported or clamped matter significantly. Also when a free edge is introduced to a stiffened shell with cutout maximum deflection and stress resultants always occur near the free edge. They observed that the superiority of a particular combination in terms of a shell action cannot be the basis of predicting the relative performances in terms of other shell actions.

Buckling Behavior of Laminated Composite Hypar Shell

Bucking analysis of laminated hypar shell (Pradyumna and Bandyopadhyay, 2010) reveals that buckling loads increase with the increase of number of layers up to 8 for all the lamination scheme except the symmetric cross-ply scheme with any of the three boundary conditions (simply supported, clamped and corner supported). Authors also showed that buckling loads increase significantly with the increase of width to thickness ratios for hypar shells for all the boundary conditions considered by them having symmetric and anti symmetric cross ply lamination scheme. Buckling load of hypar shell with the three boundary conditions also increases with increase of curvature parameters. Again they observed that buckling load of hypar shell is asymmetric with respect to the ply orientation angle. But the variation of buckling load with the increasing aspect ratio is wavy for hypar shell. The value of buckling

Figure 4. Surface of a stiffened hypar shell with cutout

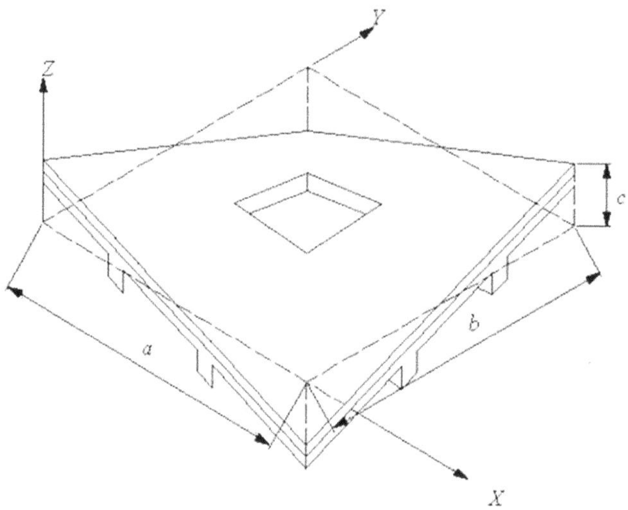

Surface equation: $z = \frac{4c}{ab}(x - a/2)(y - b/2)$

load is largest for clamped boundary condition followed by simply supported and corner supported boundary conditions.

Failure of Composite Hypar Shell

The lamina-wise failure progress mechanism with gradually increasing failure load and knowledge of failure propagation may be utilized for getting a warning in advance about the gross failure of the shell by appropriate instrumentation related to different nondestructive tests. If an engineer has a prior knowledge regarding the flow of damage then appropriate points on the shell surfaces may be tested only to detect the hidden flaws without going into an elaborate testing program. The maximum area of damage that can be allowed under a practical situation may be correlated with the nondestructive test results at predetermined points. With this aim, researchers have started paying attention to the failure aspects of laminated composite hypar shell roofs. Failure of composite hypars has been analyzed recently (Ghosh and Chakravorty, 2014). The results therein indicated that angle ply shells should be preferred compared to cross ply ones from engineering point of view. The location of the first ply failure point is extremely important to be known to a practicing engineer because any instrumentation needed for hidden flaw detection should start from that point. In other words, if it is found that the point prone to first ply failure damage is free of any hidden flaw, the shell surface is free of any damage caused due to overloading. In another study (Ghosh and Chakravorty, 2017), it is pointed out that by virtue of the geometry of a skewed hypar shell, majority of the loads can be overcome. This is why angle ply laminates which have their fibres oriented along the diagonals prove to be better than the cross ply ones in terms of first ply failure. The lamina wise failure investigation and using the information to evolve tailored laminates may be utilized as design guidelines to fabricate stiff shell surfaces for a given material consumption. They also pointed out that the damaged area of a skewed hypar shell may be explicitly correlated through an equation with a visible measurable gross response of the shell that is deflection, may be utilized by practicing civil engineers in health monitoring of the shell without going into detailed non-destructive testing programme. They also observed that the values of load factors are extremely high and indicate that the laminated composites have substantial amount of load resisting capacity beyond the working load and hence these materials are expected to exhibit a ductile failure. This virtue of these materials makes them appropriate to be used in earthquake prone zones.

Vibration Behaviour of Composite Hypar Shell

Free vibration analysis of composite hypar shell by Sahoo and Chakravorty (2005) reveals that in general fundamental frequency increases with the increase in the number of support constraints. There are, however, few departures from this general tendency when two shells of different laminations are compared. Moreover, the number of boundary constraints and their arrangement influence the mode shapes significantly. For a particular boundary condition the mode shapes sometimes change with lamination and in some other cases the lamination has almost no effect on the vibration pattern. Again fundamental frequencies and mode shapes of $(0/\theta)_n$ shells do not change significantly when n is increased beyond 3 and the relation between frequency and θ is sometimes unified and sometimes not. Representative mode shapes for laminated composite hypar shells with different stacking sequences and boundary conditions are shown in Figure 5. For shells with opposite edges clamped, fibers parallel to the clamped edges and further from the middle surface contribute more significantly to the vibration stiffness. Further, for four layered laminates $(0/\theta/0/\theta)$ and $(0/\theta/\theta/0)$ the fundamental frequency either increases or decreases but monotonically with θ. For CSCS SSSS and CFFF boundary conditions antisymmetric laminations are recommended while for FCFC, FSSF and FSFS edges symmetric laminations are better choices for any given value of θ. For CCCC shells the performances of $0/\theta/0/\theta$ and $0/\theta/\theta/0$ stacking orders are comparable and for CSSC and FCCF boundaries the relative performances of the above laminates are functions of θ. Fundamental mode shapes are symmetric if both edge conditions and laminations show biaxial symmetry. An asymmetry in fibre orientation brings about asymmetric nature of mode shapes.

Vibration of Laminated Composite Hypar Shell Roofs With Cutouts

Free vibration of composite as well as isotropic plate with cutout was studied by different researchers from time to time. Reddy (1982) investigated large amplitude flexural vibration of composite plate with cutout. Malhotra, Ganesan & Veluswamy (1989) studied free vibration of composite plate with cutout for different boundary conditions. One of the early reports on free vibration of curved panels with cutout was due to Sivasubramonian, Kulkarni, Rao & Krishnan (1997). They analysed the

Figure 5. Typical mode shapes for laminated composite hypar shells with different stacking sequences and boundary conditions

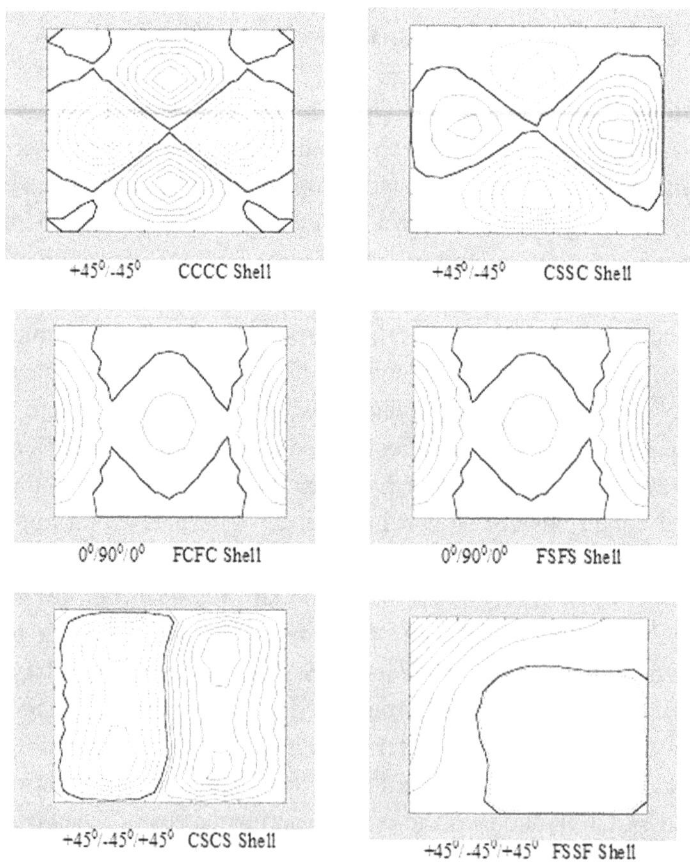

effect of cutouts on the natural frequencies of plates with some classical boundary conditions. The plate had a curvature in one direction and was straight in the other. The effect of fibre orientation and size of cutout on natural frequency on orthotropic square plates with square cutout was studied using Rayleigh-Ritz method. Later Sivakumar, Iyenger & Deb (1999), Rossi (1999), Huang & Sakiyama (1999), Hota & Padhi (2007) studied free vibration of plate with various cutout geometries. Chakravorty, Sinha & Bandyopadhyay (1998) analyzed the effect of concentric cutout on different shell options.

The free vibration of un-stiffened hypar shell with cutouts is studied considering different boundary conditions (Sahoo, 2011). The variation of fundamental frequency due to change in eccentricity of cutout along x and y direction is also considered. It is reported that concentric square cutouts may be provided safely on hypar shell surfaces for functional requirements upto the ratio of cutout size to side of shell =0.2. It is found that the arrangement of boundary constraint along the four edges is far more important than their actual number so far the free vibration stiffness is concerned. Fundamental frequency undergoes marked improvement when the edge is converted to clamped from simply supported condition.

Vibration of Stiffened Hypar Shell

Sahoo and Chakravorty (2006) in their paper analysed composite stiffened hypar shell and opined that if a designer has the limitation of providing only one central stiffener only he may preferably run the stiffeners along X direction because central x stiffeners perform either equally or better than central y stiffeners for the class of laminations considered in their study. For other laminations, however, this conclusion may not be true. Also for effective increase of the fundamental frequency the stiffeners may be provided perpendicular to the nodal lines obtained from vibrations of bare shells. In the cases where no nodal lines are exhibited by a

vibrating bare shell the performance of x and y stiffeners are comparable. It was observed that the mode shapes of angle ply shells are more complex than those of the cross ply ones with greater number of crests and troughs within shell span among shells with a single central stiffener along one or both the plan directions. Again for biaxially stiffened shells with one or more stiffener/s along each of the plan directions, the vibration modes for both the boundary conditions were found to be complicated. For some of the simply supported shells, torsional modes of vibration are identified while for clamped shells, the vibrational modes are flexural along both the plan directions. They also observed that with the increase of the number of stiffeners, either uniaxial or biaxial, the fundamental frequency increases but reaches a saturation where after there is no appreciable increase in frequency with the provision of additional stiffeners. Moreover, for any given number of stiffeners in

Figure 6. Variation of non-dimensional fundamental frequencies with stiffener depth to shell thickness ratio for orthogonally stiffened composite hypar shells

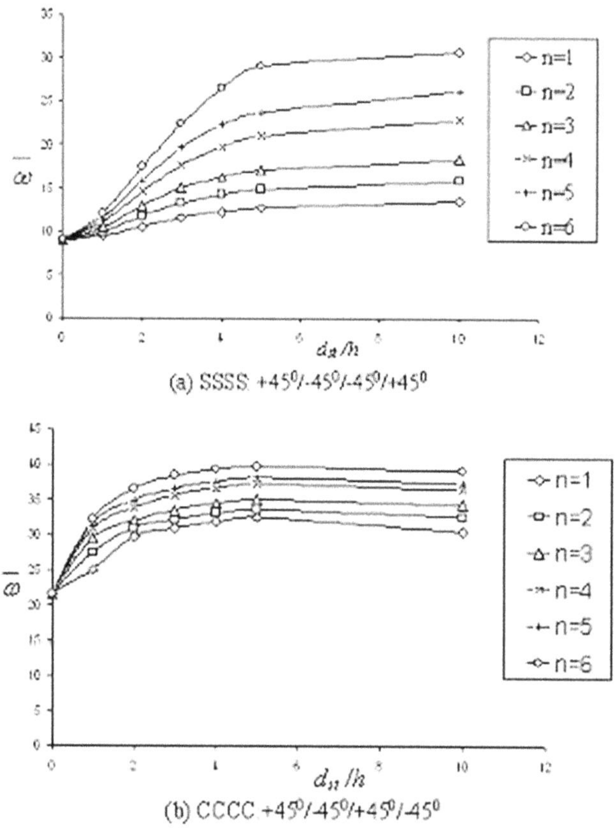

one of the directions, there is no point in providing more than five or three stiffeners for simply supported and clamped shells respectively, along the other direction. Among shells with equal number of x and y stiffeners there is no appreciable increase in frequency when the number of stiffeners in either direction exceeds 5 (Figure 6). So from their study it is proved that there are two ways of increasing the fundamental frequency of a bare shell by stiffening, either by increasing the stiffener depth keeping the number of stiffeners fixed or by increasing the number of stiffeners with a given stiffener depth.

Vibration of Stiffened Shell With Cutout

Sahoo (2012) analysed free vibration behaviour of composite stiffened hypar shell with cutout and observed that concentric cutouts may be provided safely on stiffened

Figure 7. Values of non-dimensional fundamental frequency of 0/90/0/90 stiffened hypar shell with cutout for different sizes of central cutout and boundary conditions

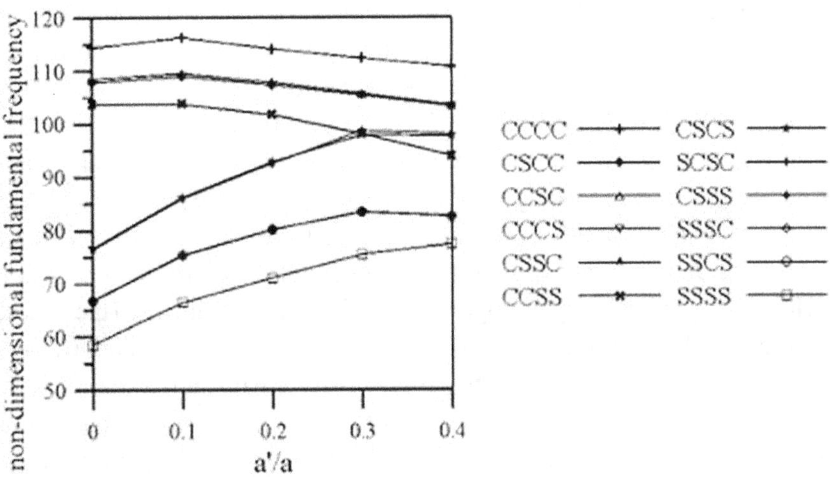

hypar shell surfaces for functional requirements upto a'/a=0.2. The arrangement of boundary constraint along the four edges is far more important than their actual number so far the free vibration stiffness is concerned (Figure 7). Fundamental frequency undergoes marked improvement when the edge is converted to clamped from simply supported condition. The author furnished the specific zones within which the cutout centre may be varied such that the loss of fundamental frequency is within 5% with respect to a shell with concentric cutout also the specific zones which should be avoided (where the frequency decrease greater than equal to 10%) will help to take a decision regarding the eccentricity of the cutout centre that can be allowed.

Hypar Shell Under Impact

Composite material has gained popularity in different industrial sectors due to its high specific weight and stiffness. With other sectors, civil engineering is no exception and the laminated composites are used as roofing material in the situation where large column-free space is required. The high specific weight also causes reduction in dead weight of structure and thus reduces seismic weight and foundation cost. In spite of these advantages, low transverse shear strength makes them vulnerable under sudden impact. This leads to research in this aspect where an accurate modeling of contact behavior is one of the most important steps to analyze an impact response problem. A scrutiny of relevant literature reveals the fact that impact response of civil engineering shell structures has not received due attention though such shells may often be subjected to impact loads. Impact forces on shell roofs are encountered due to snow fall, due to air borne debris in cyclone prone areas and in other similar situations. So the necessity to study impact response of composite shell roofs is felt. Impact response of composite hypar shell was studied by Neogi, Karmakar & Chakravorty (2011, 2013, 2017) under different boundary conditions.

Under the influence of normal low velocity impact, the contact force shows a parabolic combined loading and unloading curve with a single peak for the practical class of shells. Higher magnitude of impact velocity results in higher value of the peak contact force but due to a sharp elastic rebound, the total duration of contact force is less for higher velocity of impactor. The time instants at which the maximum contact force and the maximum dynamic displacement occur show a phase difference and interestingly in some cases, the maximum displacement and hence stresses may occur even after the contact force dies down totally. The maximum contact force, the peak dynamic displacement, and the equivalent static load are all increasing functions of impactor velocity, the relations being almost linear. However, the dynamic magnification factor shows a logarithmically decreasing tendency with increase of the velocity of impact. At some of the points, the displacement at some instant may exceed the same occurring at the point of impact. But such displacements will never govern the design as they are always less than the absolute maximum displacement occurring at the point of contact.

A comparative study of performances of simply supported cross ply and simply supported angle ply shells reveals that the nature of contact force and dynamic displacement for these two types of shells are more or less similar. One interesting difference is that for simply supported cross ply shell the peak dynamic displacement does not only show a phase lag with respect to the peak contact force but by the time displacement value reaches the peak, the contact force value dies down totally. This shows that the after effect of impact are some times more severe than the shell response during the impact and study of displacement variation even after the contact force decays to a null value is absolutely necessary. However, after passage of some more time the subsequent local maxima which are obtained do not touch the peak. The dependence of the maximum contact force, the peak dynamic displacement and the equivalent static load on the impact velocity in case of simply supported cross ply shell are similar to what is observed in case of simply supported angle ply shell.

CONCLUSION

With advancement in the field of structural engineering, hunt for smarter materials has led the research towards the application of composite materials. It is the high specific weight and specific stiffness of this material that has drawn the interest of different industrial sectors. Civil engineers also picked up composites to use it as a roofing material. Laminated composite shells can cover large column-free area and reduces dead weight of structure. Hypar shell is unique in a sense that the curvature has only the radius of cross curvature and these shells do not admit easy closed form solution particularly when the boundary conditions are complicated. Thus researchers have mostly resorted to different numerical techniques to explore the behavior of such shells under various boundary conditions and laminations along with the structural complexities including presence of stiffeners, cutouts etc. Results available in literature concerning the bending, vibration, buckling, failure, impact etc. of laminated composite hypar shells definitely lead to better understanding of these shell forms under various situations and will aid the structural engineers in proper design of such composite structures.

REFERENCES

Chakravorty, D., Sinha, P. K., & Bandyopadhyay, J. N. (1998). Applications of FEM on free and forced vibrations of laminated shells. *Journal of Engineering Mechanics*, *124*(1), 1–8. doi:10.1061/(ASCE)0733-9399(1998)124:1(1)

Chaudhury, P. B., Mitra, A., & Sahoo, S. (2016). Relative performance of antisymmetric angle-ply laminated stiffened hypar shell roofs with cutout in terms of static behavior. *Curved and Layered Structures*, *3*(1), 22–46.

Dennis, S. T., & Palazotto, A. N. (1990). Static response of a cylindrical composite panel with cutouts using a geometrically nonlinear theory. *AIAA Journal*, *28*(6), 1082–1108. doi:10.2514/3.25170

Dey, A., Bandyopadhyay, J. N., & Sinha, P. K. (1992a). Finite element analysis of laminated composite conoidal shell structures. *Computers & Structures*, *43*(3), 469–476. doi:10.1016/0045-7949(92)90281-4

Dey, A., Bandyopadhyay, J. N., & Sinha, P. K. (1992b). Finite element analysis of laminated composite paraboloid of revolution shells. *Computers & Structures*, *44*(3), 675–682. doi:10.1016/0045-7949(92)90400-T

Dey, A., Bandyopadhyay, J. N., & Sinha, P. K. (1994). Behaviour of paraboloid of revolution shell using cross-ply and antisymmetric angle-ply laminates. *Computers & Structures*, *52*(6), 1301–1308. doi:10.1016/0045-7949(94)90194-5

Ghosh, A., & Chakravorty, D. (2014). Prediction of progressive failure behaviour of composite skewed hypar shells using finite element method. *Journal of Structures, ID, 147578*, 1–8.

Ghosh, A., & Chakravorty, D. (2017). Failure analysis of civil Engineering composite shell roofs. *Procedia Engineering, 173*, 1642–1649. doi:10.1016/j.proeng.2016.12.258

Ghosh, B., & Bandyopadhyay, J. N. (1994). Bending analysis of conical shells with cutouts. *Computers & Structures, 53*(1), 9–18. doi:10.1016/0045-7949(94)90124-4

Hota, S. S., & Padhi, P. (2007). Vibration of plates with arbitrary shapes of cutouts. *Journal of Sound and Vibration, 302*(4-5), 1030–1036. doi:10.1016/j.jsv.2007.01.003

Huang, M., & Sakiyama, T. (1999). Free vibration analysis of rectangular plates with variously shaped holes. *Journal of Sound and Vibration, 226*(4), 769–786. doi:10.1006/jsvi.1999.2313

Hwang, D. Y., & Foster, W. A. Jr. (1992). Analysis of axisymmetric free vibration of isotropic shallow spherical shells with a circular hole. *Journal of Sound and Vibration, 157*(2), 331–343. doi:10.1016/0022-460X(92)90685-Q

Kielb, R. E., Leissa, A. W., & Macbain, J. C. (1985). Vibrations of twisted cantilever plates—a comparison of theoretical results. *International Journal for Numerical Methods in Engineering, 21*(8), 1365–1380. doi:10.1002/nme.1620210802

Kohnke, P. C., & Schnobrich, W. C. (1972). Analysis of eccentrically stiffened cylindrical shells. *Journal of the Structural Division, 98*(ST7), 1493–1510.

Madenci, E., & Barut, A. (1994). Pre and post-buckling response of curved thin composite panels with cutouts under compression. *International Journal for Numerical Methods in Engineering, 37*(9), 1499–1510. doi:10.1002/nme.1620370906

Malhotra, S. K., Ganesan, N., & Veluswami, M. A. (1989). Vibration of composite plates with cut-outs. *Journal of Aeronautical Society of India, 41*, 61–64.

Narita, Y., & Leissa, A. W. (1984). Vibrations of corner point supported shallow shells of rectangular planform. *Earthquake Engineering & Structural Dynamics, 12*(5), 651–661. doi:10.1002/eqe.4290120506

Neogi, S. D., Karmakar, A., & Chakravorty, D. (2011). Impact response of simply supported skewed hypar shell roofs by finite element. *Journal of Reinforced Plastics and Composites, 30*(21), 1795–1805. doi:10.1177/0731684411418865

Neogi, S. D., Karmakar, A., & Chakravorty, D. (2013). Study of dynamic behavior of multilayered clamped composite skewed hypar shell roofs under impact load. *Journal of Engineering, ID, 192176*, 1–10.

Neogi, S. D., Karmakar, A., & Chakravorty, D. (2017). Finite element analysis of laminated composite skewed hypar shell roofs under oblique impact with friction. *Procedia Engineering, 173*, 314–322. doi:10.1016/j.proeng.2016.12.023

Noor, A. K., Starnes, J. H. Jr, & Peters, J. M. (1996). Nonlinear and postbuckling responses of curved composite panels with cutouts. *Composite Structures, 34*(2), 213–240. doi:10.1016/0263-8223(95)00147-6

Pradyumna, S., & Bandyopadhyay, J. N. (2010). Buckling of laminated composite hypars and conoids using a higher-order theory. *The IES Journal Part A: Civil & Structural Engineering, 3*(2), 85–95.

Prusty, B. G., & Satsangi, S. K. (2001). Analysis of stiffened shell for ships and ocean structures by finite element method. *Ocean Engineering, 28*(6), 621–638. doi:10.1016/S0029-8018(00)00021-4

Qatu, M. S., & Leissa, A. W. (1991). Vibration studies for laminated composite twisted cantilever plates. *International Journal of Mechanical Sciences, 33*(11), 927–940. doi:10.1016/0020-7403(91)90012-R

Qatu, M. S., & Leissa, A. W. (1993). Vibration of shallow shells with two adjacent edges clamped and others free. *Mechanics of Structures and Machines, 21*(3), 285–301. doi:10.1080/08905459308905190

Rao, K. P. (1978). A rectangular laminated anisotropic shallow thin shell finite element. *Computer Methods in Applied Mechanics and Engineering, 15*(1), 13–33. doi:10.1016/0045-7825(78)90003-8

Reddy, J. N. (1982). Large amplitude flexural vibration of layered composite plates with cutouts. *Journal of Sound and Vibration, 83*(1), 1–10. doi:10.1016/S0022-460X(82)80071-0

Reddy, J. N., & Chadrashekhara, K. (1985). Geometrically nonlinear transient analysis of laminated doubly curved shells. *International Journal of Non-linear Mechanics, 20*(2), 79–90. doi:10.1016/0020-7462(85)90002-2

Rossi, R.E. (1999). Transverse vibrations of thin orthotropic rectangular plates with rectangular cutouts with fixed boundaries. *Journal of Sound and Vibration, 221*(4), 733-736.

Sahoo, S. (2011). Free vibration of laminated composite hypar shell roofs with cutouts. *Advances in Acoustics and Vibration, ID, 403684*, 1–13.

Sahoo, S. (2012). Behaviour and optimization aids of composite stiffened hypar shell roofs with cutout under free vibration. *ISRN Civil Engineering, ID, 989785*, 1–14.

Sahoo, S., & Chakravorty, D. (2004). Finite element bending behaviour of composite hyperbolic paraboloidal shells with various edge conditions. *Journal of Strain Analysis for Engineering Design, 39*(5), 499–513. doi:10.1243/0309324041896434

Sahoo, S., & Chakravorty, D. (2005). Finite element vibration characteristics of composite hypar shallow shells with various edge supports. *Journal of Vibration and Control, 11*(10), 1291–1309.

Sahoo, S., & Chakravorty, D. (2006). Stiffened composite hypar shell roofs under free vibration: Behaviour and optimization aids. *Journal of Sound and Vibration, 295*(1-2), 362–377. doi:10.1016/j.jsv.2006.01.012

Sahoo, S., & Chakravorty, D. (2008). Bending of composite stiffened hypar shell roofs under point load. *Journal of Engineering Mechanics, 134*(6), 441–454. doi:10.1061/(ASCE)0733-9399(2008)134:6(441)

Sai Ram, K. S., & Babu, T. S. (2001a). Study of bending of laminated composite shells, Part I: Shells without a Cutout. *Composite Structures, 51*(1), 103–116. doi:10.1016/S0263-8223(00)00129-X

Sai Ram, K. S., & Babu, T. S. (2001b). Study of bending of laminated composite shells, Part II: Shells with a Cutout. *Composite Structures, 51*(1), 117–126. doi:10.1016/S0263-8223(00)00130-6

Sanders, J. L. Jr. (1970). Cutouts in Shallow shells. *Journal of Applied Mechanics, 37*(2), 374–383. doi:10.1115/1.3408516

Schwarte, J. (1994). Vibrations of corner point supported rhombic hypar-shells. *Journal of Sound and Vibration, 175*(1), 105–114. doi:10.1006/jsvi.1994.1314

Seshu, P., & Ramamurti, V. (1989). Vibrations of twisted composite plates. *Journal of Aeronautical Society of India, 41*, 65–70.

Sinha, G., & Mukhopadhyay, M. (1995). Static and dynamic analysis of stiffened shells- A review. *Proceedings of Indian National Science Academy, 61A*(3&4), 195–219.

Sinha, G., Sheikh, A. H., & Mukhopadhyay, M. (1992). A new finite element model for the analysis of arbitrary stiffened shells. *Finite Elements in Analysis and Design*, *12*(3-4), 241–271. doi:10.1016/0168-874X(92)90036-C

Sivakumar, K., Iyengar, N. G. R., & Deb, K. (1999). Free vibration of laminated composite plates with cutout. *Journal of Sound and Vibration*, *221*(3), 443–470. doi:10.1006/jsvi.1998.2034

Sivasubramonian, B., Kulkarni, A. M., Rao, G. V., & Krishnan, A. (1997). Free vibration of curved panels with cutouts. *Journal of Sound and Vibration*, *200*(2), 227–234. doi:10.1006/jsvi.1996.0637

Chapter 12
Visible Light Active Nanocomposites for Photocatalytic Applications

Rohini Singh
Indian Institute of Technology Dhanbad (ISM), India

Suman Dutta
Indian Institute of Technology Dhanbad (ISM), India

ABSTRACT

This chapter explores the concept of visible light active nanocomposites for the enhanced photocatalytic hydrogen generation and dye degradation. Since the late 1960s, A. Fujishima has been involved in unfolding the fascinating characteristics of titanium dioxide (TiO_2) as semiconductor oxide. The increased growth in population and industrial development has tremendously increased the generation of waste products and consumption of energy worldwide. This situation creates an immense need of clean and sustainable alternative sources of energy. Hydrogen, having a high energy capacity, is considered as a reliable fuel for the future energy requirements. In addition to that, due to the rapid industrialisation, our water is being contaminated with various harmful industrial effluents. This chapter illustrates the significance of visible light nanocomposites for the photocatalytic application of hydrogen generation for future energy security and dye degradation for the effective effluent treatment of textile industries.

DOI: 10.4018/978-1-5225-5216-1.ch012

INTRODUCTION

The "Nanocomposite" is widely illustrated as hybrid material in which at least one of the phases exhibit dimensions in the nanometer range ($1nm=10^{-9}m$). The addition of the nanosized second material assists to achieve enhanced unique behaviour created by the synergistic between constituents. The properties of nanocomposites depend on various parameters such as extent of loading and distribution, shape, size and orientation of the nanoscale second phase and interactions within the materials forming the nanocomposites.

In this chapter, further discussion will be restricted to the visible light active nanocomposites utilized for enhanced photocatalytic hydrogen generation and dye degradation. Most of the energy that we come across today is derived from non-renewable fossil fuels such as coal and petroleum. Fossil fuels cannot be considered as ideal due to some reasons:

1. Evolution of carbon dioxide (CO_2), which is one of the major greenhouse gases
2. Risk of depletion in future and
3. Waste of time and money in the import, relocation and distribution of these fuels.

Therefore, there is an immense need of a sustainable low cost and environmental friendly energy source for future.

Hydrogen is considered as an ideal energy carrier because of the abundance, high heat energy (122 kJ/g) compared to other fuels such as gasoline (40 kJ/g) and environmental friendliness as it never produces any greenhouse gases on combustion (Dutta, 2014; Liao et al., 2012). Hydrogen generation methods can be broadly classified into two categories i.e via (i) renewable resources such as water and solar energy through photocatalytic and photoelectrochemical (PEC) water splitting, and (ii) non-renewable resources such as coal and methane through steam reforming and coal gasification. Semiconductor photocatalytic water splitting to produce large-scale clean and recyclable hydrogen (H_2) and oxygen (O_2) using solar energy is a promising technology for energy conservation (Ismail & Bahnemann, 2014; Jing et al., 2010; Maeda, 2011). Photoelectrochemical water splitting was first demonstrated by Fujishima and Honda in 1972 and since then lots of progress have been observed by the across the world in the area of photocatalytic water splitting technology (Fujishima & Honda, 1972).

In addition to that, textile dyes mainly reactive dyes are the largest source of organic compounds that possess enhanced environmental risk and approximately, 1-20% of the world's total dye production is lost as effluent during the process of dyeing. Semiconductor photocatalytic process such as advanced oxidation technology has been reported as an economic, environmental friendly and sustainable wastewater treatment method. However, the major technical barrier that restricts its commercialisation is the recovery of the photocatalyst particles after water treatment (Chong et al., 2010).

In the subsequent sections, different synthetic routes and properties of TiO_2 nanocomposite will be discussed in details. TiO_2 nanoparticles have been extensively investigated for application including dye degradation and H_2 production as a fuel using solar energy. Though TiO_2 is an efficient, chemically stable and cheap photocatalyst but its activity is restricted to UV radiation, which is only 3-5% of solar radiation. As solar radiation contains ~40% visible light, current research is directed towards the synthesis of visible light active nanocomposites in order to utilize solar energy efficiently for the photocatalyst industrial applications.

BACKGROUND

Hetrogeneous semiconductor photocatalysis is among the most promising technologies for solar energy conversions and environmental air/water purification applications. Several semiconductors have band gap energies sufficient to be used as photocatalyst including TiO_2, ZnO, Fe_2O_3, WO_3, SnO_2, ZrO_2, CdS, $SrTiO_3$ and ZnS. Having high chemical stability, enhanced photocatalytic activity, non-toxicity, corrosion resistance and low cost, TiO_2 has become most widely used photocatalyst solar hydrogen generation and photocatalytic water treatment applications (Lee & Park, 2013). Various significant crystalline properties of TiO_2 are listed in Table 1.

But, pure TiO_2 is considered as less efficient for the photocatalytic application such as-hydrogen generation and dye degradation due to 1. quick e^-/h^+ recombination in the bulk/surface of semiconductor photocatalyst 2. rapid back reaction of oxygen and hydrogen to form water on the surface of the photocatalyst, and 3. inability to

Table 1. Crystalline properties of TiO_2 (Gupta & Tripathi, 2011)

Properties	Rutile	Anatase	Brookite
Crystal structure	Tetragonal	Tetragonal	Orthorhombic
Lattice constant (Å)	a = 4.5936 c =2.9587	a = 3.784 c = 9.515	a = 9.184 b = 5.447 c = 5.154
Space group	P42/mnm	I41/amd	Pbca
Molecule (cell)	2	2	4
Volume/ molecule (Å^3)	31.2160	34.061	32.172
Density (g cm^{-3})	4.13	3.79	3.99

harness visible light. In addition to that, the conduction band and valence band levels should satisfy the energy requirement set by the redox potentials for H_2O, respectively. Therefore, in order to overcome the drawback, different modifications techniques have been developed by the researchers across the globe, to modify the pure catalyst to the effective composites. Nanoscience and nanotechnology have a great potential to contribute to the current and future energy and environmental systems (Islam & Miyazaki, 2009; Salerno, Landoni & Verganti, 2008; Uskoković, 2007; Zäch et al., 2006). Nanocomposites exhibits improved photoactivity due to large surface area and increased reaction site (Mao, Shen & Guo, 2012; Zhu & Zäch, 2009). Numerous strategies including band gap engineering, surface sensitization, electronic structure improvisation and interfacial heterostructuring are being adopted by different synthetic routes such as-metal and non-metal loading, semiconductor coupling, dye sensitization and ion doping for the nanocomposite formation (Anpo & Takeuchi, 2003; Namratha & Byrappa, 2012; Ni et al., 2007; Wen et al., 2015). Hybrid semiconductor nanocomposites have many significant applications including environmental protection, energy conversion, photonics and sensors (Li & Zhang, 2009). Some of the photoreactors used for the hydrogen production using semiconductor nanocomposite are listed in Table 2.

Further discussion will be focussed on the application of TiO_2 nanocomposite in the area of photocatalytic dye degradation and solar hydrogen generation.

The electronic structure of semiconductors plays a vital role in photocatalysis. Semiconductors consist of a valence band (VB) and conduction band (CB). Difference between these two energy states is said to be the band gap energy (E_g). Initially, in a semiconductor material both the electrons (e^-) and holes (h^+) resides in the VB and as it receives the energy greater than their E_g, the electrons in the VB gets excited to the CB (Ni et al., 2007; Preethi & Kanmani, 2013). Thus, the electrons

Table 2. Photoreactors used for the hydrogen production using semiconductor nanocomposite

Serial No.	Types of photoreactor	References
1	Batch type photoreactor made of a water-jacketed stainless vessel with a quartz window	Huang et al., 2011
2	Solar slurry photocatalytic reactors i.e., batch reactor (BR) and batch recycle reactor	Priya & Kanmani, 2009
3	A novel twin reactor	Lo et al., 2010
4	A novel bench-scale tubular photocatalytic reactor	Ruban & Sellappa, 2014
5	Photocatalytic membrane reactors(PMRs)	Molinari, Marino & Argurio, 2014

participate in the reduction reaction to produce hydrogen, whereas holes carry out oxidation reaction for dye degradation. Schematic illustration of the steps involved in a photocatalytic process is shown in Figure 1.

SYNTHETIC ROUTES FOR NANOCOMPOSITE FORMATION

Metal and Non-Metal Deposition

Different binary and ternary nanocomposites had been prepared successfully by the researchers across the globe via metal and non-metal deposition to enhance the visible light response of TiO_2 photocatalyst for its effective utilization in various photo assisted applications including dye degradation and hydrogen generation. Metal and non-metal elemental doping has been extensively utilised to insert impurity levels in the band gap of TiO_2 for the expansion of its optical activity under visible light irradiation. Transition and noble metals are most commonly used as co-catalyst to reduce the charge recombination tendency in the semiconductor photocatalysts. Noble metals, such as- Pt (Melián et al., 2013), Au (Wang et al., 2017), Pd (Xu et al., 2013), Ni (Liu et al., 2014), Cu (Yoong, Chong & Dutta, 2009) and Ag (Aysin, Ozturk & Park, 2013) acts as traps for the photogenerated electrons. Hence, the electron/hole recombination rate can be decreased such that maximum number of electrons and hole gets participated as the charge carrier in the reduction and oxidation reactions respectively. Figure 2 illustrates the photocatalytic water splitting process in the presence of noble metal as co-catalyst.

Figure 1. Schematic illustration of the steps involved in a photocatalytic process

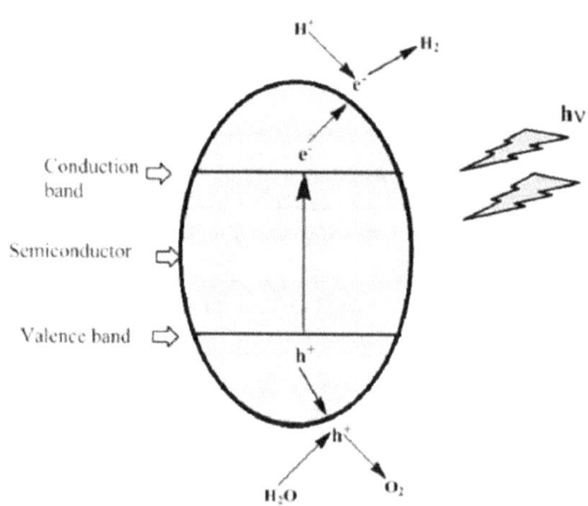

Figure 2. Photocatalytic water splitting process in the presence of noble metal as co-catalyst

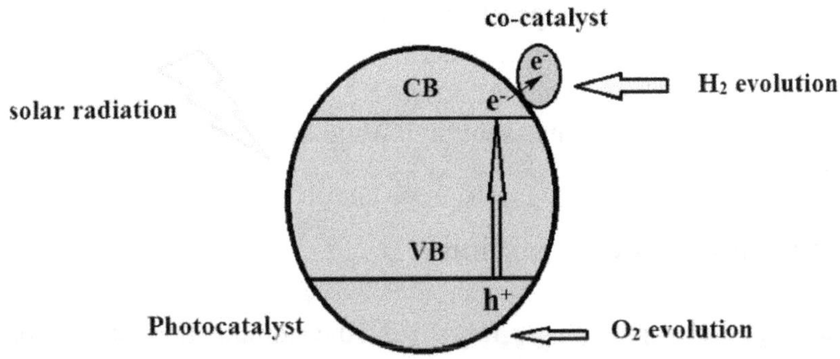

As the Fermi levels of these noble metals are lower than that of TiO_2, photo-excited electrons can be transferred from conduction (CB) of TiO_2 to the CB of noble metal particles deposited on the surface of TiO_2, while photo-generated holes reside in the valence band (VB) of TiO_2. Surface plasmon resonance effect (SPR) of the noble metal is responsible for the enhancement of the photoactivity of the prepared catalyst (Gu et al., 2016). Zhong et al. (2015) fabricated Ag/TiO_2 nanotubes by solvothermal method using ethanol and $AgNO_3$ as solvent and precursor respectively. Anatase TiO_2 nanotube arrays doped by Ce ions were synthesized by Fan et al. (2015) via the electrochemical anodic oxidation of pure titanium using NH_4F electrolyte solution containing Fe ions followed by Ag nanoparticles (AgNPs) deposition in Ce/TiO_2 nanotube arrays by microwave-assisted chemical reduction. Wang et al. (2017) prepared highly visible light active Fe-doped TiO_2 nanotube arrays by a simple solvothermal method and achieved relatively uniform doping due to high temperature and pressure in an enclosed space. Llano et al. (2010) explained that the deposition yield under similar experimental conditions significantly depends on the nature of metal and leads to a very different final properties of the deposited metal. For the metal deposited or doped samples an "enhancement factor" (Ef) is defined as the ratio between the rate of photocatalytic reactions over modified sample and the rate of same onto non-modified sample. The photoactivity enhancement highly depends on the contaminant to be degraded along with the properties and charge of the metal deposits (Maicu et al., 2011).

Non-metals most frequently utilized by the researchers to increase the photoactivity of TiO_2 using various techniques are nitrogen, carbon, boron and fluorine (Anas et al., 2016). Wu et al. (2013) successfully prepared C/TiO_2 powder by a facile calcination assisted solvothermal method. Boron doped TiO_2 was prepared by Simsek (2017) with improved visible light response by solvothermal method. The incorporation of non-metals improves the photocatalytic activity of TiO_2 under visible light irradiation most probably due to the reduced band gap energy and increased surface area due to doping. Nitrogen doped TiO_2 nanocomposites are proved to be efficient for achieving enhanced photocatalytic performance including photodegradation of organic compounds (Dolat et al., 2015; Devi, Nagaraj & Rajashekhar, 2012) and hydrogen generation from water splitting (Wang et al., 2013; Villa et al., 2012; Naik et al., 2015).

Semiconductor Heterojunction

Semiconductor heterojunction is formed by the combination of two or more semiconductor materials with different band gaps. Photocatalytic performance of resultant catalysts gets improved due to the heterostructure formed at the interface of different semiconductors have the capability to reduce the electron-hole pair recombination tendency. When a large band gap semiconductor photocatalyst is coupled with a small band gap semiconductor with a more negative CB level, electrons from CB can be injected from the small band gap semiconductor to the large band gap semiconductor (Ni et al., 2007). Thus, a wide electron/hole separation is achieved as shown in Figure 3.

An efficient charge separation and mobility plays a key role in the enhancement of photocatalytic performance via semiconductor heterostructuring strategy. Following conditions should be met to synthesise a highly active composite semiconductor photocatalyst (i) semiconductors should be free from photocorrosion, (ii) the small band gap semiconductor should have the excitation capability by visible light and (iii) electron injection should be fast as well as efficient. Liu et al. (2017) coupled TiO_2

Figure 3. Charge separation in semiconductor heterojunction

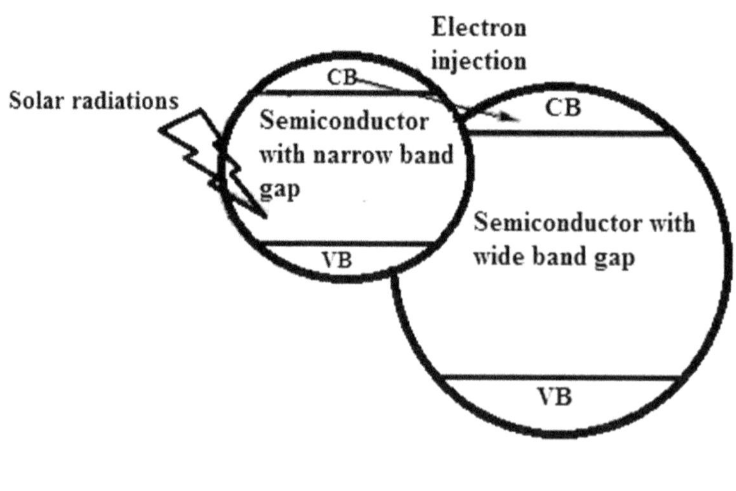

nanoparticles with Bi_2S_3, a narrow band gap (1.27 eV) semiconductor with higher conduction band positions to achieve high visible light response. Nuengmatcha et al. (2016) synthesised visible light active ZnO-graphene-TiO_2 (ZGT) composite catalyst via a simple solvothermal process and efficiently utilised for the photocatalytic degradation of rhodamine B (RhB), texbrite BAC-L (TBAC) and texbrite NFW-L (TNFW) under visible-light irradiation. Mixed silicon, aluminium and titanium oxide nanostructures was successfully prepared by Pal et al. (2016) as novel adsorbent for the removal of rhodamine 6G and methylene blue and concluded that highest adsorption capacity of the prepared adsorbent (162.96 mg MB/g) can be attributed to their high specific surface area, presence of surface Si-O groups and hydroxyl groups of amorphous titania. Novel Bi_2WO_6–TiO_2 heterostructures prepared by Lopez et al. (2011) leads to the enhanced charge carrier (electron/hole) separation, causing the excellent photocatalytic performance under solar irradiation. Titania/tungsten (VI) oxide/noble metal (gold and platinum) composites were obtained by Karacsonyi et al. (2013) for the effective utilisation in photocatalytic hydrogen production. Perez-Larios et al. (2012) prepared TiO_2–ZnO mixed oxides photoconductors with varied Zn content by the sol–gel method that was six times more active than pure TiO_2 semiconductor. Due to the narrow band gap of CuO (~1.23eV), it is most commonly coupled with TiO_2 by the researchers via different techniques (Kum et al., 2013; Xu & Sun, 2009; Chen et al., 2013; Miwa et al., 2010) for the photocatalytic hydrogen generation. In addition to that, different semiconductors such as- MnO (Ivanova & Harizanova, 2001), ZnO (Liao et al., 2004), WO_3 (Shifu et al., 2005), SnO_2 (Shifu et al., 2006), NiO (Shifu et al., 2008) and Fe_2O_3 (Ghorai, Chakraborty & Pramanik, 2011) had been coupled with TiO_2 to achieve enhanced visible light activity.

Sensitization

Another approach to modify the metal oxide is to sensitize the surface with visible light absorbing organic dyes, conducting polymers and inorganic metal complexes (Kumar & Rao, 2017). Dye sensitization is widely used to utilize visible light for energy conversion (Li et al., 2014). Some dyes having redox property and visible light sensitivity can be used in solar cells as well as photocatalytic systems. Under visible light irradiation, the excited dyes can inject electrons to CB of semiconductors to initiate the catalytic reactions. Some dyes, for example safranine, O/EDTA and T/EDTA, have the visible light absorption capability and can produce electrons as reducing agents, sufficient to generate hydrogen even without semiconductors. However, the rate of hydrogen evolution only by the visible light active dyes is very low. Increased hydrogen generation rate can be achieved by enhanced absorption of visible light and effective injection of electrons from excited dyes to the conduction band (CB) of TiO_2 photocatalyst. Dye-sensitized TiO_2 has received much attention in photovoltaic applications, such as photolysis of water to generate hydrogen and dye-sensitized cells. TiO_2 is a suitable semiconductor for dye-sensitized solar cells because its high stability even under extreme operating conditions to be utilised as stable photoelectrode in photoelectrochemical cells. In addition, the conduction band (CB) edge of TiO_2 lies slightly below the excited state energy level of many dyes for an effective electron injection. Thus, dye sensitized TiO_2 catalysts can fully utilize visible light to accelerate photocatalytic reactions. In order to accelerate the quantum yield for hydrogen production, sensitization of TiO_2 has been considered as a successful modification method. Jin et al. (2007) proposed that a eosin Y sensitized CuO incorporated TiO_2 was found beneficial for hydrogen generation under visible light irradiation accompanied by the electron donors leading to the highly improved quantum efficiency of the prepared photocatalyst towards photocatalytic water splitting. Sensitization of the TiO_2 particles for utilizing visible light has been an important subject both in hydrogen production by photocatalytic water splitting and decomposition of hazardous substances for water purification. In these sensitization reactions, it was important that the sensitizing dye molecule is adsorbed on the TiO_2 particles (Hirano et al., 2000).There are several adsorption modes: covalent attachment, electrostatic interaction, hydrogen bonding, and so on. Electrostatic interaction was reported to be more effective than the common covalent attachment in photosensitization (Pan et al., 2013). The surface functionalization can be considered as a useful tool to link the dye molecule to achieve improved characteristics of the dye coating. Numerous approaches involving chemical or electrochemical adsorption of the organic molecule on the surface of oxide have been reported in the literature in order to functionalize the electrode surface (Palmas et al., 2012). Figure 4 shows the schematic illustration for the water photosplitting process in the presence of dye sensitizer.

However, due to the ph dependency for adsorption and degradation tendency of dyes during the course of photocatalytic reactions, other sensitizers are also being preferred by the researchers such as- conducting polymers, inorganic complexes and simple chelating ligands. Metal oxide composites based on conducting polymers possess high thermal and chemical stability with remarkable charge transporting properties. Dopamine (DA) and its derivatives (norepinephrine and nitrodopamine) were chosen as monomers by Kim et al. and their polymers (pDA, pNE, and pNDA) were synthesized and subsequently complexed with TiO_2 to achieve visible light photoactivity activity (Kim et al., 2017).

The inorganic metal complexes are also utilised due to their wider absorption range under solar irradiation and ability to form stable composites. A stronger interaction between the anionic counterparts like sulfate, phosphate or carboxylic groups of the complexes with central metal ion of metal oxides plays a key role in the overall stability of the prepared composites. In addition to that, the inorganic dye

Figure 4. Water photosplitting process in the presence of dye sensitizer

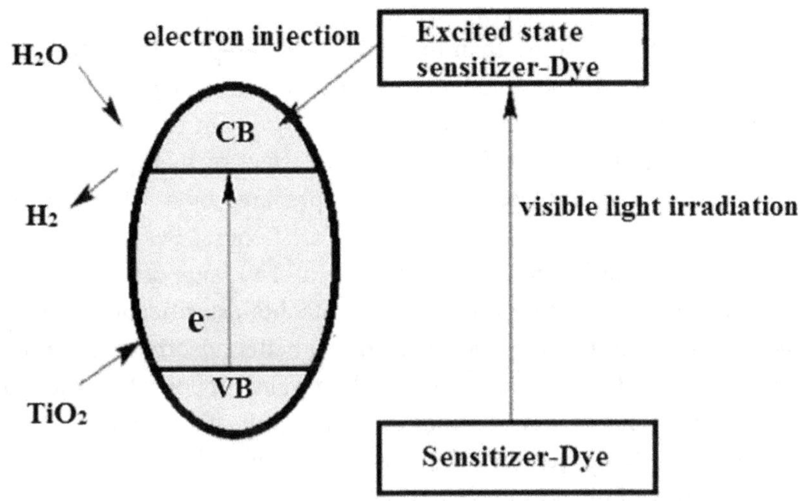

sensitized metal oxides can also decompose the organic contaminants even under anoxic conditions (Güy & Özacar, 2016).

An effective visible light response can also be achieved through surface complexation between the ligands (organic acids, phenolic com-pounds, H_2O_2, calixarenes etc.) and central metal ion of the metal oxides. Under this approach excited electrons directly migrates from the valence state of the adsorbate to the CB of metal oxide (Kumar & Rao, 2017).

Addition of Support or Co-Sorbent Material

Various methods are developed to improve photocatalytic efficiency of TiO_2 as mentioned in the earlier section. Addition of support or co-sorbent material such as silica (SiO_2), alumina (Al_2O_3), zeolite (ZSM-5) or clay and activated carbon (AC) can significantly enhance the photoactivity of prepared TiO_2. Generally, TiO_2-support composite increased photocatalytic performance is attributed to the supports' adsorption capability towards the contaminants and high specific surface area, which helps in more effective adsorption than pure TiO_2 (Naeem & Ouyang, 2013). TiO_2 pillared clays (mica, montmorillonite and saponite) were prepared by Shimizu et al. (2002) for the selective photo-oxidation of benzene. TiO_2 pillared clays can facilitate large surface area as well as pore volume, which allows the effective adsorption and desorption of the organic compounds on and from the active sites present on its surface respectively. Microporous zeolites and mesoporous molecular sieves supported TiO_2 was successfully prepared and investigated by Xu & Langford

(1997) for the effective photocatalytic degradation of acetophenone in an aqueous solution. The photocatalytic activity of the supported catalyst highly depends on the method adopted for loading than the calcination temperature of the prepared sample. Crystallinity, well defined porous structure and unusual morphology of the zeolite plays an important role in the enhancement of the photocatalytic efficiency. Anderson & Bard (1995) studied the effect of incorporation of SiO_2 on the behaviour of a TiO_2-based photocatalyst prepared by a sol-gel method for the photocatalytic degradation of Rhodamine 6G. Lepore, Persaud & Langford (1996) describes synthesis and characterization of TiO_2 supported on silica gels and modified silica gels and proposed that by alcohol esterification of the silica gel surface, photocatalyst can be transformed to a better adsorbent for removal of hazardous contaminants from an aqueous medium. The support of smaller particle size exhibit more efficiency towards photocatalytic applications (Xu, Zheng & Liu, 1999). The effect of the amount of TiO_2 loading on activated carbon support on the photodecomposition of dichloromethane was studied by Torimoto et al. (1997) and concluded that the use of mixed suspensions of TiO_2 and activated carbon showed lower activities than the TiO_2-loaded activated carbon. Yoneyama & Torimoto (2000) studied TiO_2/adsorbent hybrid photocatalysts for photodegradation of organic substances of dilute concentrations and explained that the application of adsorbents can be used as a support for TiO_2 to achieve high decomposition rates of organic substances both in air atmosphere and water phase due to the enhanced adsorption and oxidation of target contaminants and intermediates formed on the surface of loaded TiO_2. In addition to that, Singh et al. (2017) have successfully synthesised TiO_2 nanocrystals supported on ground nano egg shell waste and utilized for the effective photocatalytic dye degradation.

Black Colored Metal Oxides

Black colored metal oxides have the capability to absorb the complete visible and near infra-red light that leads to its efficient application in photocatalysis. Frequently used methods to prepare black TiO_2 involve oxidation of low valence Ti compounds, low/high pressure reduction of TiO_2 and many more. Actually, oxygen vacancies, Ti^{3+} or H atom dopants were introduced into the TiO_2 lattice structures during the synthesis of black TiO_2 photocatalyst by reduction or oxidation reactions at various atmospheres. In the reduction methods, hydrogen (H_2) gas is the most commonly used reducing agent for obtaining black TiO_2. Thus, the method adopted to synthesise black TiO_2 via reduction reaction under hydrogen atmosphere is termed as hydrogenation (Chen, Liu & Huang, 2015; Liu et al., 2016). Highly photoactive TiO_2 can be obtained is achievable by hydrogenation at constant temperature and pressure, with controlled hydrogenation duration. In 2011, black TiO_2 was initially synthesised by Chen et al. (2011) with a narrowed band gap of ~1.5eV by calcining pure white TiO_2 nanoparticles under a 20 bar pure H_2 atmosphere at about 200^{0}C for 5 days. Photocatalytic water splitting and cleaning using hydrogenated TiO_2 is believed to have high efficiency (Wang et al. 2013). Tian et al. (2015) reported the facile synthesis of hydrogenated TiO_2 (H-TiO_2) nanobelts with excellent UV and visible photocatalytic response towards methyl orange (MO) decomposition and water splitting for hydrogen generation. The improved photocatalytic property can be attributed to the Ti^{3+}ions and oxygen vacancies in TiO_2 nanobelts created by hydrogenation. Ti^{3+}ions and oxygen vacancies can accelerate visible light absorption, promote charge carrier trapping, and reduces the photogenerated electron–hole recombination. However arguments arise on the on the influence of some factors such as-morphology, size, surface active facets, surface/bulk Ti^{3+} and development of oxygen vacancies on the final physico-chemical properties of the hydrogenated catalyst (Samsudin et al., 2015). The water decontamination ability of hydrogenated TiO_2 exhibits 2.4 fold increase and its H_2 generation rate was enhanced by 1.7 times as compared to pure TiO_2. Therefore, black colored metal oxide by hydrogenation offers a cost effective method to improve the visible light activity (Zhu et al., 2015).

PROPERTIES OF NANOCOMPOSITE PHOTOCATALYSTS

The photocatalytic activity of the nanocomposite is correlated with their structural, morphological, optical and surface properties. The properties of the prepared nanocomposites are highly dependent on the method of synthesis and controlled experimental conditions. Since, photocatalysis is a surface phenomenon; therefore, the properties of nanostructured composite including particle size, crystallinity, surface area, morphology, and porosity significantly influences its photocatalytic activity (Zou et al., 2014; Lee et al., 2013).

Structural and Optical Properties

The enhanced light absorption capability, promising electronic arrangement and charge transport characteristics are the key factors for the application of nanocomposites as a photocatalyst. For the photocatalytic reactions to take place, "the VB potential must be more positive than the oxidation potential of water (1.23 V w. r. t. normal hydrogen electrode (NHE)) and CB potential must be more negative than the reduction potential of water (0.0 V w. r. t. NHE)" (Sasikala & Bharadwaj, 2012). TiO_2 is *a UV light active material* that exists in three well known crystalline phase i.e. Rutile (tetragonal), Anatase (tetragonal) and Brookite (orthorhombic) with the band gaps of 3.02, 3.2, and 2.96 eV, respectively. However, pure TiO_2 lacks visible light activity and so, to enhance its optical response towards visible radiations of solar spectrum, nanocomposites are synthesised by different methods as mentioned in the earlier sections. TiO_2 nanocomposites exhibit enhanced crystallinity, larger surface area, reduced band gap and excellent photoactivity.

Morphological Properties

TiO_2 nanocomposites can be synthesised with different morphologies such as nanospheres, nanowires, nanotubes, nanorods, nanosheet and nanoflowers (Wang et al., 2014). Morphology of TiO_2 composites depends on the method of synthesis and can also be varied by changing the synthetic parameters such as- operating temperature, pressure and precursor concentration under the same process. TiO_2 nanotubes arrays (TNAs) have three-dimensional tubular structure, large specific surface area and can be used repeatedly and conveniently (Zhang et al., 2013).

CHARACTERIZATION OF NANOCOMPOSITE

Different basic techniques generally involved in characterization of nanocomposite photocatalysts are as follows.

X-Ray Diffraction (XRD)

William Roentgen discovered X-rays in 1895 and enabled scientists to investigate crystalline structure at the atomic level. XRD is a common non-destructive technique for the study of crystal structures and atomic spacing based on constructive interference of monochromatic X-rays and a crystalline sample. X-ray diffraction has been used for (1) fingerprint characterization of crystalline materials and (2) determination of their structure. Each crystalline material has its unique characteristic X-ray powder pattern which may be used as a "fingerprint" for its identification. X-ray diffraction is one of the numerous important characterization tools used in solid state chemistry and materials science. It is very significant analytical technique for the structural characterization of nanocomposites. XRD can be utilised for the phase identification of a crystalline material along with unit cell dimensions. Thus, XRD is primarily used for the following listed purposes.

1. Identification of unknown crystalline materials and crystal structure of identified materials
2. Distinguishing between amorphous and crystalline material and quantification of the percent crystallinity of a sample

3. Quantitative determination of amounts of different phases in multi-phase mixtures by peak-ratio calculations
4. Determination of crystallite size from analysis of peak broadening and determine crystallite shape from study of peak symmetry.

Ultraviolet–Visible (UV-Vis) Spectroscopy

Ultraviolet–visible spectroscopy (UV-Vis) refers to the absorption/reflectance spectroscopy in the ultraviolet-visible spectral region. Optical properties of the synthesized nanocomposites are determined by the UV-Visible absorption/reflectance spectroscopy. UV-Vis absorption spectroscopy is the measurement of the attenuation of a beam of light after it passes through a sample or after reflection from a sample surface. Absorption measurements can be at a single wavelength or over an extended spectral range. Ultraviolet absorption spectra arise from transition of electron within a molecule from a lower level to a higher level. To absorb the visible light, photocatalysts should be of narrow band gap energy so that the electrons in the CB can easily get excited to VB even at lower intensities of the incident radiations.

In UV-Vis spectroscopy, molecules undergo electronic transition involving σ, π and n electrons.

Following four types of electronic transitions are possible:

1. $\sigma \rightarrow \sigma^*$ transition
2. $\mathrm{n} \rightarrow \sigma^*$ transition
3. $\mathrm{n} \rightarrow \pi^*$ transition
4. $\pi \rightarrow \pi^*$ transition

Field Emission Scanning Electron Microscopy (FESEM)

FESEM uses a focused beam of electrons to generate an image or to analyse the specimen. A field emission source releases these electrons and the object is scanned by electrons according to a zig-zag pattern. An FESEM is used to visualize very small topographic details on the surface or entire or fractioned objects. The morphology, agglomeration tendency and size distribution of the nanocomposites are confirmed by observing FESEM images. Different 1D (wire), 2D (rod) and 3D (sphere) particle morphology can be observed via FESEM. Compared with convention scanning electron microscopy (SEM), field emission SEM (FESEM) produces clearer, less electrostatically distorted images with three to six times better spatial resolution. Transmission electron microscopy (TEM, also sometimes conventional transmission electron microscopy or CTEM) is a microscopy technique in which a beam of electrons is transmitted through an ultrathin (less than 100 nm thick) specimen section to form a high resolution image.

Energy Dispersive X-Ray Analysis (EDX)

EDX is a widely used technique to analyse the chemical composition. EDX systems are attachments to Electron Microscopy instruments (Scanning Electron Microscopy (SEM) or Transmission Electron Microscopy (TEM)) instruments where the imaging capability of the microscope identifies the specimen of interest. This method detects the X-rays generated as the result of the electron beam interactions with the analyte sample. Elemental mapping of the distribution of the different chemical components constituting the specimen can also be obtained. Its characterization capabilities are due the fundamental principle that each element has a unique atomic structure allowing a unique set of peaks on its electromagnetic emission spectrum.

Fourier Transform Infrared Spectroscopy (FTIR)

The FT-IR spectrum provides the information about the nature of functional group and detects the residual organic compound present in the prepared nanocomposites. The information gathered from FT-IR spectra can assist in better understanding of the mechanism involved in the formation of the photocatalyst by various methods. The prepared samples are pressed into thin pallets and used to obtain an infrared spectrum of absorption or emission. An FTIR spectrometer collects high-resolution data over a wide spectral range. FTIR spectrometers are the 3rd generation infrared spectrometer. FTIR spectrometers possess numerous significant advantages: (1) The signal-to-noise ratio of spectrum is significantly higher than the previous generation infrared spectrometers. (2) The accuracy of wavenumber is high. (3) The resolution is high having wide scan range._

Brunauer-Emmett-Teller (BET) /Barrett-Joyner-Halenda (BJH) Analysis

Brunauer–Emmett–Teller (BET) and Barrett-Joyner-Halenda (BJH) theory serves as the basis for an important analytical technique for the measurement of the specific surface area and porosity of a photocatalyst respectively. BET analysis facilitates precise specific surface area evaluation of materials by nitrogen multilayer adsorption measured as a function of relative pressure using a fully automated analyzer. BJH analysis can also be employed to determine pore area and specific pore volume using adsorption and desorption techniques. This technique characterizes pore size distribution independent of external area due to particle size of the sample. The redox reactions involved in the photocatalytic water splitting occurs on the surface of the selected photocatalyst. Thus, a large surface area and porosity are always desired for increased number of reaction site and to enhance the hydrogen production and dye degradation efficiency of synthesized photocatalysts.

FUTURE RESEARCH DIRECTIONS

In future research on visible light active nanocomposites for photocatalytic applications, it is expected to deal with the following limitations:

- **Reduction in the Agglomeration Tendency of Nanocomposite Photocatalysts:** Agglomeration in nanoparticles is a serious issue that needs to overcome in order to realize the maximum benefits of nanotechnology. Due to agglomeration, the active surface reaction sites for the photocatalytic redox reaction decreases which in turn leads to the lower efficiency of nano sized photocatalysts. Thus, sincere efforts should be made to synthesise highly efficient photocatalyst nanoparticles with reduced agglomeration tendency.
- **Synthesis of Cost Effective and Highly Efficient Noble Photocatalyst and Cocatalyst:** Presently, nanotechnology is very expensive and involves complex procedures. Therefore, synthesis of noble visible light active nanocomposites through simple and economical methods is essential to meet the target of maximum solar energy conversion.
- **Separation of Photocatalyst Nanoparticles Post Treatment:** If photoactive nanoparticles are applied in suspension, after the treatment the particles have to be separated. But due to very small size of the nanoparticle (1-100 nm), it is very difficult to carry out the separation. Filtration is expensive as it involves high energy demands (Dey, 2012). So, for the industrial acceptability of nanocomposites for the photocatalytic applications, the development of cost effective separation technique is required.
- **Exploration of the Interfacial Properties of Nanocomposite:** A better understanding of the surface/interfacial chemistry is essential for the synthesis of noble nanocomposites with enhanced photoactivity. The interfacial surface energy and absorption capacity plays a key role in the transfer of electron and energy within the substances constituting the nanocomposites.

CONCLUSION

Current material science research in the area of heterogeneous photocatalysis is focused on the development of novel cost effective techniques to achieve maximum utilization of solar energy. Thus, TiO_2 nanocomposites formed via different methods allow the effective absorption of visible light that constitutes about 40-45% of solar spectrum. In conclusion, the research advancement in the era of nanocomposites formation had really increased the hope of industrial acceptance of photocatalytic process for various applications such as solar hydrogen generation through photocatalytic water splitting and industrial wastewater treatment. However, researchers across the globe are continuously attempting furthermore facile and cost effective methods to enhance the visible light activity and hence the photoactivity of TiO_2 as more as possible.

ACKNOWLEDGMENT

The authors thank Indian Institute of Technology (Indian School of Mines), Dhanbad for encouraging this research programme.

REFERENCES

Anas, M., Han, D. S., Mahmoud, K., Park, H., & Abdel-Wahab, A. (2016). Photocatalytic degradation of organic dye using titanium dioxide modified with metal and non-metal deposition. *Materials Science in Semiconductor Processing*, *41*, 209–218. doi:10.1016/j.mssp.2015.08.041

Anderson, C., & Bard, A. J. (1995). An improved photocatalyst of TiO_2/SiO_2 prepared by a sol-gel synthesis. *Journal of Physical Chemistry*, *99*(24), 9882–9885. doi:10.1021/j100024a033

Anpo, M., & Takeuchi, M. (2003). The design and development of highly reactive titanium oxide photocatalysts operating under visible light irradiation. *Journal of Catalysis*, *216*(1), 505–516. doi:10.1016/S0021-9517(02)00104-5

Aysin, B., Ozturk, A., & Park, J. (2013). Silver-loaded TiO_2 powders prepared through mechanical ball milling. *Ceramics International*, *39*(6), 7119–7126. doi:10.1016/j.ceramint.2013.02.054

Chen, W. T., Jovic, V., Sun-Waterhouse, D., Idriss, H., & Waterhouse, G. I. (2013). The role of CuO in promoting photocatalytic hydrogen production over TiO_2. *International Journal of Hydrogen Energy*, *38*(35), 15036–15048. doi:10.1016/j.ijhydene.2013.09.101

Chen, X., Liu, L., & Huang, F. (2015). Black titanium dioxide (TiO_2) nanomaterials. *Chemical Society Reviews*, *44*(7), 1861–1885. doi:10.1039/C4CS00330F PMID:25590565

Chen, X., Liu, L., Peter, Y. Y., & Mao, S. S. (2011). Increasing solar absorption for photocatalysis with black hydrogenated titanium dioxide nanocrystals. *Science*, *331*(6018), 746–750. doi:10.1126/science.1200448 PMID:21252313

Chong, M. N., Jin, B., Chow, C. W., & Saint, C. (2010). Recent developments in photocatalytic water treatment technology: A review. *Water Research*, *44*(10), 2997–3027. doi:10.1016/j.watres.2010.02.039 PMID:20378145

Devi, L. G., Nagaraj, B., & Rajashekhar, K. E. (2012). Synergistic effect of Ag deposition and nitrogen doping in TiO_2 for the degradation of phenol under solar irradiation in presence of electron acceptor. *Chemical Engineering Journal*, *181–182*, 259–266. doi:10.1016/j.cej.2011.11.076

Dey, T. (2012). *Nanotechnology for water purification*. Universal-Publishers.

Dolat, D., Mozia, S., Wróbel, R. J., Moszyński, D., Ohtani, B., Guskos, N., & Morawski, A. W. (2015). Nitrogen-doped, metal-modified rutile titanium dioxide as photocatalysts for water remediation. *Applied Catalysis B: Environmental*, *162*, 310–318. doi:10.1016/j.apcatb.2014.07.001

Dutta, S. (2014). A review on production, storage of hydrogen and its utilization as an energy resource. *Journal of Industrial and Engineering Chemistry*, *20*(4), 1148–1156. doi:10.1016/j.jiec.2013.07.037

Fan, X., Wan, J., Liu, E., Sun, L., Hu, Y., Li, H., ... Fan, J. (2015). High-efficiency photoelectrocatalytic hydrogen generation enabled by Ag deposited and Ce doped TiO_2 nanotube arrays. *Ceramics International*, *41*(3), 5107–5116. doi:10.1016/j.ceramint.2014.12.083

Fujishima, A., & Honda, K. (1972). Electrochemical photolysis of water at a semiconductor electrode. *Nature*, *238*(5358), 37–38. doi:10.1038/238037a0 PMID:12635268

Ghorai, T. K., Chakraborty, M., & Pramanik, P. (2011). Photocatalytic performance of nano-photocatalyst from TiO_2 and Fe_2O_3 by mechanochemical synthesis. *Journal of Alloys and Compounds*, *509*(32), 8158–8164. doi:10.1016/j.jallcom.2011.05.069

Gu, Y., Li, C., Bai, J., Wang, J., & Ma, T. (2016). One-step solvothermal synthesis of Au-TiO_2 loaded electrospun carbon fibers to enhance photocatalytic activity. *Vacuum*, *130*, 1–6. doi:10.1016/j.vacuum.2016.04.031

Gupta, S. M., & Tripathi, M. (2011). A review of TiO_2 nanoparticles. *Chinese Science Bulletin*, *56*(16), 1639–1657. doi:10.1007/s11434-011-4476-1

Güy, N., & Özacar, M. (2016). The influence of noble metals on photocatalytic activity of ZnO for Congo red degradation. *International Journal of Hydrogen Energy*, *41*(44), 20100–20112. doi:10.1016/j.ijhydene.2016.07.063

Hirano, K., Suzuki, E., Ishikawa, A., Moroi, T., Shiroishi, H., & Kaneko, M. (2000). Sensitization of TiO_2 particles by dyes to achieve H_2 evolution by visible light. *Journal of Photochemistry and Photobiology A Chemistry*, *136*(3), 157–161. doi:10.1016/S1010-6030(00)00342-7

Huang, C., Yao, W., Ali, T., & Muradov, N. (2011). Development of efficient photoreactors for solar hydrogen production. *Solar Energy*, *85*(1), 19–27. doi:10.1016/j.solener.2010.11.004

Islam, N., & Miyazaki, K. (2009). Nanotechnology innovation system: Understanding hidden dynamics of nanoscience fusion trajectories. *Technological Forecasting and Social Change*, *76*(1), 128–140. doi:10.1016/j.techfore.2008.03.021

Ismail, A. A., & Bahnemann, D. W. (2014). Photochemical splitting of water for hydrogen production by photocatalysis: A review. *Solar Energy Materials and Solar Cells*, *128*, 85–101. doi:10.1016/j.solmat.2014.04.037

Ivanova, T., & Harizanova, A. (2001). Characterization of TiO_2 and TiO_2–MnO oxides prepared by sol–gel method. *Solid State Ionics*, *138*(3), 227–232. doi:10.1016/S0167-2738(00)00798-0

Jin, Z., Zhang, X., Li, Y., Li, S., & Lu, G. (2007). 5.1% Apparent quantum efficiency for stable hydrogen generation over eosin-sensitized CuO/TiO_2 photocatalyst under visible light irradiation. *Catalysis Communications*, *8*(8), 1267–1273. doi:10.1016/j.catcom.2006.11.019

Jing, D., Guo, L., Zhao, L., Zhang, X., Liu, H., Li, M., ... Zhang, K. (2010). Efficient solar hydrogen production by photocatalytic water splitting: From fundamental study to pilot demonstration. *International Journal of Hydrogen Energy*, *35*(13), 7087–7097. doi:10.1016/j.ijhydene.2010.01.030

Karácsonyi, É., Baia, L., Dombi, A., Danciu, V., Mogyorósi, K., Pop, L. C., ... Pap, Z. (2013). The photocatalytic activity of TiO_2/WO_3/noble metal (Au or Pt) nanoarchitectures obtained by selective photodeposition. *Catalysis Today*, *208*, 19–27. doi:10.1016/j.cattod.2012.09.038

Kim, S., Moon, G. H., Kim, G., Kang, U., Park, H., & Choi, W. (2017). TiO_2 complexed with dopamine-derived polymers and the visible light photocatalytic activities for water pollutants. *Journal of Catalysis*, *346*, 92–100. doi:10.1016/j.jcat.2016.11.027

Kum, J. M., Yoo, S. H., Ali, G., & Cho, S. O. (2013). Photocatalytic hydrogen production over CuO and TiO_2 nanoparticles mixture. *International Journal of Hydrogen Energy*, *38*(31), 13541–13546. doi:10.1016/j.ijhydene.2013.08.004

Kumar, S. G., & Rao, K. K. (2017). Comparison of modification strategies towards enhanced charge carrier separation and photocatalytic degradation activity of metal oxide semiconductors (TiO_2, WO_3 and ZnO). *Applied Surface Science*, *391*, 124–148. doi:10.1016/j.apsusc.2016.07.081

Lee, S. S., Bai, H., Liu, Z., & Sun, D. D. (2013). Optimization and an insightful properties—Activity study of electrospun TiO_2/CuO composite nanofibers for efficient photocatalytic H_2 generation. *Applied Catalysis B: Environmental*, *140*, 68–81. doi:10.1016/j.apcatb.2013.03.033

Lee, S. Y., & Park, S. J. (2013). TiO_2 photocatalyst for water treatment applications. *Journal of Industrial and Engineering Chemistry*, *19*(6), 1761–1769. doi:10.1016/j.jiec.2013.07.012

Lepore, G. P., Persaud, L., & Langford, C. H. (1996). Supporting titanium dioxide photocatalysts on silica gel and hydrophobically modified silica gel. *Journal of Photochemistry and Photobiology A Chemistry*, *98*(1-2), 103–111. doi:10.1016/1010-6030(95)04242-3

Li, J., & Zhang, J. Z. (2009). Optical properties and applications of hybrid semiconductor nanomaterials. *Coordination Chemistry Reviews*, *253*(23), 3015–3041. doi:10.1016/j.ccr.2009.07.017

Li, X., Zhang, J., Kang, S., Li, G., & Mu, J. (2014). Visible light photocatalytic activity of CuO/Cr_2O_3 co-loaded multiwalled carbon nanotubes sensitized with eosin Y for hydrogen evolution from water. *Ceramics International, 40*(7), 10171–10176. doi:10.1016/j.ceramint.2014.02.055

Liao, C. H., Huang, C. W., & Wu, J. (2012). Hydrogen production from semiconductor-based photocatalysis via water splitting. *Catalysts, 2*(4), 490–516. doi:10.3390/catal2040490

Liao, S., Donggen, H., Yu, D., Su, Y., & Yuan, G. (2004). Preparation and characterization of ZnO/TiO_2, $SO_4^{2-}/ZnO/TiO_2$ photocatalyst and their photocatalysis. *Journal of Photochemistry and Photobiology A Chemistry, 168*(1), 7–13. doi:10.1016/j.jphotochem.2004.05.010

Liu, X., Zhu, G., Wang, X., Yuan, X., Lin, T., & Huang, F. (2016). Progress in Black Titania: A New Material for Advanced Photocatalysis. *Advanced Energy Materials, 6*(17), 1600452. doi:10.1002/aenm.201600452

Liu, Y., Shi, Y., Liu, X., & Li, H. (2017). A facile solvothermal approach of novel Bi_2S_3/TiO_2/RGO composites with excellent visible light degradation activity for methylene blue. *Applied Surface Science, 396*, 58–66. doi:10.1016/j.apsusc.2016.11.028

Liu, Y., Wang, Z., Fan, W., Geng, Z., & Feng, L. (2014). Enhancement of the photocatalytic performance of Ni-loaded TiO_2 photocatalyst under sunlight. *Ceramics International, 40*(3), 3887–3893. doi:10.1016/j.ceramint.2013.08.030

Llano, B., Restrepo, G., Marín, J. M., Navío, J. A., & Hidalgo, M. C. (2010). Characterisation and photocatalytic properties of titania–silica mixed oxides doped with Ag and Pt. *Applied Catalysis A, General, 387*(1), 135–140. doi:10.1016/j.apcata.2010.08.021

Lo, C. C., Huang, C. W., Liao, C. H., & Wu, J. C. (2010). Novel twin reactor for separate evolution of hydrogen and oxygen in photocatalytic water splitting. *International Journal of Hydrogen Energy, 35*(4), 1523–1529. doi:10.1016/j.ijhydene.2009.12.032

López, S. M., Hidalgo, M. C., Navío, J. A., & Colón, G. (2011). Novel $Bi_2WO_6–TiO_2$ heterostructures for Rhodamine B degradation under sunlike irradiation. *Journal of Hazardous Materials, 185*(2), 1425–1434. doi:10.1016/j.jhazmat.2010.10.065 PMID:21074938

Maeda, K. (2011). Photocatalytic water splitting using semiconductor particles: History and recent developments. *Journal of Photochemistry and Photobiology C, Photochemistry Reviews*, *12*(4), 237–268. doi:10.1016/j.jphotochemrev.2011.07.001

Maicu, M., Hidalgo, M. C., Colón, G., & Navío, J. A. (2011). Comparative study of the photodeposition of Pt, Au and Pd on pre-sulphated TiO_2 for the photocatalytic decomposition of phenol. *Journal of Photochemistry and Photobiology A Chemistry*, *217*(2), 275–283. doi:10.1016/j.jphotochem.2010.10.020

Mao, S. S., Shen, S., & Guo, L. (2012). Nanomaterials for renewable hydrogen production, storage and utilization. *Progress in Natural Science: Materials International*, *22*(6), 522–534. doi:10.1016/j.pnsc.2012.12.003

Melián, E. P., López, C. R., Méndez, A. O., Díaz, O. G., Suárez, M. N., Rodríguez, J. D., ... Hevia, D. F. (2013). Hydrogen production using Pt-loaded TiO_2 photocatalysts. *International Journal of Hydrogen Energy*, *38*(27), 11737–11748. doi:10.1016/j.ijhydene.2013.07.006

Miwa, T., Kaneco, S., Katsumata, H., Suzuki, T., Ohta, K., Verma, S. C., & Sugihara, K. (2010). Photocatalytic hydrogen production from aqueous methanol solution with $CuO/Al_2O_3/TiO_2$ nanocomposite. *International Journal of Hydrogen Energy*, *35*(13), 6554–6560. doi:10.1016/j.ijhydene.2010.03.128

Molinari, R., Marino, T., & Argurio, P. (2014). Photocatalytic membrane reactors for hydrogen production from water. *International Journal of Hydrogen Energy*, *39*(14), 7247–7261. doi:10.1016/j.ijhydene.2014.02.174

Naeem, K., & Ouyang, F. (2013). Influence of supports on photocatalytic degradation of phenol and 4-chlorophenol in aqueous suspensions of titanium dioxide. *Journal of Environmental Sciences (China)*, *25*(2), 399–404. doi:10.1016/S1001-0742(12)60055-2 PMID:23596962

Naik, B., Moon, S. Y., Kim, S. H., & Park, J. Y. (2015). Enhanced photocatalytic generation of hydrogen by Pt-deposited nitrogen-doped TiO_2 hierarchical nanostructures. *Applied Surface Science*, *354*, 347–352. doi:10.1016/j.apsusc.2015.05.102

Namratha, K., & Byrappa, K. (2012). Novel solution routes of synthesis of metal oxide and hybrid metal oxide nanocrystals. *Progress in Crystal Growth and Characterization of Materials*, *58*(1), 14–42. doi:10.1016/j.pcrysgrow.2011.10.005

Ni, M., Leung, M. K., Leung, D. Y., & Sumathy, K. (2007). A review and recent developments in photocatalytic water-splitting using TiO_2 for hydrogen production. *Renewable & Sustainable Energy Reviews*, *11*(3), 401–425. doi:10.1016/j.rser.2005.01.009

Ni, M., Leung, M. K., Leung, D. Y., & Sumathy, K. (2007). A review and recent developments in photocatalytic water-splitting using TiO_2 for hydrogen production. *Renewable & Sustainable Energy Reviews*, *11*(3), 401–425. doi:10.1016/j.rser.2005.01.009

Nuengmatcha, P., Chanthai, S., Mahachai, R., & Oh, W. C. (2016). Visible light-driven photocatalytic degradation of rhodamine B and industrial dyes (texbrite BAC-L and texbrite NFW-L) by ZnO-graphene-TiO_2 composite. *Journal of Environmental Chemical Engineering*, *4*(2), 2170–2177. doi:10.1016/j.jece.2016.03.045

Pal, U., Sandoval, A., Madrid, S. I. U., Corro, G., Sharma, V., & Mohanty, P. (2016). Mixed titanium, silicon, and aluminum oxide nanostructures as novel adsorbent for removal of rhodamine 6G and methylene blue as cationic dyes from aqueous solution. *Chemosphere*, *163*, 142–152. doi:10.1016/j.chemosphere.2016.08.020 PMID:27529381

Palmas, S., Da Pozzo, A., Mascia, M., Vacca, A., & Ricci, P. C. (2012). Sensitization of TiO_2 nanostructures with coumarin 343. *Chemical Engineering Journal*, *211*, 285–292. doi:10.1016/j.cej.2012.09.093

Pan, L., Zou, J. J., Wang, S., Huang, Z. F., Zhang, X., & Wang, L. (2013). Enhancement of visible-light-induced photodegradation over hierarchical porous TiO_2 by nonmetal doping and water-mediated dye sensitization. *Applied Surface Science*, *268*, 252–258. doi:10.1016/j.apsusc.2012.12.074

Pérez-Larios, A., Lopez, R., Hernandez-Gordillo, A., Tzompantzi, F., Gómez, R., & Torres-Guerra, L. M. (2012). Improved hydrogen production from water splitting using TiO_2–ZnO mixed oxides photocatalysts. *Fuel*, *100*, 139–143. doi:10.1016/j.fuel.2012.02.026

Preethi, V., & Kanmani, S. (2013). Photocatalytic hydrogen production. *Materials Science in Semiconductor Processing*, *16*(3), 561–575. doi:10.1016/j.mssp.2013.02.001

Priya, R., & Kanmani, S. (2009). Batch slurry photocatalytic reactors for the generation of hydrogen from sulfide and sulfite waste streams under solar irradiation. *Solar Energy*, *83*(10), 1802–1805. doi:10.1016/j.solener.2009.06.012

Ruban, P., & Sellappa, K. (2014). Development and performance of bench-scale reactor for the photocatalytic generation of hydrogen. *Energy*, *73*, 926–932. doi:10.1016/j.energy.2014.06.107

Salerno, M., Landoni, P., & Verganti, R. (2008). Designing foresight studies for Nanoscience and Nanotechnology (NST) future developments. *Technological Forecasting and Social Change*, *75*(8), 1202–1223. doi:10.1016/j.techfore.2007.11.011

Samsudin, E. M., Hamid, S. B. A., Juan, J. C., Basirun, W. J., & Kandjani, A. E. (2015). Surface modification of mixed-phase hydrogenated TiO_2 and corresponding photocatalytic response. *Applied Surface Science*, *359*, 883–896. doi:10.1016/j.apsusc.2015.10.194

Sasikala, R., & Bharadwaj, S. R. (2012). Photocatalytic hydrogen generation from water using solar radiation. *Chemistry Division. BARC Newsletter*, (325), 10-15.

Shifu, C., Lei, C., Shen, G., & Gengyu, C. (2005). The preparation of coupled WO_3/TiO_2 photocatalyst by ball milling. *Powder Technology*, *160*(3), 198–202. doi:10.1016/j.powtec.2005.08.012

Shifu, C., Lei, C., Shen, G., & Gengyu, C. (2006). The preparation of coupled SnO_2/TiO_2 photocatalyst by ball milling. *Materials Chemistry and Physics*, *98*(1), 116–120. doi:10.1016/j.matchemphys.2005.08.073

Shifu, C., Sujuan, Z., Wei, L., & Wei, Z. (2008). Preparation and activity evaluation of p–n junction photocatalyst NiO/TiO_2. *Journal of Hazardous Materials*, *155*(1), 320–326. doi:10.1016/j.jhazmat.2007.11.063 PMID:18166268

Shimizu, K. I., Kaneko, T., Fujishima, T., Kodama, T., Yoshida, H., & Kitayama, Y. (2002). Selective oxidation of liquid hydrocarbons over photoirradiated TiO_2 pillared clays. *Applied Catalysis A, General*, *225*(1), 185–191. doi:10.1016/S0926-860X(01)00863-8

Simsek, E. B. (2017). Solvothermal synthesized boron doped TiO_2 catalysts: Photocatalytic degradation of endocrine disrupting compounds and pharmaceuticals under visible light irradiation. *Applied Catalysis B: Environmental*, *200*, 309–322. doi:10.1016/j.apcatb.2016.07.016

Singh, R., Kumari, P., Chavan, P., Datta, S., & Dutta, S. (2017). Synthesis of solvothermal derived TiO_2 nanocrystals supported on ground nano egg shell waste and its utilization for the photocatalytic dye degradation. *Optical Materials*, *73*, 377–383. doi:10.1016/j.optmat.2017.08.040

Tian, J., Leng, Y., Cui, H., & Liu, H. (2015). Hydrogenated TiO_2 nanobelts as highly efficient photocatalytic organic dye degradation and hydrogen evolution photocatalyst. *Journal of Hazardous Materials*, *299*, 165–173. doi:10.1016/j.jhazmat.2015.06.019 PMID:26118828

Torimoto, T., Okawa, Y., Takeda, N., & Yoneyama, H. (1997). Effect of activated carbon content in TiO_2-loaded activated carbon on photodegradation behaviors of dichloromethane. *Journal of Photochemistry and Photobiology A Chemistry*, *103*(1-2), 153–157. doi:10.1016/S1010-6030(96)04503-0

Uskoković, V. (2007). Nanotechnologies: What we do not know. *Technology in Society*, *29*(1), 43–61. doi:10.1016/j.techsoc.2006.10.005

Villa, K., Black, A., Domènech, X., & Peral, J. (2012). Nitrogen doped TiO_2 for hydrogen production under visible light irradiation. *Solar Energy*, *86*(1), 558–566. doi:10.1016/j.solener.2011.10.029

Wang, C., Hu, Q., Huang, J., Wu, L., Deng, Z., Liu, Z., & Cao, Y. (2013). Efficient hydrogen production by photocatalytic water splitting using N-doped TiO_2 film. *Applied Surface Science*, *283*, 188–192. doi:10.1016/j.apsusc.2013.06.080

Wang, Q., Jin, R., Zhang, M., & Gao, S. (2017). Solvothermal preparation of Fe-doped TiO_2 nanotube arrays for enhancement in visible light induced photoelectrochemical performance. *Journal of Alloys and Compounds*, *690*, 139–144. doi:10.1016/j.jallcom.2016.07.281

Wang, X., Dong, H., Hu, Z., Qi, Z., & Li, L. (2017). Fabrication of a Cu_2O/Au/TiO_2 composite film for efficient photocatalytic hydrogen production from aqueous solution of methanol and glucose. *Materials Science and Engineering B*, *219*, 10–19. doi:10.1016/j.mseb.2017.02.011

Wang, Y., He, Y., Lai, Q., & Fan, M. (2014). Review of the progress in preparing nano TiO_2: An important environmental engineering material. *Journal of Environmental Sciences (China)*, *26*(11), 2139–2177. doi:10.1016/j.jes.2014.09.023 PMID:25458670

Wang, Z., Yang, C., Lin, T., Yin, H., Chen, P., Wan, D., ... Jiang, M. (2013). H-Doped Black Titania with Very High Solar Absorption and Excellent Photocatalysis Enhanced by Localized Surface Plasmon Resonance. *Advanced Functional Materials*, *23*(43), 5444–5450. doi:10.1002/adfm.201300486

Wen, J., Li, X., Liu, W., Fang, Y., Xie, J., & Xu, Y. (2015). Photocatalysis fundamentals and surface modification of TiO_2 nanomaterials. *Chinese Journal of Catalysis*, *36*(12), 2049–2070. doi:10.1016/S1872-2067(15)60999-8

Wu, X., Yin, S., Dong, Q., Guo, C., Li, H., Kimura, T., & Sato, T. (2013). Synthesis of high visible light active carbon doped TiO_2 photocatalyst by a facile calcination assisted solvothermal method. *Applied Catalysis B: Environmental, 142*, 450–457. doi:10.1016/j.apcatb.2013.05.052

Xu, S., & Sun, D. D. (2009). Significant improvement of photocatalytic hydrogen generation rate over TiO_2 with deposited CuO. *International Journal of Hydrogen Energy, 34*(15), 6096–6104. doi:10.1016/j.ijhydene.2009.05.119

Xu, W., Zhu, S., Li, Z., Cui, Z., & Yang, X. (2013). Synthesis and catalytic properties of Pd nanoparticles loaded nanoporous TiO_2 material. *Electrochimica Acta, 114*, 35–41. doi:10.1016/j.electacta.2013.09.084

Xu, Y., & Langford, C. H. (1997). Photoactivity of titanium dioxide supported on MCM41, zeolite X, and zeolite Y. *The Journal of Physical Chemistry B, 101*(16), 3115–3121. doi:10.1021/jp962494l

Xu, Y., Zheng, W., & Liu, W. (1999). Enhanced photocatalytic activity of supported TiO_2: Dispersing effect of SiO_2. *Journal of Photochemistry and Photobiology A Chemistry, 122*(1), 57–60. doi:10.1016/S1010-6030(98)00470-5

Yoneyama, H., & Torimoto, T. (2000). Titanium dioxide/adsorbent hybrid photocatalysts for photodestruction of organic substances of dilute concentrations. *Catalysis Today, 58*(2), 133–140. doi:10.1016/S0920-5861(00)00248-0

Yoong, L. S., Chong, F. K., & Dutta, B. K. (2009). Development of copper-doped TiO_2 photocatalyst for hydrogen production under visible light. *Energy, 34*(10), 1652–1661. doi:10.1016/j.energy.2009.07.024

Zäch, M., Hägglund, C., Chakarov, D., & Kasemo, B. (2006). Nanoscience and nanotechnology for advanced energy systems. *Current Opinion in Solid State and Materials Science, 10*(3), 132–143. doi:10.1016/j.cossms.2007.04.004

Zhang, S., Wang, H., Yeung, M., Fang, Y., Yu, H., & Peng, F. (2013). $Cu(OH)_2$-modified TiO_2 nanotube arrays for efficient photocatalytic hydrogen production. *International Journal of Hydrogen Energy, 38*(18), 7241–7245. doi:10.1016/j.ijhydene.2013.04.027

Zhong, J. S., Wang, Q. Y., & Yu, Y. F. (2015). Solvothermal preparation of Ag nanoparticles sensitized TiO_2 nanotube arrays with enhanced photoelectrochemical performance. *Journal of Alloys and Compounds, 620*, 168–171. doi:10.1016/j.jallcom.2014.08.205

Zhu, G., Yin, H., Yang, C., Cui, H., Wang, Z., Xu, J., ... Huang, F. (2015). Black Titania for Superior Photocatalytic Hydrogen Production and Photoelectrochemical Water Splitting. *ChemCatChem*, *7*(17), 2614–2619. doi:10.1002/cctc.201500488

Zhu, J., & Zäch, M. (2009). Nanostructured materials for photocatalytic hydrogen production. *Current Opinion in Colloid & Interface Science*, *14*(4), 260–269. doi:10.1016/j.cocis.2009.05.003

Zou, Y., Kang, S. Z., Li, X., Qin, L., & Mu, J. (2014). TiO_2 nanosheets loaded with Cu: A low-cost efficient photocatalytic system for hydrogen evolution from water. *International Journal of Hydrogen Energy*, *39*(28), 15403–15410. doi:10.1016/j.ijhydene.2014.07.174

Chapter 13
Electroless Nickel Coatings for High Temperature Applications

Arkadeb Mukhopadhyay
Jadavpur University, India

Tapan Kumar Barman
Jadavpur University, India

Prasanta Sahoo
Jadavpur University, India

ABSTRACT

This chapter aims to discuss the evolution of electroless nickel coatings with respect to their tribological behavior with special emphasis on their applicability at high-temperature-based applications. Electroless nickel coatings have tremendous potential as anti-wear and anti-friction coatings under ambient condition. The investigation of their tribological properties at high temperatures is relatively new. At demanding conditions, most conventional lubricants lose their properties and hence the use of self-lubricating coatings is unavoidable. Due to high melting point of nickel and oxidation resistance, electroless nickel-based coatings may prove to be suitable candidate at high temperatures. A review and analysis of the tribological characteristics of electroless nickel coatings especially at high temperatures is therefore necessary and has been reported in this chapter. Future research directions for the improvement of coating properties at high temperature are also identified from the review.

DOI: 10.4018/978-1-5225-5216-1.ch013

INTRODUCTION

Reduction of friction and wear of machine parts where tribological contact exists by the application of surface coatings is being widely investigated by researchers since excellent surface properties may be achieved without altering the bulk properties of the material. Amongst several popular coating techniques, the electroless method has gained immense importance since overwhelming results have been achieved by the appropriate control of coating bath parameters. High hardness, wear resistance, corrosion resistance, low coefficient of friction (COF) and uniform deposition are the key properties of electroless nickel (EN) coatings (Sahoo & Das, 2011). Since it is an autocatalytic process and avoids the use of electricity a wide variety of substrates may be coated with ease. In general, an electroless bath comprises of a source of nickel ions, reducing agent, complexants, buffers, stabilizers and surfactants (Loto, 2016). Appropriate control of the bath parameters lead to enhanced corrosion resistance as well as tribological behavior. EN coatings finds use in aerospace, automobile, chemical, electrical, electronics, textile, food processing, mining machinery, etc. industries due to their outstanding properties compared to the electro-deposited ones (Sudagar et al., 2013).

EN coatings have evolved over the last few decades as an advanced material with tremendous potential especially in cases where tribological contact takes place under aggressive environments. The coatings may be classified as pure nickel, alloy and composite coatings (Sudagar et al., 2013; Sahoo & Das, 2011; Agarwala & Agarwala, 2003; Mukhopadhyay et al., 2016a). Pure nickel coatings were manufactured by using hydrazine as reducing agent which has been discontinued lately due to the associated hazards (Sudagar et al., 2013). EN alloys are obtained mainly from sodium hypophosphite and sodium borohydride based baths resulting in formation of Ni-P (Sahoo, 2009; Mukhopadhyay et al., 2016b, 2016c; 2017a; Sahoo and Roy, 2017) and Ni-B (Krishnaveni et al., 2005; Vitry et al., 2010, 2012a, 2012b; Vitry & Bonin, 2017a, 2017b; Mukhopadhyay et al., 2016a, 2016d; Duari et al., 2016a) coatings respectively. Several metals could be alloyed with the binary coatings such as tungsten (Balaraju & Rajam, 2009; Balaraju et al., 2012; Palaniappa & Seshadri, 2008; Mukhopadhyay et al., 2016e, 2016f; Duari et al., 2017), copper (Balaraju & Rajam, 2005; Yu et al., 2001; Liu et al., 2015; Sahoo & Roy, 2017; Duari et al., 2016b), molybdenum (Vargas Mendoza et al., 2006; Balaraju et al., 2014), tin (Popoola et al., 2016; Balaraju et al., 2007), etc. Depending upon the alloying element, the properties of EN coatings such as hardness, corrosion/wear resistance, COF, magnetic properties etc. may be improved. EN composite coatings are obtained by the incorporation of hard ceramic or soft self lubricating particles along with the binary variants. These particles include Al_2O_3, TiO_2, SiC, diamond, ZnO, SiO_2, hBN, Si_3N_4, B_4C, PTFE, MoS_2, WS_2, etc (Agarwala & Agarwala, 2003; Gadhari

& Sahoo, 2016). Recent developments include the development and exploration of properties of carbon nanotubes (CNTs) co-deposited in EN coatings (Sudagar et al., 2013). With such a wide range of the deposits and its flexibility to coat even non-conducting substrates, EN coatings have provided an effective solution to a variety of engineering problems starting right from reduction of friction, wear and corrosion to decorative purposes. Although sodium hypophosphite based coatings enjoy a dominant share in the market, recent investigations are exploring the properties of sodium borohydride based coatings due to their higher hardness, wear resistance and anti-friction properties. Needless to say, with such a wide range of applicability, EN coatings are increasingly replacing chromium due to its environmental hazards.

With rapid advancements in aerospace, automotive and manufacturing sectors mechanical components are being subjected to friction and wear at high temperatures. This requires materials with excellent high temperature stability and oxidation resistance. EN alloy and composite coatings are now being investigated to cater into the high temperature domain due to the high melting point and oxidation resistance of nickel. Excellent self lubricating properties at elevated temperatures have been reported by researchers (Kundu et al., 2016; Masoumi et al., 2012; Ghaderi et al., 2016a; Alirezaei et al., 2013a, 2013b; Staia et al., 2002; Franco et al., 2016; León et al., 2003). Formation of oxide glazes and microstructural changes due to in-situ heat treatment effect during the high temperature sliding resulted in enhanced tribological properties of the EN coatings (Li et al., 2013; Alirezaei et al., 2013a, 2013b; Mukhopadhyay et al., 2016g). Inclusion of refractory materials is also seen to improve the high temperature stability of EN coatings (Mukhopadhyay et al., 2017b, 2017c). Recent studies have been directed towards the investigation of high temperature tribological characteristics of Ni-B coatings and its variants (Mukhopadhyay et al., 2016g, 2017b, 2017d). From the ongoing discussion, it may be concluded that EN coatings have gained immense importance as an advanced surface coating catering into a wide variety of industrial needs. Keeping in mind the increasingly aggressive environments, the present chapter aims to discuss the advancements of EN coatings and its tribological properties with special emphasis on its high temperature applicability. From high temperature studies the applicability and suitability of EN coatings at demanding conditions will be discussed and its viability as a potential high temperature candidate will be assessed. The scope of improving the tribological characteristics and thermal stability of EN coatings will be discussed along with recent developments in high temperature tribological characteristics of sodium borohydride based EN coatings. The tribological characteristics of the coatings under dry and lubricated environments are also discussed stating the recent progresses and development of newer variants of the coatings.

FUNDAMENTALS OF ELECTROLESS NICKEL DEPOSITION

Electroless nickel deposition is an autocatalytic process which involves the reduction of metallic ions in a solution 'bath' and the deposition of coating takes place through the oxidation of a reducing agent which supplies electric current. In this process, catalytic reduction of nickel ions present in the solution bath takes place leading to deposition of nickel and thereby eliminating the use of electricity. The cations of the metal to be deposited receive electrons from the substrate and get reduced while the reductant delivers electrons to this substrate and it gets oxidized (Sudagar et al., 2013). Once reaction starts, it continues due to the autocatalytic activity and evolution of hydrogen takes place in the process. In comparison with the conventional electroplating process, EN coatings have higher corrosion resistance, wear resistance, low COF and high hardness. Another key advantage of the electroless deposition is the deposit homogeneity. This is due to the fact that the driving force for reduction of nickel ions is provided by a reducing agent and this driving potential is essentially constant at all points on the surface of substrate (Loto, 2016). Therefore, a uniform deposit thickness is achieved with low porosities even on intricate parts. This leads to a high corrosion resistance of the EN coatings compared to the electroplated ones.

EN coatings may be broadly classified as pure nickel, alloy/poly-alloy and composite coatings. Pure nickel coatings are generally obtained from hydrazine bath but have been discontinued since they are costly, difficult to control and tend to be unstable at high temperatures. The alloy/poly alloy coatings are obtained from sodium hypophosphite, sodium borhydride, N-dimethyl amine borane (DMAB) and N-diethyalmine borane (DEAB). Depending on the reducing agent, the EN alloy coatings may be Ni-P or Ni-B and their variants. Electroless composite coatings are obtained by the inclusion of soft lubricious or hard ceramic particles to the EN alloy matrix. The variant and coating type to be used depends on the application for which it is required. The electroless bath is generally composed of an aqueous solution consisting of a source of nickel ions, reducing agent, complexing agent, buffer, stabilizers, wetting agents, etc. that affects the quality of the deposits in several ways and their proper control is required. Depending on the bath pH the Ni-P or Ni-B alloys are obtained from acid or alkaline baths.

The deposition mechanism of EN acid/alkaline bath coatings (Loto, 2016) has several controversies. But the electrochemical and atomic hydrogen mechanisms are widely accepted for Ni-P coatings. The electrochemical mechanism involves the catalytic oxidation of the hypophosphite. This yields electrons at the catalytic surface and reduces nickel and hydrogen ions (Sahoo & Das, 2011). The mechanism may be explained by the following reactions as mentioned by Sahoo & Das (2011):

$$H_2PO_2^- + H_2O \rightarrow H_2PO_3^- + 2H + 2e^- \quad (1)$$

$$Ni^{2+} + 2e^- \rightarrow Ni \quad (2)$$

$$2H^+ + 2e^- \rightarrow H_2 \quad (3)$$

$$H_2PO_2^- + 2H^+ + e^- \rightarrow P + 2H_2O \quad (4)$$

On the other hand, in atomic hydrogen mechanism, due to the catalytic dehydrogenation of hypophosphite molecule adsorbed at the surface, release of atomic hydrogen takes place (Sahoo & Das, 2011). Further, this nascent hydrogen reduces nickel at the catalyst surface (Loto, 2016). The atomic hydrogen mechanism may be represented as follows (Sahoo & Das, 2011):

$$H_2PO_2^- + H_2O \rightarrow HPO_3^{--} + H^+ + 2H_{ads} \quad (5)$$

$$2H_{ads} + Ni^{++} \rightarrow Ni + 2H^+ \quad (6)$$

$$H_2PO_2 + H_{ads} \rightarrow H_2O + OH^- + P \quad (7)$$

$$\left(H_2PO_2\right)^{2-} + H_2O \rightarrow H^+ + \left(HPO_3\right)^{2-} + H_2 \quad (8)$$

On the other hand, borohydride oxidizes according to the equation as follows (Anik et al., 2008):

$$BH_4^- + 4OH^- = BO_2^- + 2H_2O + 2H_2(g) + 4e^- \quad (9)$$

$$BH_4^- = B + 2H_2(g) + e^- \quad (10)$$

The reduction reactions in the coating bath may be expressed as (Anik et al., 2008):

$$Ni^{2+} + 2e^- = Ni \quad (11)$$

$$2H_2O + 2e^- = 2OH^- + H_2(g) \quad (12)$$

The reactions mentioned above are affected by the bath parameters such as the concentration of nickel, reducing agent, complexing agent, accelerator, surfactants, bath agitation and bath temperature. These parameters in turn affect coating characteristics such as structure, density, electrical resistivity, magnetic property, internal stress, deposit thickness, surface roughness, microhardness, wear resistance, COF, corrosion resistance, etc. The mechanical properties of the coatings are also affected by heat treatment, laser treatment, nitridation and heat treatment in active atmosphere. Several studies have been carried out to achieve enhanced characteristics of EN coatings, but the tribological properties and factors that affect them will be addressed mainly in this chapter in the subsequent sections.

SURFACE MORPHOLOGY AND MICROSTRUCTURE

It is well known that friction and wear process is a surface phenomenon. Surface characteristics of a material have influence on the tribological behavior. The excellent tribological characteristics of EN coatings have been attributed to the surface morphology, microstructure and particularly in case of Ni-B coatings and its variants to their columnar growths. The surface of Ni-P coatings and its poly alloys exhibit well defined nodulated structures in as-deposited condition. On heat treatment, these nodules grow in size with the grain boundaries becoming more prominent. Typical scanning electron micrographs of as-deposited and heat treated (400°C for 1 hour) Ni-P coatings derived from hypophosphite based bath is shown in Fig. 1. Ni-B coating derived from borohydride based bath is shown in Fig. 2 in as-deposited and heat treated (350°C for 1 hour) condition. Ni-B coatings present a dense nodulated structure in as-deposited condition while the nodules appear inflated on heat treatment. The surface typically resembles a cauliflower. Such morphology leads to the excellent anti-friction properties of Ni-B coatings. The initiation and growth of Ni-B coatings was studied elaborately by Vitry et al. (2010).The phases include a short induction period, formation of a thin continuous layer, formation of nodules, densification of the coating followed by several nucleation/growth phases. The growth of the nodules takes place in a succession of germination phases (formation of small nodules) and growth phases where densification of nodules take place along with their growth (Vitry et al., 2012b). Another aspect that imparts superior tribological properties to Ni-B coatings in comparison with Ni-P coatings is the columnar structures. A micrograph of cross-section of as-deposited Ni-B coating is shown in Fig. 3. This columnar morphology begins to form as the growth of the nodules of Ni-B coating in lateral direction stops (Vitry et al. 2012b). These columnar growths reduce the actual area of contact and thereby the adhesive forces between coating and counterface material and hence wear resistance improves and COF decreases. They also act as pockets for retaining lubricants under adhesive wear situations as well as leads to lower consumption of lubricants. SEM micrographs of EN poly-alloy coatings obtained from hypophosphite and borohydride based baths in as-deposited condition is shown in Fig. 4 and Fig. 5 respectively. Each poly-alloy coating has unique properties compared to their binary variants such as high hardness, wear resistance, corrosion resistance, low COF, higher thermal stability, oxidation resistance, magnetic properties, etc.

Another important aspect of the coatings is their crystallinity. Generally as-deposited EN coatings are considered amorphous. Nevertheless, the content of phosphorus or boron in EN coatings determines their crystallinity in as-deposited state. Based on the phosphorus content, Ni-P coatings may be categorized as low-P (3-5 wt. %), medium-P (6-9 wt. %) and high-P (10-12 wt. %) coatings. As the phosphorus content decreases, the crystallinity of the deposits changes from amorphous to amorphous/nanocrystalline and finally to nanocrystalline structures. Similar observations have been also made for Ni-B coatings. Inclusion of alloying

Figure 1. SEM micrograph of (a) as-deposited and (b) heat treated Ni-P coating

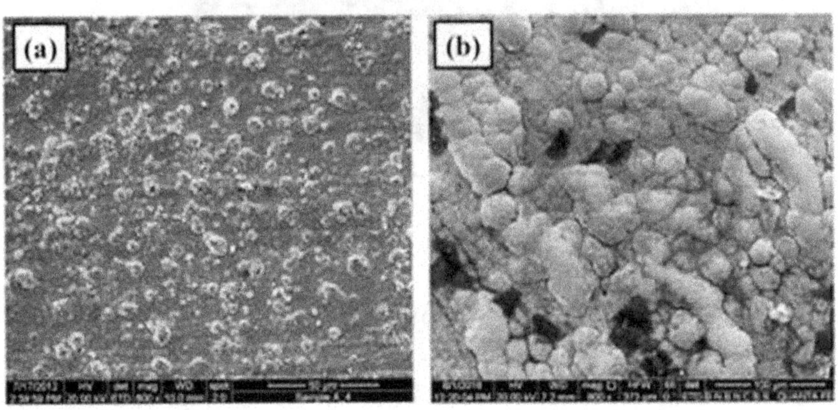

Figure 2. SEM micrograph of (a) as-deposited and (b) heat treated Ni-B coating

Figure 3. Cross-section SEM image of Ni-B coating

elements also influences the crystallinity of EN coatings mainly by affecting the phosphorus or boron content of the coatings. The crystallinity of the coatings may be observed from X-ray diffraction (XRD) analysis. X-ray patterns of as-deposited EN coatings are shown in Fig. 6. A broad hump in the XRD generally denotes amorphous nature (disorder in arrangement) of the coatings and crystallinity

Figure 4. SEM micrographs of EN poly alloy coatings from hypophosphite based bath (a) Ni-P-W and (b) Ni-P-Cu in as-deposited condition

Figure 5. SEM micrographs of EN poly alloy coatings from borohydride based bath (a) Ni-B-W and (b) Ni-B-Mo in as-deposited condition

improves as the full width at half maximum (FWHM) decreases. On heat treatment precipitation of phosphides is observed for Ni-P coatings while borides are seen in case of Ni-B coatings. Due to precipitation of these hard phases on heat treatment, the microhardness of EN coatings improves. Apart from precipitation hardening, the improvement in microhardness of poly-alloy coatings is attributed to the solid solution strengthening phenomenon. XRD patterns of heat treated EN coatings are shown in Fig. 7. Compared to the as-deposited state, the reflection from Ni (111) becomes sharp and intense on heat treatment.

Figure 6. XRD results of as-deposited EN coatings (a) Ni-P and (b) Ni-B

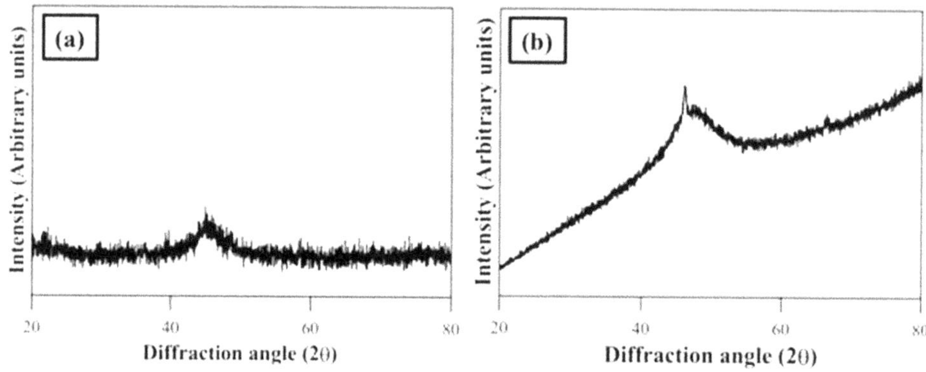

Figure 7. XRD results of heat treated EN coatings (a) Ni-P and (b) Ni-B

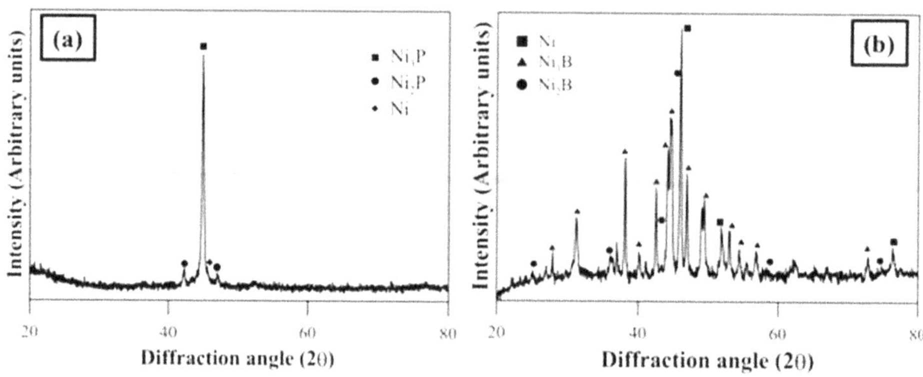

TRIBOLOGICAL CHARACTERISTICS OF ELECTROLESS NICKEL COATINGS

The tribological characteristics of EN coatings have long been investigated since its inception by Brenner and Riddell (1946). The high hardness, wear resistance and self lubricating capabilities make them suitable candidates for wear reduction of mating components. In this section the tribological properties of EN alloy/ composite coatings and their evolution over time i.e. how the family of EN coatings have expanded with a quest to achieve improved properties are discussed along with recent progresses in the high temperature tribological domain.

Tribology Under Dry Condition at Ambient Temperature

The wear behavior and mechanisms governing the same for Ni-P coatings have gained immense importance from researchers over the years. Ni-P coatings are susceptible to severe adhesion in as-deposited condition when sliding against plain carbon steel under dry condition (Gawne & Ma, 1987). Under controlled heat treatment precipitation of crystalline nickel and Ni_3P takes place which is incompatible with steel and prevents severe wear. In comparison with the as-deposited coatings, the heat treated ones show higher wear resistance due to high hardness (Staia et al., 1996) achieved from the precipitated crystalline phases. Taheri et al. (2001) observed that the microhardness of Ni-P coatings depend on the phosphorus content as well as aging time. But the kinetic coefficient of friction is reduced irrespective of heat treatment condition. Yan et al. (2008) could achieve record high hardness of $910HV_{100}$ for as-deposited coatings and consequently a higher wear resistance by controlling the phosphorus content. Sahoo (2009) indicated in his work that the tribological behavior of Ni-P coatings can be significantly improved by controlling electroless bath parameters. Ni-P coatings have also successfully reduced the wear of AZ91 magnesium alloy and show excellent sliding property due to formation of oxide glazes (Novák et al., 2011). A high performance Ni-P coating has been obtained by Ansari and Thakur (2017) on AZ91 magnesium alloy by maintaining the bath pH at 6.5. Alloying with tungsten improves the microhardness and wear resistance (Palaniappa & Seshadri, 2008) while inclusion of copper produces deposits with lower hardness and wear resistance (Sahoo & Roy, 2017) in as-deposited condition though the corrosion resistance improves. Heat treated Ni-P-W coatings undergo abrasive wear with wear debris cutting through the coating producing micro-grooves and micro-furrows (Duari et al., 2017; Mukhopadhyay et al., 2016e, 2016f). While heat treated Ni-P-Cu coatings (500°C for 1 hour) undergo failure by fracture of the nodules due to embrittlement (Duari et al., 2016b). Worn specimens of Ni-P-W and Ni-P-Cu coatings are shown in Fig. 8. Inclusion of tin also improves the wear resistance of Ni-P coatings (Popoola et al., 2016). Another excellent way of achieving high wear resistance is by the deposition of functionally graded Ni-P coatings and the sequence of low P/medium P/high P from the substrate exhibit the best wear resistance (Hadipour et al., 2016). Such deposits with varying phosphorus through the thickness have better surface finish and low porosity. As a result of gradual variation of phosphorus, higher wear resistance is achieved and formation of cracks is prevented (Anvari et al., 2015). Composite particles have been in general found to improve the tribological properties of Ni-P coatings. Hard particles such as Al_2O_3, TiO_2, Si_3N_4, B_4C, BN, SiC, ZrO_2, SiO_2, WC and diamond causes a remarkable improvement in microhardness and wear resistance. Soft particles such as PTFE, MoS_2, WS_2, Gr, IF-MoS_2, etc. improve the friction behavior of Ni-P coatings. In most of the cases, it has been observed that there exists an optimum heat treatment temperature and particle concentration up to which the microhardness and tribological behavior of Ni-P coatings improve above which it deteriorates. The detailed tribological behavior along with configuration of test setups for electroless composite coatings has been discussed elaborately by Gadhari and Sahoo (2016). Recent studies report the co-deposition of calcinated sea shell particles, halloysite nanotubes and multiwalled carbon nanotubes with high performances (Jagatheeshwaran et al., 2015; 2016; Ranganatha et al., 2012).

Significant attention is now being focused on the development of borohydride reduced coatings due to their higher hardness in as-deposited condition in comparison with hypophosphite reduced Ni-P alloy. As observed earlier in Fig. 2(b) and other researchers (Sahoo & Das, 2011), Ni-B coatings have a cauliflower like surface morphology which gives them self lubricating properties. Moreover, the columnar growths reduce COF and wear rate of the coatings by reducing actual contact area. It has been observed that Ni-B coatings deposited on a variety of substrates such as magnesium, AZ91D alloy (Correa et al., 2013), 316L stainless steel (Bülbül et al., 2013) and pure titanium (Çelik et al., 2016) have tremendously improved their friction and wear characteristics. Krishnaveni et al. (2005) observed the tribological behavior and the effect of heat treatment of Ni-B coatings. It was revealed that as-deposited Ni-B coatings underwent higher wear and COF than heat treated ones. The heat treated Ni-B coatings show better friction and wear characteristics compared to as-deposited ones due to the high hardness achieved by precipitation hardening phenomenon. The as-deposited coatings undergo adhesive wear while abrasive wear is marked for heat treated Ni-B coatings. Similar observations have been made by Mukhopadhyay et al. (2017d). SEM micrograph of worn specimen of as-deposited and heat treated (350°C for 1 hour) Ni-B coating is shown in Fig. 9 where the adhesive and abrasive wear phenomenon has been observed (Mukhopadhyay et al. 2017d).

Figure 8. SEM micrograph of worn specimen of heat treated (a) Ni-P-W and (b) Ni-P-Cu coating under dry sliding condition

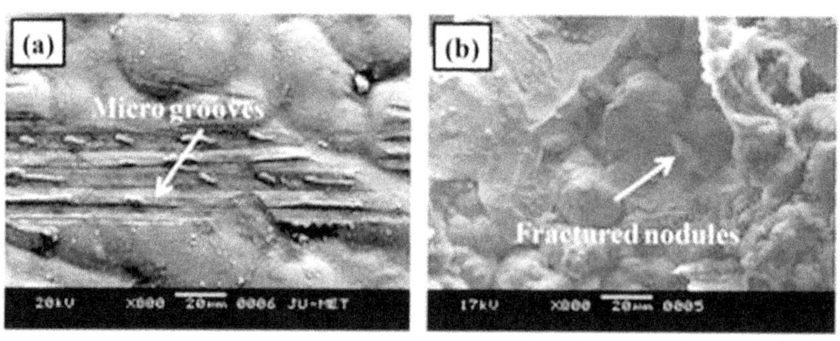

Similar observations have been also reported by Balaraju et al. (2016) in a recent study. Das & Sahoo (2011) observed the effect of coating bath parameters on the tribological characteristics of Ni-B coating under dry condition and used design of experiments coupled with grey relational analysis for optimization. It is observed in the same study that the reducer concentration (sodium borohydride) has the highest significance followed by bath temperature in controlling the wear depth and COF of Ni-B coatings. The sodium borohydride concentration and bath temperature affects the boron content in the coatings. Increasing sodium borohydride increases the boron content (Anik et al., 2008). While Vitry & Bonin (2017b) observed that a very high boron deposition could be achieved by suitably controlling the bath temperature which increases the wear resistance of Ni-B coatings (Vitry & Bonin, 2017b).

The tribological behavior of Ni-B coatings also depend on the test parameters such as applied normal load, sliding speed, sliding distance, sliding duration, etc. Krishnaveni et al. (2005) observed an increase in the wear rate and COF with an increase in applied normal load. Interaction effects of the tribo-testing parameters such as normal load, sliding speed and sliding time on the wear depth and COF could be observed clearly from 3D surface and contour plots (Mukhopadhyay et al., 2016a). Madah et al. (2015) observed the wear mechanism of Ni-B coating at different applied normal load and sliding distance. At 5 and 60N normal load and sliding distance of 100 and 300m, a combination of adhesive and abrasive wear occurs. At the same normal loads, if the sliding distance is increased to 300 and 500m, the dominant wear mechanism is abrasive and fatigue. At 300 and 500m, when the applied normal loads are 20 and 40N, the wear mechanism is fatigue with severe oxidation. Ni-B coatings completely failed at 700m at all applied normal load. For 5 and 20N load, the lifetime is 500m whereas for 40 and 60N load, the lifetime decreases to 300m. In another work by Madah et al. (2016), phase transformations

Figure 9. SEM micrograph of worn specimen of Ni-B coating tested under dry sliding condition at ambient temperature (a) as-deposited and (b) heat treated (350°C for 1 hour)

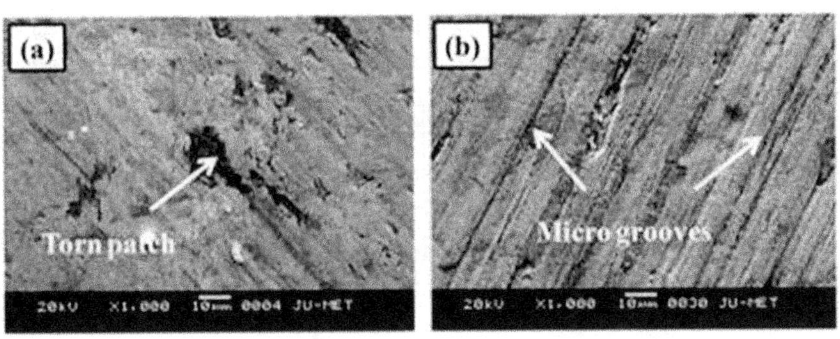

of the coatings could be observed after dry sliding which had a strong influence on supporting the tribo-patches and improving the tribological behavior of the deposits. Ni-B coatings in as-deposited and heat treated state respond well to scratch tests where load has been applied at a rate of 100Nmin^{-1} allowing it to reach 150N (Vitry et al., 2011). The scratch mark is observed at different loads and no spalling or delamination with overall ductile behavior throughout the test is concluded.

The tribological behavior of Ni-B coatings is also influenced by alloying with tungsten or molybdenum. A noteworthy improvement in wear resistance is obtained on inclusion of tungsten to Ni-B coatings (Aydeniz et al., 2013). The COF and relative wear resistance is also good for Ni-B-Mo heat treated at 300°C for 1 hour (Serin & Göksenli, 2015). An increase in microhardness from 570HV_{100} to 860HV_{100} has been reported by Ghaderi et al. (2016b) when composite particles of Al_2O_3 are incorporated into Ni-B matrix. The wear loss and COF of Ni-B-Al_2O_3 composite coatings are therefore significantly lower in comparison with Ni-B alloy. On impregnating Ni-B coatings with 40% PTFE, a COF as low as < 0.1 is achieved and rubbing test duration of 3600s could be endured by the coatings (Wan et al., 2016).

Apart from the polyalloy or composite variants of Ni-P or Ni-B coatings, an effective way of achieving properties intermediate to two is by the deposition of duplex Ni-P/Ni-B coating or the properties may be even better than the two binary alloy coatings. The wear resistance and microhardness of coatings with Ni-B as outer layer while Ni-P as inner layer is higher than Ni-P or Ni-B binary alloys (Sankara Narayanan et al., 2003). Duplex NiP/NiB coatings deposited on 2024 aluminium alloy show high wear resistance when heat treated at 180°C for 4 hour in an atmosphere of 95% argon and 5% hydrogen (Vitry et al., 2012c). The deposition and tribological characterization of duplex coatings with high boron (7-9 wt. %) content and mid phosphorus (7-9 wt. %) content has been reported by Bonin and Vitry (2016). The coatings exhibit properties intermediate to Ni-P and Ni-B but better properties (abrasive wear resistance) are achieved with Ni-B as top layer. Duplex Ni-P/Ni-P-W coatings with high phosphorus intermediate layer show enhanced wear resistance and characterized by mild adhesive wear (Liu et al., 2012). The tribological behavior of the Ni-P/Ni-P-W system further improves on laser treatment. Electroless Ni-P/Ni-P-Mo coatings also show better tribological properties than Ni-P coatings which is further improved on laser annealing (Liu et al., 2016). The wear of outer Ni-P-Mo layer is characterized by micro furrows and micro-fracture indicative of abrasive wear. As the inner layer is reached, the tribological characteristics are dominated by the properties of Ni-P coating and mostly by the peeled off Ni-P-Mo outer layer (Liu et al., 2016). In a recent investigation by Vitry and Bonin (2017a), multi-layered i.e. 10 layers with alternating Ni-P and Ni-B or Ni-B and Ni-P were seen to exhibit higher wear resistance than Ni-B (which is considered to have higher hardness and wear resistance than most EN alloy coatings) or Ni-P mono layered coatings. The multi-layered coatings helped in combining the high corrosion resistance of Ni-P and high wear resistance of Ni-B.

Tribology Under Lubricated Condition at Ambient Temperature

EN coatings have tremendous potential under lubricated condition due to the nodulated surface morphology which allows retaining the lubricants during sliding wear process. Eventually, the lubricant consumption may be reduced due to such behavior. Several investigations have been therefore carried out to assess the suitability of the coatings under different lubrication situations. There are several mechanical components that work under lubricated sliding condition. With the increasing environmental concerns, lubricants of mineral origin are being increasingly replaced with bio-lubricants especially in areas where they come in contact with water, food or people (Ramalho & Miranda, 2007). One of the major concerns with bio-lubricants is their corrosive action that accelerates the wear of cylinder liners and affects automobile engines (Xu et al., 2014). Since EN coatings exhibit high corrosion resistance, they can be a potential candidate in such contact situations. Ramalho & Miranda (2007) observed the tribological characteristics of Ni-P and Ni-P-PTFE (polytetrafluoroethylene) coatings under lubrication of gearbox oil, chain way bio oil, SAFE LUBE™ gear lube oil, emulsion containing 20% by volume of vegetable base oil in water and water. The addition of PTFE to Ni-P is helpful in boundary lubrication condition due to the self lubricity provided by PTFE. The wear occurs under lubrication mainly by abrasion (Ramalho & Miranda, 2007). Ni-P-PTFE coatings also exhibit decent tribological characteristics under water lubricated condition due to its highly hydrophobic nature provided the counterface is highly hydrophilic (Ger et al., 2002). Xu et al. (2014) observed that Ni-P and Ni-P-Cu coatings can effectively prevent cylinders from corrosive action of bio-oil as well as plays an anti-wear and anti-friction role. In comparison to Ni-P coatings, Ni-P-Cu coating provides better tribological performance due to a tribo-film of CuO formed on worn surface. To improve the tribological characteristics of piston ring and cylinder liner friction pair under bio-oil and alleviate its corrosive action, electroless Ni-P, Ni-P-MoS_2, Ni-P-GO and Ni-P-MoS_2-GO were deposited on the rings (Peng et al., 2017). Adhesive wear, scratching, spalling and mild wear are associated with Ni-P, Ni-P-MoS_2, Ni-P-GO, and Ni-P-MoS_2-GO coatings. Ni-P-MoS_2-GO coated piston rings were concluded to exhibit best performance and have tremendous potential for accelerating the application of bio-oil in IC engines (Peng et al., 2017).

Electroless Ni-P-CNT (carbon nanotube) coatings show tremendous potential under oil lubricated condition in comparison with Ni-P-SiC or Ni-P-Gr coatings (Chen et al., 2002). This happens due to improved mechanical properties and the topological structure of CNTs which can easily slide or roll within surfaces of tribo-pairs. The tribological behavior of sodium hypophosphite as well as borohydride based EN coatings under engine oil lubricated condition has been extensively studied by Mukhopadhyay et al. (2016b, 2016c, 2016d, 2016e) and Duari et al. (2016a, 2017). Design of experiments (DOE) and a hybrid grey fuzzy reasoning analysis has been employed by Mukhopadhyay et al. (2016c, 2016d) to assess the tribological behavior of Ni-P and Ni-B coating under engine oil (SERVO PRIDE, a product of Indian Oil) lubricated condition with varying tribological test parameters (applied normal load, sliding speed and sliding duration). Applied normal load and sliding duration is observed to be the most significant parameters affecting the wear depth and COF of the Ni-P and Ni-B coatings respectively under lubrication. Formation of a protective oxide layer is seen in both Ni-P as well as Ni-B coatings and the wear surfaces are characterized by micro-ploughing effects indicating mild-abrasive wear. The use of artificial intelligence was successfully demonstrated in modeling the tribological behavior of Ni-P and Ni-P-W coating under lubricated condition compared to conventional regression techniques (Mukhopadhyay et al. 2016b, 2016e). Electroless Ni-P-W coatings show enhanced tribological behavior and the wear depth as well as COF decreases significantly when sliding under lubrication in comparison with dry condition. Worn surface of Ni-P-W coating appears to be smooth with mild scratches (Duari et al., 2017; Mukhopadhyay et al., 2016e). Worn surface morphology of Ni-P, Ni-B and Ni-P-W coating after tribological tests under lubricated condition is shown in Fig. 10.

Tribology Under Corrosive Environment at Ambient Temperature

EN coatings are preferred in chemical as well as petrochemical industries due to their high corrosion resistance. Since the coatings also possess high hardness and wear resistance, they may prove to be excellent candidates for tribological contacts working under corrosive environment. Electroless Ni-P coatings may provide effective shielding against chemical attack that can aggravate the wear of mechanical components especially in mining and marine industries. Panja et al. (2016) evaluated the friction and wear characteristics of Ni-P coatings under brine, acidic and alkaline environment and the effect of varying applied normal load and sliding speed on the properties is also investigated. The COF is seen to decrease with an increase in applied normal load for all the three environments. While the wear rate of Ni-P coatings gradually increases with load for brine and alkaline environments up to

Figure 10. Worn surface of (a) Ni-P (b) Ni-B and (c) Ni-P-W coatings under lubrication

40N after which a decreasing trend is observed. In acidic environment, the wear rate shows an increasing trend with applied normal load. It is also concluded in the same study that acidic environment has the most degrading effect in comparison with alkaline or brine environment. Several investigations has been also reported on tailoring the bath components to achieve enhanced tribological behavior under alkaline, brine and acidic environment using Taguchi's experimental design and grey relational analysis (Panja et al., 2015; Panja & Sahoo, 2014, 2015).

Carbon fiber reinforced plastic (CFRP) composites are widely used in aeronautical and astronautical applications. The properties of such composites and their corrosion wear characteristics could be improved by the deposition of electroless Ni-P coatings. Lee (2009) investigated the corrosion wear characteristics of Ni-P coated CFRP composites in 3.5% NaCl solution. It is observed that the Ni-P coatings deposited on the composites having roughness > 0.3μm showed improved corrosion resistance. As the roughness of substrate increased, the micro-porosity, coating thickness and a smooth surface morphology of Ni-P are obtained. Ni-P coatings also improve the wear corrosion resistance of the coatings especially when phosphorus content and coating thickness increased while polishing condition of CFRP substrates decreased. The strong passivity of Ni-P coating acts as a lubricator and reduces the friction coefficient. Similar results have been also observed by Lee (2008) for Ni-P coatings deposited on glass fiber reinforced plastic (GFRP) composites that are used frequently in wind turbine blades.

With the advancements in technology, exploration of sea based resources are becoming extremely important which has led to the development of a new research field of tribology namely ocean tribology. Sea water has corrosive effects on mechanical components such as hydraulic drives which utilize sea water directly as working fluid. The sea water hydraulic drive is considered to be a key technology for deep-sea submergence vehicle, underwater robot and submarine oil production system (Wang et al., 2009). Hence, selection of corrosion and wear resistant friction pairs is necessary. Electroless Ni-P coatings have been reported to improve the wear corrosion resistance of such components due to its high hardness and corrosion resistance properties (Wang et al., 2009).

Tribology Under Dry Condition at Elevated Temperature

From literature and ongoing discussion, it is clear that EN coatings have tremendous potential under dry, lubricated or corrosive environment. The coatings have been suitably tailored to sustain such environments. Recent investigations have now been directed towards the tribological behavior of EN coatings at elevated temperature. It is expected that the mechanical properties of the coatings would deteriorate at high temperatures. But it has been observed that several factors control the tribological behavior of EN coatings at high temperature which improves the wear resistance and COF. Electroless Ni-P coatings show excellent tribological behavior at 550°C (Masoumi et al., 2012). The as-deposited coatings show better friction and wear characteristics at 550°C in comparison with heat treated ones. This happens due to crystallization. It is well known that heat treated EN coatings undergo lower wear in comparison with the as-deposited ones. But the contrary is observed at elevated temperatures. There exists an optimum structure for better wear resistance of Ni-P alloys (Masoumi et al., 2012). Optimum amount and size of granular nickel particles dispersed in Ni_3P matrix prevent dislocations from looping. When a crack propagates into such a particle, enough plastic deformation is allowed to slow down the crack. The stress concentration is reduced to a high degree and crack growth is slowed down thereby reducing wear volume. This optimum particle size is achieved in case of as-deposited Ni-P coatings while for heat treated coatings, the nickel particles grow further in size thereby increasing the mean free path for crack propagation. Moreover, since complete crystallization of the as-deposited coatings does not occur, there is high adhesive tendency which supports an adhesive oxide layer on the coatings surface. This again helps in reducing wear rate and COF.

The effect of load and speed at elevated temperatures on the tribological characteristics of Ni-P coatings has been reported by Kundu et al. (2016). The wear rate and COF of the coatings increases with an increase in applied normal load while it decreases with sliding speed at 100-500°C for both as-deposited and heat treated condition. The COF is observed to be significantly low at 500°C due to the formation of protective oxide layer. The as-deposited coating shows enhanced wear and friction in comparison with heat treated ones at 500°C due to the microstructural changes on heat treatment at high temperature. The percentage change in hardness post wear test indicated the maximum rise for the as-deposited coatings.

Tribological behavior of Ni-P coatings with different phosphorus content in between 3.5 and 12% has been observed by Ghaderi et al. (2016a) at 250°C test temperature. Highest and lowest wear resistance has been observed for coatings with 3 and 12% phosphorus respectively. This has been attributed to the decrease in hardness with an increase in phosphorus content and occurrence of adhesive wear. The low phosphorus coatings were observed to be protected by an oxide film while the high phosphorus deposits lacked the same. An increase in hardness post wear test has been also reported similar to that observed by Kundu et al. (2016) for high phosphorus coatings due to growth and crystallization of nickel grains from amorphous phase.

Electroless composite coatings containing self lubricating particles may prove to be better candidates at high temperatures compared to the binary Ni-P alloy. Co-deposition of nanocrystalline silver particles to the binary Ni-P alloy improves the tribological behavior of the deposits by manifolds at a test temperature of 500°C (Alirezaei et al., 2013a). The wear rate of binary Ni-P coating increased by ~10 times at 500°C in comparison with room temperature test while for Ni-P-Ag coatings, the wear rate increased by about 2 times. The COF of Ni-P-Ag coatings at elevated temperatures have been also observed to be lower than Ni-P coatings at 500°C. Silver particles could be easily deformed due to its low shear strength and diffusion of silver from inside of the composite bulk to the surface leads to its self lubricating characteristics. Because of self lubricating characteristics, this type of coating is considered to be 'chameleon' due to their excellent self lubricating properties. Inclusion of Al_2O_3 along with Ni-P-Ag coatings did not cause any significant improvement in tribological behavior at high temperature (Alirezaei et al., 2013b). But Ni-P-Ag-Al_2O_3 hybrid nano composite coatings could be a suitable candidate to achieve optimum hardness and wear characteristics at high temperatures.

The effect of temperature on the tribological behavior of Ni-P-MoS_2 coating heat treated at 300°C for 2 hour has been observed due to their self lubricating characteristics (Li et al., 2013). The temperature was varied between ambient to 600°C. The COF of Ni-P-MoS_2 coatings decreases with an increase in test temperature up to 500°C. Best tribological properties are marked at 400°C. The highest wear rate is observed at 200°C after which the wear rate decreases at 300-500°C temperature range and deteriorates again at 600°C due to softening of the coatings. Similar observations are also noted for COF. The enhanced tribological properties of the coatings at elevated temperatures have been attributed to the formation of lubricious oxide glaze composed of NiO and MoO_3. The worn specimens are mainly characterized by deformation and mild scuffing.

Inclusion of SiC nano particles improves the tribological behavior of EN conditions under dry ambient sliding condition. The tribological behavior of Ni-P-SiC coatings at an elevated temperature of 200°C has been observed by Franco et al. (2016). The coatings heat treated at 400°C for 1 hour showed lower wear rates compared to the as-deposited ones. The wear rate also decreases with an increase in the SiC content from 0-42% in the EN matrix. The wear morphology of particles free coating is characterized by adhesive wear with formation of surface as well as sub-surface micro cracks due to applied mechanical and thermal stresses. The crack propagation is interrupted on the inclusion of SiC and also acts as load bearing sites. The wear mechanism of composite and heat treated coating is mainly observed to be abrasive with tribo-oxidation and formation of protective oxide layers. Phase transformations post wear test at 200°C was not observed to be profound though the onset of evolution of crystallization could be observed in XRD results with reflection from Ni (111) becoming sharper. On the contrary, Staia et al. (2002) observed a deterioration in the wear resistance of Ni-P-(24.5%)SiC heat treated in argon at 400°C for 1 hour when tested at an elevated temperature of 100 and 300°C. An over aging effect due to the heat treatment and high operating temperature led to the softening of the coatings and loosening of SiC from the EN matrix and consequently a higher wear and COF has been marked. Although, in comparison with tests conducted at 100°C; better wear performance has been reported at 300°C. This has been attributed to the formation of a more stable and uniform film of iron oxide which acts as a protective layer.

The tribological performance of electroless Ni-P-BN (h) coatings deteriorated as the test temperature is increased to 100-400°C from ambient condition (25°C) (León et al., 2003). Adhesive wear mechanism was observed to be dominant at high temperature. The ploughing action is increased at high temperatures by the presence of BN (h) particles which are unable to support the load due to their low hardness and is detached due to the considerable softening of the coating matrix. This coupled with the cyclic nature of load of ball-on-disc geometry led to the formation of surface and sub-surface crack. High accumulation of strain and adhesive forces cause the opening of these cracks and formation of surfaces with high roughness takes place. At 300 and 400°C, the wear surfaces are characterized by tearing of the coating, gross plastic deformation and high amount of material transfer to the counterface.

By far, the high temperature studies of EN coatings has been restricted to sodium hypophosphite reduced alloy and composite coatings. Recently the high temperature tribology of borohydride reduced Ni-B coating has been reported by Mukhopadhyay et al. (2016g). For the as-plated coatings, it has been observed that the wear rate and COF increases significantly on increasing the operating temperature to 100°C. This is attributed to the hardened debris cutting through softer coating matrix forming deep groves and adding to the ploughing component of friction. The wear rate and COF improves on further increasing the temperature to 300°C which is expected due to initiation of phase transformations in the coatings that support the tribo-patches and load bearing sites. At 500°C, deterioration of wear resistance takes place due to oxidation of the coatings and loss of boron content though the COF improves due to softening of the coatings. The wear mechanism at room temperature has been observed to be adhesive following which it changes to abrasive on increasing the temperature to 100/300°C and finally to oxidative with formation of blackish patches of oxides of nickel at 500°C. The worn surfaces of as-plated Ni-B coatings as observed by Mukhopadhyay et al. (2016g) are shown in Fig. 11.

Thereafter, the effect of inclusion of W and Mo to Ni-B coatings on their high temperature tribology in their as-plated state has been reported by Mukhopadhyay et al. (2017b). At room temperature, the wear rate and COF of Ni-B-W coatings is reported to be the lowest while Ni-B-Mo coating shows the worst performance due to its lower microhardness, a cracked surface morphology and severe adhesion

Figure 11. SEM micrograph of worn specimen of as-deposited Ni-B coating at different operating temperatures (a) 100 (b) 300 and (c) 500°C

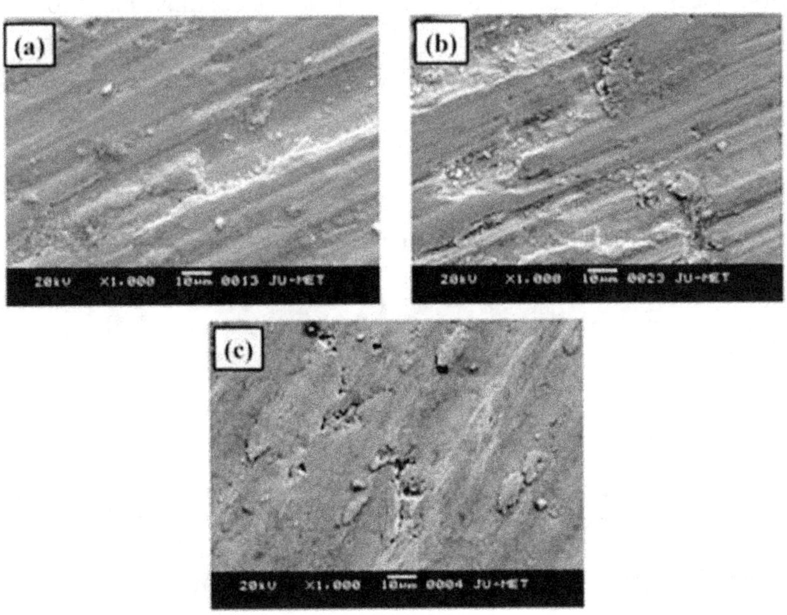

patches. At 100°C, the wear rate of both Ni-B-W and Ni-B-Mo coatings increase in comparison to room temperature but is lower than Ni-B coatings. At 300°C, the best tribological behavior is exhibited by Ni-B coatings while at 500°C, clearly Ni-B-W and Ni-B-Mo coatings exhibit the best friction and wear characteristics. At high temperatures, such enhanced tribological behavior is observed for Ni-B-Mo coatings due to self lubricious and protective tribo-patches of oxides of nickel and molybdenum. Formation of a smooth surface and a tribo-oxidative patch for Ni-B-W and Ni-B-Mo coatings at 500°C test temperature could be clearly observed through SEM and the micrograph has been shown in Fig. 12. The occurrence of phase transformation and microstructural changes due to in-situ heat treatment effect during sliding wear at elevated temperatures have been clearly observed through XRD analysis of the worn specimens.

Heat treatment leads to a noteworthy improvement in friction and wear characteristics of Ni-B characteristics as discussed earlier. The effect of heat treatment on the tribological characteristics at elevated temperatures has been investigated by Mukhopadhyay et al. (2017d). Electroless Ni-B coatings were heat treated at 350, 400 and 450°C for 1 hour in a muffle furnace and subjected to high temperature tribological tests at 100, 300 and 500°C. It has been observed that enhanced tribological characteristic at all the operating temperatures is shown by Ni-B coatings heat treated at 350°C. This has been attributed to the achievement of optimum nickel particle size and complete crystallization of the coatings. Whereas, an over-aging phenomenon occurs for the coatings heat treated at 400/450°C. At a high temperature of 500°C, the property of coatings heat treated at 400/450°C degrade severely and the worn surfaces are characterized by delamination of coatings due to formation of cracks. Due to the repeated nature of loading, these surface and sub-surface cracks propagate and fatigue failure has been observed by Mukhopadhyay et al. (2017d).

Figure 12. SEM micrograph of worn specimen of as-deposited (a) Ni-B-W and (b) Ni-B-Mo coating at 500°C operating temperature

The corresponding SEM micrographs of coatings heat treated at 400 and 450°C and tested at 500°C operating temperature are shown in Fig. 13.

In comparison with Ni-B coating, the tribological behavior is significantly improved for Ni-B-Mo coatings when heat treated at 450°C for 1 hour at 100, 300 and 500°C operating temperature (Mukhopadhyay et al. 2017c). The wear morphology of Ni-B-Mo coatings heat treated at 450°C indicates a smooth surface with formation of oxides of nickel and molybdenum. Moreover, the surface also reveals formation of flattened debris patches which act as load bearing areas. Consequently the friction and wear characteristics are improved. The SEM micrograph of Ni-B-Mo coating heat treated at 450°C and subjected to wear tests at 500°C is shown in Fig. 14 along with the debris particles formed. Another important aspect observed for Ni-B-Mo coatings is their high thermal stability at elevated temperatures. XRD results of worn specimen of as-deposited Ni-B-Mo coating reveal the sharpening of the peak of nickel with an increase in operating temperature indicating possible microstructural changes.

APPLICATIONS AT AMBIENT AND HIGH TEMPERATURES

Due to the high hardness, wear resistance, low COF and corrosion resistance EN plating has gained immense importance. Ability to coat over a variety of substrates and intricate parts uniformly are added advantages. Due to their uniform deposition, EN finds usage in valves, bores, internal surfaces, deep recesses, blind holes and threaded parts. Some typical applications of EN include aerospace hardware, cylinder liners, automotive parts, chemical and textiles industry, turbine blades working in extreme conditions, fluid power components, printed circuit boards, etc. (Loto, 2016). Petrochemical applications of EN include rod pumps, couplings,

Figure 13. SEM micrograph of worn specimen of Ni-B coating heat treated at (a) 400°C and (b) 450°C tested at 500°C operating temperature

Figure 14. SEM micrograph of (a) worn specimen and (b) wear debris of Ni-B-Mo coating heat treated at 450°C and subjected to 500°C operating temperature

mud pumps, extruder and blenders, fire tubes, tanks and vessels (Loto, 2016). Small arm weapons used in dusty regions are also protected by EN coatings. Steel compressors used in chlor-alkali plant are effectively protected from wear and corrosion by Ni-P coatings containing high percentage of phosphorus. Due to high corrosion resistance and excellent tribological behavior of Ni-P coatings in brine environment, they are widely used for transportation of brine in chlor-alkali plants. Other applications in the chemical process industry include nuts and bolts, cooling tower water pumps, components of rail car carrying sodium hydroxide such as safety vents, valve assemblies and cover plate, flow meters for handling aqueous solutions, etc. High phosphorus EN is generally found to be suitable for prevention of stress corrosion cracking and reducing corrosion rate. Applications in food industry include hydraulics and gears in meat processing, grain processing, breweries and poultry processing machines. Likewise, due to excellent properties of EN, they find wide usage in automotive and aerospace industries such as pistons, carburetor parts, cylinder liners, fuel injection systems, cast iron slip yokes used in vehicle power trains, engine valves, gas turbine vanes, high pressure compressor spacers, etc. (Parkinson, 1997). The enormous research carried out as discussed previously therefore suffice and widen the usage of EN coatings further in the fast growing industrial scenario.

EN coatings have high oxidation resistance up to 500°C and nickel has a high melting point due to which the high temperature tribological property is receiving attention recently. At very high temperatures most conventional lubricants lose their properties leading to the deposition and development of multi-layer and nano-composite coatings which can cater into the demands of such applications. EN coatings due to their high oxidation resistance, hardness, wear resistance, low COF and high thermal stability of some of its alloys may prove to be suitable candidates for tribological contact applications at high temperature and demanding conditions. Some typical applications that involve tribological contact at elevated temperature are crushers, hammer bars, nuclear valves sealing surface, hot metal working and extrusion tools and dies, rolls for rolling mills, machine tools, etc. Greaseless guns and small arm weapons have been found to be effectively protected by electroless Ni-B coating produced by UltraCem process (Riddle & Bailer, 2005). The barrel bores of such weapons are exposed to some of the most severe environment when controlled explosion occurs as a round is fired. Chromium coatings were found to be severely cracked due to high temperature fatigue in the barrel bores. Ni-B coatings have proven to be effective for such lubrication free or 'greaseless' guns. The hot drawing of sheets in automotive industries is quite common for manufacturing bumper beams, hood reinforcements, rocker panels or door beams. The applicability of EN coating to the tools may be studied to enhance their working life. High strength steels are used in automotive industries as structural reinforcements and they are generally formed at high temperatures. The wear and oxidation of the tool may be improved by the use of EN coatings. Lifetime and efficiency of highly loaded turbine vanes used in aerospace industries and gas turbines are reduced due to the severe operating environment. The applicability of EN coatings may be investigated in those sectors.

CONCLUSION

The present chapter discusses the deposition and tribological characterization of EN coatings. The evolution of the deposits over time into a large family of functional coatings i.e. various alloys, poly-alloys and composites is evident from the discussion. EN coatings may be tailor made to suit any particular application and there is a wide variety now available to cater into the demanding needs of industries. The coatings have received significant attention from researchers with the quest still on to further develop the coating properties, deposition on a wide variety of substrates and deposition of newer composite variants with enhanced properties. Most importantly, the EN variants are environment friendly and hence are a suitable substitute to chromium. EN coatings have already proven their excellent corrosion resistance and tribological properties at ambient temperature condition though still there is scope for further improvement and research is ongoing as evident from the review. Recent investigations on EN coatings are trying to assess and improve the thermal stability and high temperature tribological properties especially for the borohydride reduced variants. Investigations clearly indicate that there is still large scope for the development of the coatings so that they may cater into the high temperature domain. The coating properties may be further improved by tailoring the EN bath components. The effect of different constituents of the EN bath on high temperature stability and tribological properties may be an interesting area of research. More emphasis needs to be laid on improving the mechanical properties of EN alloy matrix irrespective of the co-deposited composite particles.

ACKNOWLEDGMENT

The authors gratefully acknowledge the support of DST, PURSE – II and COE, TEQIP – II program of Jadavpur University, Kolkata, India – 700032.

REFERENCES

Agarwala, R. C., & Agarwala, V. (2003). Electroless alloy/composite coatings: A review. *Sadhana*, *28*(3-4), 475–493. doi:10.1007/BF02706445

Alirezaei, S., Monirvaghefi, S. M., Saatchi, A., Ürgen, M., & Motallebzadeh, A. (2013a). High temperature tribology of nanocrystalline Ni–P–Ag coating. *Surface Engineering*, *29*(4), 306–311. doi:10.1179/1743294412Y.0000000085

Alirezaei, S., Monirvaghefi, S. M., Saatchi, A., Ürgen, M., & Motallebzadeh, A. (2013b). High temperature friction and wear behaviour of Ni–P–Ag–Al2O3 hybrid nanocomposite coating. *Transactions of the IMF*, *91*(4), 207-213.

Anik, M., Körpe, E., & Şen, E. (2008). Effect of coating bath composition on the properties of electroless nickel–boron films. *Surface and Coatings Technology*, *202*(9), 1718–1727. doi:10.1016/j.surfcoat.2007.07.031

Ansari, M. I., & Thakur, D. G. (2017). A high performance of ENi–P coatings on mechanical and tribological properties with influence of bath pH on AZ91 magnesium alloy. *Journal of Failure Analysis and Prevention*, *17*(1), 100–106. doi:10.1007/s11668-016-0210-2

Anvari, S. R., Monirvaghefi, S. M., & Enayati, M. H. (2015). Wear characteristics of functionally graded nanocrystalline Ni–P coatings. *Surface Engineering*, *31*(9), 693–700. doi:10.1179/1743294415Y.0000000023

Aydeniz, A. I., Göksenli, A., Dil, G., Muhaffel, F., Calli, C., & Yüksel, B. (2013). Electroless Ni-B-W coatings for improving hardness, wear and corrosion resistance. *Materiali in Tehnologije*, *47*(6), 803–806.

Balaraju, J. N., Jahan, S. M., Jain, A., & Rajam, K. S. (2007). Structure and phase transformation behavior of electroless Ni–P alloys containing tin and tungsten. *Journal of Alloys and Compounds*, *436*(1), 319–327. doi:10.1016/j.jallcom.2006.07.045

Balaraju, J. N., Manikandanath, N. T., & Grips, V. W. (2012). Phase transformation behavior of nanocrystalline Ni–W–P alloys containing various W and P contents. *Surface and Coatings Technology*, *206*(10), 2682–2689. doi:10.1016/j.surfcoat.2011.11.024

Balaraju, J. N., Priyadarshi, A., Kumar, V., Manikandanath, N. T., Kumar, P. P., & Ravisankar, B. (2016). Hardness and wear behaviour of electroless Ni–B coatings. *Materials Science and Technology*, *32*(16), 1654–1665. doi:10.1080/02670836.2015.1137683

Balaraju, J. N., & Rajam, K. S. (2005). Electroless deposition of Ni–Cu–P, Ni–W–P and Ni–W–Cu–P alloys. *Surface and Coatings Technology*, *195*(2), 154–161. doi:10.1016/j.surfcoat.2004.07.068

Balaraju, J. N., & Rajam, K. S. (2009). Surface morphology and structure of electroless ternary NiWP deposits with various W and P contents. *Journal of Alloys and Compounds*, *486*(1), 468–473. doi:10.1016/j.jallcom.2009.06.173

Balaraju, J. N., Raman, N., & Manikandanath, N. T. (2014). Nanocrystalline electroless nickel poly-alloy deposition: incorporation of W and Mo. *Transactions of the IMF*, *92*(3), 169-176. doi:10.1179/0020296713Z.000000000123

Bonin, L., & Vitry, V. (2016). Mechanical and wear characterization of electroless nickel mono and bilayers and high boron-mid phosphorus electroless nickel duplex coatings. *Surface and Coatings Technology*, *307*, 957–962. doi:10.1016/j.surfcoat.2016.10.021

Brenner, A., & Riddell, G. E. (1946). Nickel plating on steel by chemical reduction. *Journal of Research of the National Bureau of Standards*, *37*(1), 31–34. doi:10.6028/jres.037.019

Bülbül, F., Altun, H., Ezirmik, V., & Küçük, Ö. (2013). Investigation of structural, tribological and corrosion properties of electroless Ni–B coating deposited on 316L stainless steel. *Proceedings of the Institution of Mechanical Engineers. Part J, Journal of Engineering Tribology*, *227*(6), 629–639. doi:10.1177/1350650112464928

Çelik, İ., Karakan, M., & Bülbül, F. (2016). Investigation of structural and tribological properties of electroless Ni–B coated pure titanium. *Proceedings of the Institution of Mechanical Engineers. Part J, Journal of Engineering Tribology*, *230*(1), 57–63. doi:10.1177/1350650115588568

Chen, W. X., Tu, J. P., Gan, H. Y., Xu, Z. D., Wang, Q. G., Lee, J. Y., ... Zhang, X. B. (2002). Electroless preparation and tribological properties of Ni-P-Carbon nanotube composite coatings under lubricated condition. *Surface and Coatings Technology*, *160*(1), 68–73. doi:10.1016/S0257-8972(02)00408-5

Correa, E., Zuleta, A. A., Guerra, L., Gómez, M. A., Castaño, J. G., Echeverría, F., ... Thompson, G. E. (2013). Tribological behavior of electroless Ni–B coatings on magnesium and AZ91D alloy. *Wear*, *305*(1), 115–123. doi:10.1016/j.wear.2013.06.004

Das, S. K., & Sahoo, P. (2011). Tribological characteristics of electroless Ni–B coating and optimization of coating parameters using Taguchi based grey relational analysis. *Materials & Design*, *32*(4), 2228–2238. doi:10.1016/j.matdes.2010.11.028

Duari, S., Mukhopadhyay, A., Barman, T. K., & Sahoo, P. (2016a). Optimization of wear performance of electroless Ni-B coating under lubrication. *International Journal of Engineering and Technologies*, *7*, 94–103. doi:10.18052/www.scipress.com/IJET.7.94

Duari, S., Mukhopadhyay, A., Barman, T. K., & Sahoo, P. (2016b). Investigation of friction and wear properties of electroless Ni–P–Cu coating under dry condition. *Journal of Molecular and Engineering Materials*, *04*(04), 1640013. doi:10.1142/S225123731640013X

Duari, S., Mukhopadhyay, A., Barman, T. K., & Sahoo, P. (2017). Study of wear and friction of chemically deposited Ni-P-W coating under dry and lubricated condition. *Surfaces and Interfaces*, *6*, 177–189. doi:10.1016/j.surfin.2017.01.009

Franco, M., Sha, W., Aldic, G., Malinov, S., & Çimenoğlu, H. (2016). Effect of reinforcement and heat treatment on elevated temperature sliding of electroless Ni–P/SiC composite coatings. *Tribology International*, *97*, 265–271. doi:10.1016/j.triboint.2016.01.047

Gadhari, P., & Sahoo, P. (2016). Electroless nickel-phosphorus composite coatings: A Review. *International Journal of Manufacturing, Materials, and Mechanical Engineering*, *6*(1), 14–50. doi:10.4018/IJMMME.2016010102

Gawne, D. T., & Ma, U. (1987). Wear mechanisms in electroless nickel coatings. *Wear*, *120*(2), 125–149. doi:10.1016/0043-1648(87)90063-9

Ger, M. D., Hou, K. H., Wang, L. M., & Hwang, B. J. (2002). The friction and wear of Ni–P–PTFE composite deposits under water lubrication. *Materials Chemistry and Physics, 77*(3), 755–764. doi:10.1016/S0254-0584(02)00153-0

Ghaderi, M., Rezagholizadeh, M., Heidary, A., & Monirvaghefi, S. M. (2016b). The effect of Al_2O_3 nanoparticles on tribological and corrosion behavior of electroless Ni–B–Al_2O_3 composite coating. *Protection of Metals and Physical Chemistry of Surfaces, 52*(5), 854–858. doi:10.1134/S2070205116050087

Ghaderi, M., Rezagholizadeh, M., Vaghefi, S. M., & Heidary, A. (2016a). Investigation of high temperature wear resistance of electroless nickel coating with different contents of phosphorous. *Protection of Metals and Physical Chemistry of Surfaces, 52*(3), 538–542. doi:10.1134/S2070205116030126

Hadipour, A., Bahrololoom, M. E., Monirvaghefi, S. M., & Bahramkia, A. R. (2016). Evaluation of wear properties for Ni–P electroless coatings. *Transactions of the Indian Institute of Metals, 69*(9), 1733–1743. doi:10.1007/s12666-016-0833-8

Jagatheeshwaran, M. S., Elayaperumal, A., & Arulvel, S. (2015). Wear characteristics of electroless NiP/bio-composite coatings on En8 steel. *Journal of Manufacturing Processes, 20*, 206–214. doi:10.1016/j.jmapro.2015.08.002

Jagatheeshwaran, M. S., Elayaperumal, A., & Arulvel, S. (2016). The role of calcinated sea shell particles on friction-wear behavior of electroless NiP coating: Fabrication and characterization. *Surface and Coatings Technology, 304*, 492–501. doi:10.1016/j.surfcoat.2016.07.053

Krishnaveni, K., Narayanan, T. S., & Seshadri, S. K. (2005). Electroless Ni–B coatings: Preparation and evaluation of hardness and wear resistance. *Surface and Coatings Technology, 190*(1), 115–121. doi:10.1016/j.surfcoat.2004.01.038

Kundu, S., Das, S. K., & Sahoo, P. (2016). Tribological behaviour of electroless Ni-P deposits under elevated temperature. Silicon. doi:10.1007/s12633-016-9450-8

Lee, C. K. (2008). Corrosion and wear-corrosion resistance properties of electroless Ni–P coatings on GFRP composite in wind turbine blades. *Surface and Coatings Technology, 202*(19), 4868–4874. doi:10.1016/j.surfcoat.2008.04.079

Lee, C. K. (2009). Structure, electrochemical and wear-corrosion properties of electroless nickel–phosphorus deposition on CFRP composites. *Materials Chemistry and Physics, 114*(1), 125–133. doi:10.1016/j.matchemphys.2008.08.088

León, O. A., Staia, M. H., & Hintermann, H. E. (2003). High temperature wear of an electroless Ni–P–BN (h) composite coating. *Surface and Coatings Technology*, *163*, 578–584. doi:10.1016/S0257-8972(02)00663-1

Li, Z., Wang, J., Lu, J., & Meng, J. (2013). Tribological characteristics of electroless Ni–P–MoS_2 composite coatings at elevated temperatures. *Applied Surface Science*, *264*, 516–521. doi:10.1016/j.apsusc.2012.10.055

Liu, H., Guo, R. X., & Liu, Z. (2012). Characteristics of microstructure and performance of laser-treated electroless Ni–P/Ni–W–P duplex coatings. *Transactions of Nonferrous Metals Society of China*, *22*(12), 3012–3020. doi:10.1016/S1003-6326(11)61564-1

Liu, H., Lv, Y. Y., Liu, Z., & Thompson, G. E. (2016). Dry sliding wear behaviour and structural characteristics of laser-annealed electroless Ni–P/Ni–Mo–P duplex coatings. *Tribology International*, *103*, 343–351. doi:10.1016/j.triboint.2016.07.014

Liu, J., Wang, X., Tian, Z., Yuan, M., & Ma, X. (2015). Effect of copper content on the properties of electroless Ni–Cu–P coatings prepared on magnesium alloys. *Applied Surface Science*, *356*, 289–293. doi:10.1016/j.apsusc.2015.08.072

Loto, C. A. (2016). Electroless nickel plating–A review. *Silicon*, *8*(2), 177–186. doi:10.1007/s12633-015-9367-7

Madah, F., Amadeh, A. A., & Dehghanian, C. (2016). Investigation on the phase transformation of electroless Ni–B coating after dry sliding against alumina ball. *Journal of Alloys and Compounds*, *658*, 272–279. doi:10.1016/j.jallcom.2015.10.156

Madah, F., Dehghanian, C., & Amadeh, A. A. (2015). Investigations on the wear mechanisms of electroless Ni–B coating during dry sliding and endurance life of the worn surfaces. *Surface and Coatings Technology*, *282*, 6–15. doi:10.1016/j.surfcoat.2015.09.003

Masoumi, F., Ghasemi, H. R., Ziaei, A. A., & Shahriari, D. (2012). Tribological characterization of electroless Ni–10% P coatings at elevated test temperature under dry conditions. *The International Journal of Advanced Manufacturing Technology*, *62*(9-12), 1063–1070. doi:10.1007/s00170-011-3832-2

Mukhopadhyay, A., Barman, T. K., & Sahoo, P. (2016g). Effect of operating temperature on tribological behavior of as-plated Ni-B coating deposited by electroless method. *Tribology Transactions*, 1–12. doi:10.1080/10402004.2016.1271929

Mukhopadhyay, A., Barman, T. K., & Sahoo, P. (2017b). Tribological behavior of sodium borohydride reduced electroless nickel alloy coatings at room and elevated temperatures. *Surface and Coatings Technology*, *321*, 464–476. doi:10.1016/j.surfcoat.2017.05.015

Mukhopadhyay, A., Barman, T. K., & Sahoo, P. (2017c). Effect of heat treatment on tribological behavior of electroless Ni-B-Mo coatings at different operating temperatures. Silicon. doi:10.1007/s12633-017-9594-1

Mukhopadhyay, A., Barman, T. K., & Sahoo, P. (2017d). Effects of heat treatment on tribological behavior of electroless Ni–B coating at elevated temperatures. *Surface Review and Letters*, 24(Supp01), 1850014. doi:10.1142/S0218625X18500142

Mukhopadhyay, A., Duari, S., Barman, T. K., & Sahoo, P. (2016a). Application of grey fuzzy logic for the optimization of tribological performance of electroless Ni-B coating. *Journal of Manufacturing Technology Research*, *7*(1/2), 1–24.

Mukhopadhyay, A., Duari, S., Barman, T. K., & Sahoo, P. (2016b). Wear analysis of electroless Ni-P coating under lubricated condition using fuzzy logic. *Portugaliae Electrochimica Acta*, *34*(1), 63–82. doi:10.4152/pea.201601063

Mukhopadhyay, A., Duari, S., Barman, T. K., & Sahoo, P. (2016c). Optimization of friction and wear properties of electroless Ni–P coatings under lubrication using grey fuzzy logic. *Journal of the Institution of Engineers*, *98*(2), 255–268. doi:10.1007/s40033-016-0133-9

Mukhopadhyay, A., Duari, S., Barman, T. K., & Sahoo, P. (2016d). Tribological performance optimization of electroless Ni–B coating under lubricated condition using hybrid grey fuzzy logic. *Journal of the Institution of Engineers (India): Series D*, *97*(2), 215-231.

Mukhopadhyay, A., Duari, S., Barman, T. K., & Sahoo, P. (2016e). Investigation of Wear Behavior of Electroless Ni-P-W Coating under Dry and Lubricated Conditions Using RSM and Fuzzy Logic. *Portugaliae Electrochimica Acta*, *34*(4), 231–255. doi:10.4152/pea.201604231

Mukhopadhyay, A., Duari, S., Barman, T. K., & Sahoo, P. (2016f). Optimization of wear behavior of electroless Ni-P-W coating under dry and lubricated conditions using genetic algorithm (GA). *Jurnal Tribologi*, *11*, 24–48.

Mukhopadhyay, A., Duari, S., Barman, T. K., & Sahoo, P. (2017a). Tribology of electroless Ni-P coating under lubricated condition: An RSM and GA approach. *International Journal of Surface Engineering and Interdisciplinary Materials Science*, *5*(1), 37–57. doi:10.4018/IJSEIMS.2017010103

Novak, M., Vojtěch, D., Novak, P., & Vítů, T. (2011). Tribological properties of heat-treated electroless Ni–P coatings on AZ91 alloy. *Applied Surface Science*, *257*(23), 9982–9985. doi:10.1016/j.apsusc.2011.06.119

Palaniappa, M., & Seshadri, S. K. (2008). Friction and wear behavior of electroless Ni–P and Ni–W–P alloy coatings. *Wear*, *265*(5), 735–740. doi:10.1016/j.wear.2008.01.002

Panja, B., Das, S. K., & Sahoo, P. (2015). Tribological behaviour of electroless Ni–P coatings in brine solution and optimization of coating parameters using Taguchi based grey relation analysis. *Journal of the Institution of Engineers (India): Series C, 96*(3), 299-309.

Panja, B., Das, S. K., & Sahoo, P. (2016). Tribological behavior of electroless Ni–P coatings in various corrosive environments. *Surface Review and Letters, 23*(5) 1650040. doi: 10.1142/S0218625X16500402

Panja, B., & Sahoo, P. (2014). Taguchi-grey based optimization of friction and wear of electroless Ni-P coatings in acidic environment. *International Journal of Surface Engineering and Interdisciplinary Materials Science*, *2*(2), 53–69. doi:10.4018/IJSEIMS.2014070104

Panja, B., & Sahoo, P. (2015). Tribological behavior of electroless Ni-P coatings in alkaline environment and optimization of coating parameters using Taguchi based grey relational analysis. *Indian Journal of Engineering and Materials Sciences*, *22*(5), 503–512.

Parkinson, R. (1997). Properties and applications of electroless nickel. *Nickel Development Institute Technical Series*, (10081), 1-37.

Peng, Y., Xu, Y., Geng, J., Dearn, K. D., & Hu, X. (2017). Tribological assessment of coated piston ring-cylinder liner contacts under bio-oil lubricated conditions. *Tribology International*, *107*, 283–293. doi:10.1016/j.triboint.2016.12.004

Popoola, A. P. I., Loto, C. A., Osifuye, C. O., Aigbodion, V. S., & Popoola, O. M. (2016). Corrosion and wear properties of Ni-Sn-P ternary deposits on mild steel via electroless method. *Alexandria Engineering Journal*, *55*(3), 2901–2908. doi:10.1016/j.aej.2016.06.018

Ramalho, A., & Miranda, J. C. (2007). Tribological characterization of electroless NiP coatings lubricated with biolubricants. *Wear*, *263*(1), 592–597. doi:10.1016/j.wear.2006.10.027

Ranganatha, S., Venkatesha, T. V., & Vathsala, K. (2012). Development of high performance electroless Ni–P–HNT composite coatings. *Applied Surface Science*, *263*, 149–156. doi:10.1016/j.apsusc.2012.09.020

Riddle, Y. W., & Bailer, T. O. (2005). Friction and wear reduction via an Ni-B electroless bath coating for metal alloys. *JOM*, *57*(4), 40–45. doi:10.1007/s11837-005-0080-7

Sahoo, P. (2009). Wear behaviour of electroless Ni–P coatings and optimization of process parameters using Taguchi method. *Materials & Design*, *30*(4), 1341–1349. doi:10.1016/j.matdes.2008.06.031

Sahoo, P., & Das, S. K. (2011). Tribology of electroless nickel coatings–a review. *Materials & Design*, *32*(4), 1760–1775. doi:10.1016/j.matdes.2010.11.013

Sahoo, P., & Roy, S. (2017). Tribological behavior of electroless Ni-P, Ni-P-W and Ni-P-Cu coatings: A comparison. *International Journal of Surface Engineering and Interdisciplinary Materials Science*, *5*(1), 1–15. doi:10.4018/IJSEIMS.2017010101

Sankara Narayanan, T. S. N., Krishnaveni, K., & Seshadri, S. K. (2003). Electroless Ni–P/Ni–B duplex coatings: Preparation and evaluation of microhardness, wear and corrosion resistance. *Materials Chemistry and Physics*, *82*(3), 771–779. doi:10.1016/S0254-0584(03)00390-0

Serin, I. G., & Göksenli, A. (2015). Effect of annealing temperature on hardness and wear resistance of electroless Ni–B–Mo coatings. *Surface Review and Letters*, *22*(5) 1550058. doi: 10.1142/S0218625X15500584

Staia, M. H., Castillo, E. J., Puchi, E. S., Lewis, B., & Hintermann, H. E. (1996). Wear performance and mechanism of electroless Ni-P coating. *Surface and Coatings Technology*, *86*, 598–602. doi:10.1016/S0257-8972(96)03086-1

Staia, M. H., Conzono, A., Cruz, M. R., Roman, A., Lesage, J., Chicot, D., & Mesmacque, G. (2002). Wear behaviour of silicon carbide/electroless nickel composite coatings at high temperature. *Surface Engineering*, *18*(4), 265–269. doi:10.1179/026708401225005359

Sudagar, J., Lian, J., & Sha, W. (2013). Electroless nickel, alloy, composite and nano coatings–A critical review. *Journal of Alloys and Compounds*, *571*, 183–204. doi:10.1016/j.jallcom.2013.03.107

Taheri, R., Oguocha, I. N. A., & Yannacopoulos, S. (2001). The tribological characteristics of electroless NiP coatings. *Wear, 249*(5), 389–396. doi:10.1016/S0043-1648(01)00539-7

Vargas Mendoza, L., Barba, A., Bolarin, A., & Sanchez, F. (2006). Age hardening of Ni–P–Mo electroless deposit. *Surface Engineering, 22*(1), 58–62. doi:10.1179/174329406X84976

Vitry, V., & Bonin, L. (2017a). Formation and characterization of multilayers borohydride and hypophosphite reduced electroless nickel deposits. *Electrochimica Acta, 243*, 7–17. doi:10.1016/j.electacta.2017.04.152

Vitry, V., & Bonin, L. (2017b). Increase of boron content in electroless nickel-boron coating by modification of plating conditions. *Surface and Coatings Technology, 311*, 164–171. doi:10.1016/j.surfcoat.2017.01.009

Vitry, V., Kanta, A. F., & Delaunois, F. (2010). Initiation and formation of electroless nickel–boron coatings on mild steel: Effect of substrate roughness. *Materials Science and Engineering B, 175*(3), 266–273. doi:10.1016/j.mseb.2010.08.003

Vitry, V., Kanta, A. F., & Delaunois, F. (2011). Mechanical and wear characterization of electroless nickel-boron coatings. *Surface and Coatings Technology, 206*(7), 1879–1885. doi:10.1016/j.surfcoat.2011.08.008

Vitry, V., Kanta, A. F., Dille, J., & Delaunois, F. (2012a). Structural state of electroless nickel–boron deposits (5wt.% B): Characterization by XRD and TEM. *Surface and Coatings Technology, 206*(16), 3444–3449. doi:10.1016/j.surfcoat.2012.02.003

Vitry, V., Sens, A., Kanta, A. F., & Delaunois, F. (2012b). Experimental study on the formation and growth of electroless nickel–boron coatings from borohydride-reduced bath on mild steel. *Applied Surface Science, 263*, 640–647. doi:10.1016/j.apsusc.2012.09.126

Vitry, V., Sens, A., Kanta, A. F., & Delaunois, F. (2012c). Wear and corrosion resistance of heat treated and as-plated Duplex NiP/NiB coatings on 2024 aluminum alloys. *Surface and Coatings Technology, 206*(16), 3421–3427. doi:10.1016/j.surfcoat.2012.01.049

Wan, Y., Yu, Y., Cao, L., Zhang, M., Gao, J., & Qi, C. (2016). Corrosion and tribological performance of PTFE-coated electroless nickel boron coatings. *Surface and Coatings Technology, 307*, 316–323. doi:10.1016/j.surfcoat.2016.09.001

Wang, J., Yan, F., & Xue, Q. (2009). Friction and wear behavior of ultra-high molecular weight polyethylene sliding against GCr15 steel and electroless Ni–P alloy coating under the lubrication of seawater. *Tribology Letters*, *35*(2), 85–95. doi:10.1007/s11249-009-9435-5

Xu, Y., Zheng, X., Hu, X., Yin, Y., & Lei, T. (2014). Preparation of the electroless Ni–P and Ni–Cu–P coatings on engine cylinder and their tribological behaviors under bio-oil lubricated conditions. *Surface and Coatings Technology*, *258*, 790–796. doi:10.1016/j.surfcoat.2014.07.079

Yan, M., Ying, H. G., & Ma, T. Y. (2008). Improved microhardness and wear resistance of the as-deposited electroless Ni–P coating. *Surface and Coatings Technology*, *202*(24), 5909–5913. doi:10.1016/j.surfcoat.2008.06.180

Yu, H. S., Luo, S. F., & Wang, Y. R. (2001). A comparative study on the crystallization behavior of electroless Ni-P and Ni-Cu-P deposits. *Surface and Coatings Technology*, *148*(2), 143–148. doi:10.1016/S0257-8972(01)01345-7

Compilation of References

Abdel-Azim, A. N., Shash, Y., Mostafa, S. F., & Younan, A. (1995). Casting of 2024-A1 alloy reinforced with A12O3 particles. *Journal of Materials Processing Technology*, *55*(3-4), 199–205. doi:10.1016/0924-0136(95)01954-5

Abdelmouleh, M., Boufi, S., Belgacem, M. N., Duarte, A. P., Salah, A. B., & Gandini, A. (2004). Modification of cellulosic fibres with functionalised silanes: Development of surface properties. *International Journal of Adhesion and Adhesives*, *24*(1), 43–54. doi:10.1016/S0143-7496(03)00099-X

Abdelmouleh, M., Boufi, S., Belgacem, M. N., & Dufresne, A. (2007). Short natural-fibre reinforced polyethylene and natural rubber composites: Effect of silane coupling agents and fibres loading. *Composites Science and Technology*, *67*(7), 1627–1639. doi:10.1016/j.compscitech.2006.07.003

Abu Talib, A. R., Ali, A., Badie, M. A., Azida Che Lah, N., & Golestaneh, A. F. (2010). Developing a hybrid, carbon/glass fiber-reinforced, epoxy composite automotive drive shaft. *Materials & Design*, *31*(1), 514–521. doi:10.1016/j.matdes.2009.06.015

Agari, Y., Ueda, A., Omura, Y., & Nagai, S. (1997). Thermal diffusivity and conductivity of PMMA–PC blends. *Polymer*, *38*(4), 801–807. doi:10.1016/S0032-3861(96)00577-0

Agari, Y., & Uno, T. (1986). Estimation on Thermal Conductivities of Filled Polymers. *Journal of Applied Polymer Science*, *32*(7), 5705–5712. doi:10.1002/app.1986.070320702

Agarwala, R. C., & Agarwala, V. (2003). Electroless alloy/composite coatings: A review. *Sadhana*, *28*(3-4), 475–493. doi:10.1007/BF02706445

Akhshik, M., Panthapulakkal, S., Tjong, J., & Sain, M. (2017). Life cycle assessment and cost analysis of hybrid fiber-reinforced engine beauty cover in comparison with glass fiber-reinforced counterpart. *Environmental Impact Assessment Review*, *65*, 111–117. doi:10.1016/j.eiar.2017.04.005

Alajmi, M., & Shalwan, A. (2015). Correlation between Mechanical Properties with Specific Wear Rate and the Coefficient of Friction of Graphite/Epoxy Composites. *Materials (Basel)*, *8*(7), 4162–4175. doi:10.3390/ma8074162 PMID:28793431

Alavudeen, A., Rajini, N., Karthikeyan, S., Thiruchitrambalam, M., & Venkateshwaren, N. (2015). Mechanical properties of banana/kenaf fiber-reinforced hybrid polyester composites: Effect of woven fabric and random orientation. *Materials & Design (1980-2015), 66*, 246-257.

Alirezaei, S., Monirvaghefi, S. M., Saatchi, A., Ürgen, M., & Motallebzadeh, A. (2013a). High temperature tribology of nanocrystalline Ni–P–Ag coating. *Surface Engineering, 29*(4), 306–311. doi:10.1179/1743294412Y.0000000085

Alirezaei, S., Monirvaghefi, S. M., Saatchi, A., Ürgen, M., & Motallebzadeh, A. (2013b). High temperature friction and wear behaviour of Ni–P–Ag–Al2O3 hybrid nanocomposite coating. *Transactions of the IMF, 91*(4), 207-213.

AL-Oqla, F. M., Sapuan, S. M., Ishak, M. R., & Nuraini, A. A. (2016). A decision-making model for selecting the most appropriate natural fiber – Polypropylene-based composites for automotive applications. *Journal of Composite Materials, 50*(4), 543–556. doi:10.1177/0021998315577233

Al-Qureshi, H. A. (2001). Automobile leaf springs from composite materials. *Journal of Materials Processing Technology, 118*(1–3), 58–61. doi:10.1016/S0924-0136(01)00863-9

Ambare, R., & Mishra, H. (2016). Design and performance testing of master leaf for Mahindra Pick-up. *International Research Journal of Engineering and Technology, 3*(7), 1140–1143.

Anas, M., Han, D. S., Mahmoud, K., Park, H., & Abdel-Wahab, A. (2016). Photocatalytic degradation of organic dye using titanium dioxide modified with metal and non-metal deposition. *Materials Science in Semiconductor Processing, 41*, 209–218. doi:10.1016/j.mssp.2015.08.041

Anderson, C., & Bard, A. J. (1995). An improved photocatalyst of TiO_2/SiO_2 prepared by a sol-gel synthesis. *Journal of Physical Chemistry, 99*(24), 9882–9885. doi:10.1021/j100024a033

Anik, M., Körpe, E., & Şen, E. (2008). Effect of coating bath composition on the properties of electroless nickel–boron films. *Surface and Coatings Technology, 202*(9), 1718–1727. doi:10.1016/j.surfcoat.2007.07.031

Anpo, M., & Takeuchi, M. (2003). The design and development of highly reactive titanium oxide photocatalysts operating under visible light irradiation. *Journal of Catalysis, 216*(1), 505–516. doi:10.1016/S0021-9517(02)00104-5

Ansari, M. I., & Thakur, D. G. (2017). A high performance of ENi–P coatings on mechanical and tribological properties with influence of bath pH on AZ91 magnesium alloy. *Journal of Failure Analysis and Prevention, 17*(1), 100–106. doi:10.1007/s11668-016-0210-2

Anvari, S. R., Monirvaghefi, S. M., & Enayati, M. H. (2015). Wear characteristics of functionally graded nanocrystalline Ni–P coatings. *Surface Engineering, 31*(9), 693–700. doi:10.1179/1743294415Y.0000000023

Arbelaiz, A., Fernández, B., Cantero, G., Llano-Ponte, R., Valea, A., & Mondragon, I. (2005). Mechanical properties of flax fibre/polypropylene composites. Influence of fibre/matrix modification and glass fibre hybridization. *Composites. Part A, Applied Science and Manufacturing, 36*(12), 1637–1644. doi:10.1016/j.compositesa.2005.03.021

Arsha, A. G., Jayakumar, E., Rajan, T. P. D., Antony, V., & Pai, B. C. (2015). Design and fabrication of functionally graded in-situ aluminium composites for automotive pistons. *Materials & Design*, *88*, 1201–1209. doi:10.1016/j.matdes.2015.09.099

Asim, M., Jawaid, M., Abdan, K., & Ishak, M. R. (2017). The Effect of Silane Treated Fibre Loading on Mechanical Properties of Pineapple Leaf/Kenaf Fibre Filler Phenolic Composites. *Journal of Polymers and the Environment*, 1–8.

Astrom, B. T. (1997). *Manufacturing of Polymer Composites*. CRC Press.

Athijayamani, A., Thiruchitrambalam, M., Natarajan, U., & Pazhanivel, B. (2010). Influence of alkali-treated fibers on the mechanical properties and machinability of roselle and sisal fiber hybrid polyester composite. *Polymer Composites*, *31*(4), 723–731.

Au, C., & Buyukozturk, O. (2005). Effect of fiber orientation and ply mix on fiber reinforced polymer-confined concrete. *Journal of Composites for Construction*, *9*(5), 397–407. doi:10.1061/(ASCE)1090-0268(2005)9:5(397)

Aydeniz, A. I., Göksenli, A., Dil, G., Muhaffel, F., Calli, C., & Yüksel, B. (2013). Electroless Ni-B-W coatings for improving hardness, wear and corrosion resistance. *Materiali in Tehnologije*, *47*(6), 803–806.

Ayrilmis, N., Jarusombuti, S., Fueangvivat, V., Bauchongkol, P., & White, R. H. (2011). Coir fiber reinforced polypropylene composite panel for automotive interior applications. *Fibers and Polymers*, *12*(7), 919–926. doi:10.1007/s12221-011-0919-1

Aysin, B., Ozturk, A., & Park, J. (2013). Silver-loaded TiO_2 powders prepared through mechanical ball milling. *Ceramics International*, *39*(6), 7119–7126. doi:10.1016/j.ceramint.2013.02.054

Babu, N. R., & Sangeethkumar, E. (2015). Improving the mechanical properties of the propeller shaft by using composite materials. *Applied Mechanics and Materials*, *813–814*, 1090–1097. doi:10.4028/www.scientific.net/AMM.813-814.1090

Badie, M. A., Mahdi, E., & Hamouda, A. M. S. (2011). An investigation into hybrid carbon/glass fiber reinforced epoxy composite automotive drive shaft. *Materials & Design*, *32*(3), 1485–1500. doi:10.1016/j.matdes.2010.08.042

Baetens, R., Jelle, B. P., & Gustavsen, A. (2011). Aerogel insulation for building applications: A state-of-the-art review. *Energy and Building*, *43*(4), 761–769. doi:10.1016/j.enbuild.2010.12.012

Bakis, C. E., Bank, L. C., Brown, V., Cosenza, E., Davalos, J. F., Lesko, J. J., ... Triantafillou, T. C. (2002). Fiber-reinforced polymer composites for construction—state of the art review. *Journal of Composites for Construction*, *6*(2), 73–87. doi:10.1061/(ASCE)1090-0268(2002)6:2(73)

Balaji Thattaiparthasarathy, K., Pillay, S., Ning, H., & Vaidya, U. K. (2008). Process simulation, design and manufacturing of a long fiber thermoplastic composite for mass transit application. *Composites. Part A, Applied Science and Manufacturing*, *39*(9), 1512–1521. doi:10.1016/j.compositesa.2008.05.017

Balaraju, J. N., Jahan, S. M., Jain, A., & Rajam, K. S. (2007). Structure and phase transformation behavior of electroless Ni–P alloys containing tin and tungsten. *Journal of Alloys and Compounds*, *436*(1), 319–327. doi:10.1016/j.jallcom.2006.07.045

Balaraju, J. N., Manikandanath, N. T., & Grips, V. W. (2012). Phase transformation behavior of nanocrystalline Ni–W–P alloys containing various W and P contents. *Surface and Coatings Technology*, *206*(10), 2682–2689. doi:10.1016/j.surfcoat.2011.11.024

Balaraju, J. N., Priyadarshi, A., Kumar, V., Manikandanath, N. T., Kumar, P. P., & Ravisankar, B. (2016). Hardness and wear behaviour of electroless Ni–B coatings. *Materials Science and Technology*, *32*(16), 1654–1665. doi:10.1080/02670836.2015.1137683

Balaraju, J. N., & Rajam, K. S. (2005). Electroless deposition of Ni–Cu–P, Ni–W–P and Ni–W–Cu–P alloys. *Surface and Coatings Technology*, *195*(2), 154–161. doi:10.1016/j.surfcoat.2004.07.068

Balaraju, J. N., & Rajam, K. S. (2009). Surface morphology and structure of electroless ternary NiWP deposits with various W and P contents. *Journal of Alloys and Compounds*, *486*(1), 468–473. doi:10.1016/j.jallcom.2009.06.173

Balaraju, J. N., Raman, N., & Manikandanath, N. T. (2014). Nanocrystalline electroless nickel poly-alloy deposition: incorporation of W and Mo. *Transactions of the IMF*, *92*(3), 169-176. do i:10.1179/0020296713Z.000000000123

Balcombe, R., Mackett, R., Paulley, N., Preston, J., Shires, J., Titheridge, H., … White, P. (2004). *The demand for public transport: a practical guide* (No. TRL593). Academic Press.

Banga, H., Singh, V. K., & Choudhary, S. K. (2015). *Fabrication and Study of Mechanical Properties of Bamboo Fibre Reinforced Bio-Composites*. Academic Press.

Bednarcyk, B. A. (2003). An inelastic micro/macro theory for hybrid smart/metal composites. *Composites. Part B, Engineering*, *34*(2), 175–197. doi:10.1016/S1359-8368(02)00067-7

Begum, S. N., & Murthy, S. B. (2016). Modelling and structural analysis of vehicle chassis frame made of polymeric composite material. *International Research Journal of Engineering and Technology*, *3*(8), 574–582.

Belingardi, G., Beyene, A. T., Koricho, E. G., & Martorana, B. (2015). Alternative lightweight materials and component manufacturing technologies for vehicle frontal bumper beam. *Composite Structures*, *120*, 483–495. doi:10.1016/j.compstruct.2014.10.007

Bell, S., & Arcuri, M. (2012). A speedboat with a vacuum-infused and bonded foam sandwich design. *JEC composites*, (75), 30-31.

Beyenea, A. T., Korichob, E. G., Belingardib, G., & Martoranac, B. (2014). Design and manufacturing issues in the development of lightweight solution for a vehicle frontal bumper. *Procedia Engineering*, *88*, 77–84. doi:10.1016/j.proeng.2014.11.129

Bhagat, V. K., Biswas, S., & Dehury, J. (2014). Physical, mechanical, and water absorption behavior of coir/glass fiber reinforced epoxy based hybrid composites. *Polymer Composites*, *35*(5), 925–930. doi:10.1002/pc.22736

Bhajantri, V. S., Bajantri, S. C., Shindolkar, A. M., & Amarapure, S. S. (2014). Design and Analysis of Composite Drive Shaft. *International Journal of Research in Engineering and Technology*, *3*(15), 738–745. doi:10.15623/ijret.2014.0315138

Bhaskar, J., & Singh, V. K. (2012). Physical and Mechanical Properties of Coconut Shell Particle Reinforced-Epoxy Composite. *Journal of Material Environmental Science, 4*(2), 227-232.

Bhaskar, J., & Singh, V. K. (2013). Water Absorption and Compressive Properties of Coconut Shell Particle Reinforced-Epoxy Composite. *Journal of Material Environmental Sciences (Ruse)*, *4*(1), 113–118.

Bielawski, R. (2014). Konstrukcje lotnictwa wojskowego. *Materiały Kompozytowe*, 11-14.

Bigg, D. M. (1995). Thermal conductivity of heterophase polymer compositions. *Advances in Polymer Science*, *119*, 1–30. doi:10.1007/BFb0021279

Bindal, A., Singh, S., Batra, N. K., & Khanna, R. (2013). Development of glass/jute fibers reinforced polyester composite. *Indian Journal of Marine Sciences*, *2013*, 1–6.

Binder, A. K., & Rae, J. B. (n.d.). Automotive industry. In *Encyclopedia Britannica*. Retrieved from https://www.britannica.com/topic/automotive-industry

Biswas, S. (2012). Mechanical properties of bamboo-epoxy composites a structural application. *Advanced Materials Research*, *1*(3), 221–231. doi:10.12989/amr.2012.1.3.221

Blais, P. Y., & Couture, M. (1999). Precast, prestressed pedestrian bridge-world's first reactive powder concrete structure. *PCI Journal*, *44*(5), 60–71. doi:10.15554/pcij.09011999.60.71

Bledzki, A. K., & Gassan, J. (1999). Composites reinforced with cellulose based fibres. *Progress in Polymer Science*, *24*(2), 221–274. doi:10.1016/S0079-6700(98)00018-5

Bledzki, A. K., Mamun, A. A., Lucka-Gabor, M., & Gutowski, V. S. (2008). The effects of acetylation on properties of flax fibre and its polypropylene composites. *Express Polymer Letters*, *2*(6), 413–422. doi:10.3144/expresspolymlett.2008.50

Bonin, L., & Vitry, V. (2016). Mechanical and wear characterization of electroless nickel mono and bilayers and high boron-mid phosphorus electroless nickel duplex coatings. *Surface and Coatings Technology*, *307*, 957–962. doi:10.1016/j.surfcoat.2016.10.021

Bosser, L., Poisson, E., & Davies, P. (2012). Bio-sourced sandwich materials for marine applications. *JEC Composites*, (75), 25-27.

Boudenne, A., Ibos, L., Fois, M., Majeste, J. C., & Géhin, E. (2005). Electrical and thermal behavior of polypropylene filled with copper particles. *Composites. Part A, Applied Science and Manufacturing*, *36*(11), 1545–1554. doi:10.1016/j.compositesa.2005.02.005

Brenner, A., & Riddell, G. E. (1946). Nickel plating on steel by chemical reduction. *Journal of Research of the National Bureau of Standards*, *37*(1), 31–34. doi:10.6028/jres.037.019

Brylawski, M. M., & Lovins, A. B. (1998). Advanced composites: The car is at the crossroads. *Society for the Advancement of Material and Process Engineering. Materials and Process Affordability: Keys to the Future.*, *43*, 1135–1148.

Bülbül, F., Altun, H., Ezirmik, V., & Küçük, Ö. (2013). Investigation of structural, tribological and corrosion properties of electroless Ni–B coating deposited on 316L stainless steel. *Proceedings of the Institution of Mechanical Engineers. Part J, Journal of Engineering Tribology*, *227*(6), 629–639. doi:10.1177/1350650112464928

Bunsell, A. R., & Harris, B. (1974). Hybrid carbon and glass fibre composites. *Composites*, *5*(4), 157–164. doi:10.1016/0010-4361(74)90107-4

Bunsell, A. R., & Renard, J. (2005). *Fundamentals of fibre reinforced composite materials*. CRC Press. doi:10.1201/9781420056969

Burton, D. S., Gao, X., & Brinson, L. C. (2006). Finite element simulation of a self-healing shape memory alloy composite. *Mechanics of Materials*, *38*(5-6), 525–537. doi:10.1016/j.mechmat.2005.05.021

Cabeza, L. F., Castellón, C., Nogués, M., Medrano, M., Leppers, R., & Zubillaga, O. (2007). Use of microencapsulated PCM in concrete walls for energy savings. *Energy and Building*, *39*(2), 113–119. doi:10.1016/j.enbuild.2006.03.030

Cai, J., Liu, S., Feng, J., Kimura, S., Wada, M., Kuga, S., & Zhang, L. (2012). Cellulose-silica nanocomposite aerogels by in situ formation of silica in cellulose gel. *Angewandte Chemie*, *124*(9), 2118–2121. doi:10.1002/ange.201105730 PMID:22275132

Car chassis frame. (n.d.). Retrieved August 4, 2017, from https://www.shutterstock.com/tr/image-vector/plan-car-chassis-showing-wheels-transmission-84399883

Castejon, L., Miravete, A., & Cuartero, J. (2006). Composite bus rollover simulation and testing. *International Journal of Heavy Vehicle Systems*, *13*(4), 281–297. doi:10.1504/IJHVS.2006.010584

Castellón, C., Medrano, M., Roca, J., Nogués, M., Castell, A., & Cabeza, L. F. (2007). *Use of microencapsulated phase change materials in building applications*. Retrieved July 30, 2017, from http://celsius.co.kr/phase_change_materials/download/energy/Use_of_Microencapsulated_PCM_in_Building_Applications.pdf

Castleman, L. S., Motzkin, S. M., Alicandri, F. P., Bonawit, V. L., & Johnson, A. A. (1976). Biocompatibility of nitinol alloy as an implant material. *Journal of Biomedical Materials Research*, *10*(5), 695–731. doi:10.1002/jbm.820100505 PMID:977602

Catauro, M., Raucci, M. G., Continenza, M. A., & Marotta, A. (2004). Biocompatibility Tests with Fibroblasts of Cao Rich Calcium Silicate Glasses. *Journal of Materials Science*, *39*(1), 373–375. doi:10.1023/B:JMSC.0000008093.18719.0b

Causin, V., Marega, C., Marigo, A., Ferrara, G., & Ferraro, A. (2006). Morphological and structural characterization of polypropylene/conductive graphite nanocomposites. *European Polymer Journal*, *42*(12), 3153–3161. doi:10.1016/j.eurpolymj.2006.08.017

Çelik, İ., Karakan, M., & Bülbül, F. (2016). Investigation of structural and tribological properties of electroless Ni–B coated pure titanium. *Proceedings of the Institution of Mechanical Engineers. Part J, Journal of Engineering Tribology*, *230*(1), 57–63. doi:10.1177/1350650115588568

Chakrabarti, S. K. (2005). *Handbook of Offshore Engineering*. Amsterdam: Elsevier.

Chakravorty, D., Sinha, P. K., & Bandyopadhyay, J. N. (1998). Applications of FEM on free and forced vibrations of laminated shells. *Journal of Engineering Mechanics*, *124*(1), 1–8. doi:10.1061/(ASCE)0733-9399(1998)124:1(1)

Chandrappa, A. K., & Biligiri, K. P. (2016). Pervious concrete as a sustainable pavement material-research findings and future prospects: A state-of-the-art review. *Construction & Building Materials*, *111*, 262–274. doi:10.1016/j.conbuildmat.2016.02.054

Chan, Y. W., Chen, Y. S., & Liu, Y. S. (2003). Development of bond strength of reinforcement steel in self-consolidating concrete. *ACI Structural Journal*, *100*(4), 490–498.

Chan, Y. W., & Chu, S. H. (2004). Effect of silica fume on steel fiber bond characteristics in reactive powder concrete. *Cement and Concrete Research*, *34*(7), 1167–1172. doi:10.1016/j.cemconres.2003.12.023

Chapman, G. B. (1999). *A Thermoplastic Approach to a Composite Automotive Body*. SAE Technical Paper. Retrieved from http://papers.sae.org/1999-01-3222/

Chaudhary, V., Bajpai, P. K., & Maheshwari, S. (2017). Studies on Mechanical and Morphological Characterization of Developed Jute/Hemp/Flax Reinforced Hybrid Composites for Structural Applications. *Journal of Natural Fibers*, 1–18.

Chaudhury, P. B., Mitra, A., & Sahoo, S. (2016). Relative performance of antisymmetric angle-ply laminated stiffened hypar shell roofs with cutout in terms of static behavior. *Curved and Layered Structures*, *3*(1), 22–46.

Chavhan, G. R., Burande, S. W., & Dhole, L. P. (2014). Analysis of shock absorber using different material of spring. *International Journal of Advances in Engineering and Technology*, *5*(4), 19–21.

Chen, P. W., & Chung, D. D. L. (1993). Carbon-fiber-reinforced concrete smart structures capable of nondestructive flaw detection. In *Proceedings of North American Conference on Smart Structures and Materials* (*vol. 1*, pp. 22). Academic Press. doi:10.1117/12.148502

Cheng, S. C., & Vachon, R. I. (1969). The Prediction of the Thermal Conductivity of Two and Three Phase Solid Heterogeneous Mixtures. *International Journal of Heat and Mass Transfer*, *12*(3), 249–264. doi:10.1016/0017-9310(69)90009-X

Chen, H., Ginzburg, V. V., Yang, J., Yang, Y., Liu, W., Huang, Y., ... Chen, B. (2016). Thermal conductivity of polymer-based composites: Fundamentals and applications. *Progress in Polymer Science*, *59*, 41–85. doi:10.1016/j.progpolymsci.2016.03.001

Chen, W. T., Jovic, V., Sun-Waterhouse, D., Idriss, H., & Waterhouse, G. I. (2013). The role of CuO in promoting photocatalytic hydrogen production over TiO_2. *International Journal of Hydrogen Energy*, *38*(35), 15036–15048. doi:10.1016/j.ijhydene.2013.09.101

Chen, W. X., Tu, J. P., Gan, H. Y., Xu, Z. D., Wang, Q. G., Lee, J. Y., ... Zhang, X. B. (2002). Electroless preparation and tribological properties of Ni-P-Carbon nanotube composite coatings under lubricated condition. *Surface and Coatings Technology*, *160*(1), 68–73. doi:10.1016/S0257-8972(02)00408-5

Chen, W., Yu, H., Li, Q., Liu, Y., & Li, J. (2011). Ultralight and highly flexible aerogels with long cellulose I nanofibers. *Soft Matter*, *7*(21), 10360–10368. doi:10.1039/c1sm06179h

Chen, X., Liu, L., & Huang, F. (2015). Black titanium dioxide (TiO_2) nanomaterials. *Chemical Society Reviews*, *44*(7), 1861–1885. doi:10.1039/C4CS00330F PMID:25590565

Chen, X., Liu, L., Peter, Y. Y., & Mao, S. S. (2011). Increasing solar absorption for photocatalysis with black hydrogenated titanium dioxide nanocrystals. *Science*, *331*(6018), 746–750. doi:10.1126/science.1200448 PMID:21252313

Chen, Y. M., & Ting, J. M. (2002). Ultra-high thermal conductivity polymer composites. *Carbon*, *40*(3), 359–362. doi:10.1016/S0008-6223(01)00112-9

Cheon, S., Choi, J. H., & Lee, D. G. (1995). Development of the composite bumper beam for passenger cars. *Composite Structures*, *32*(1-4), 491–499. doi:10.1016/0263-8223(95)00078-X

Cheyrezy, M., Maret, V., & Frouin, L. (1995). Microstructural analysis of RPC (reactive powder concrete). *Cement and Concrete Research*, *25*(7), 1491–1500. doi:10.1016/0008-8846(95)00143-Z

Chiang, K. T., & Chang, F. P. (2006). Application of grey-fuzzy logic on the optimal design of an injection mould part with a thin shell feature. *International Communications in Heat and Mass Transfer*, *33*(1), 94–101. doi:10.1016/j.icheatmasstransfer.2005.08.006

Choi, S., & Kim, J. (2013). Thermal conductivity of epoxy composites with a binary-particle system of aluminum oxide and aluminum nitride fillers. *Composites. Part B, Engineering*, *51*, 140–147. doi:10.1016/j.compositesb.2013.03.002

Chong, M. N., Jin, B., Chow, C. W., & Saint, C. (2010). Recent developments in photocatalytic water treatment technology: A review. *Water Research*, *44*(10), 2997–3027. doi:10.1016/j.watres.2010.02.039 PMID:20378145

Choqueuse, D., Davies, P., & Bigourdan, B. (2012). Composite pressure vessels for deep-sea applications. *JEC Composites Magazine*, 75.

Chris, W. (2015). *This 'thirsty' concrete absorbs 880 gallons of water a minute- here's how it works*. Retrieved July 30, 2017, from http://www.techinsider.io/how-magical-concrete-absorbs-water-2015-9/

Christine, E., Corbin, S. F. J., & Apte, S. P. (1998). *Porous metal structures and processes for their production.* US Patent, 5846664 A.

Cicala, G., Cristaldi, G., Recca, G., Ziegmann, G., El-Sabbagh, A., & Dickert, M. (2009). Properties and performances of various hybrid glass/natural fibre composites for curved pipes. *Materials & Design, 30*(7), 2538–2542. doi:10.1016/j.matdes.2008.09.044

Clark, G. L., Bals, C. K., & Layson, M. A. (1991). Effects of fibre and property orientation on 'C' shaped cross sections. *SAE Technical Paper*, 910049.

Cole, G. S., & Sherman, A. M. (1995). Light weight materials for automotive applications. *Materials Characterization, 35*(1), 3–9. doi:10.1016/1044-5803(95)00063-1

Compton, B. G., & Lewis, J. A. (2014). 3D-printing of lightweight cellular composites. *Advanced Materials, 26*(34), 5930–5935. doi:10.1002/adma.201401804 PMID:24942232

Corinaldesi, V., Letelier, V., & Moriconi, G. (2011). Behaviour of beam-column joints made of recycled-aggregate concrete under cyclic loading. *Construction & Building Materials, 25*(4), 1877–1882. doi:10.1016/j.conbuildmat.2010.11.072

Corinaldesi, V., & Moriconi, G. (2009). Influence of mineral additions on the performance of 100% recycled aggregate concrete. *Construction & Building Materials, 23*(8), 2869–2876. doi:10.1016/j.conbuildmat.2009.02.004

Correa, E., Zuleta, A. A., Guerra, L., Gómez, M. A., Castaño, J. G., Echeverría, F., ... Thompson, G. E. (2013). Tribological behavior of electroless Ni–B coatings on magnesium and AZ91D alloy. *Wear, 305*(1), 115–123. doi:10.1016/j.wear.2013.06.004

Ćosić, K., Korat, L., Ducman, V., & Netinger, I. (2015). Influence of aggregate type and size on properties of pervious concrete. *Construction & Building Materials, 78*, 69–76. doi:10.1016/j.conbuildmat.2014.12.073

Cui, W., Du, F., Zhao, J., Zhang, W., Yang, Y., Xie, X., & Mai, Y. W. (2011). Improving thermal conductivity while retaining high electrical resistivity of epoxy composites by incorporating silica-coated multi-walled carbon nanotubes. *Carbon, 49*(2), 495–500. doi:10.1016/j.carbon.2010.09.047

Damper. (n.d.). Retrieved August 4, 2017, from https://www.shutterstock.com/tr/image-illustration/shock-absorber-car-done-3d-isolated-78229843

Dang, H. N., Sebaibi, N., Boutouil, M., Leleyter, L., & Baraud, F. (2014). A modified method for the design of pervious concrete mix. *Construction & Building Materials, 73*(73), 271–282.

Das, S. (2001). *The cost of automotive polymer composites: a review and assessment of DOE's lightweight materials composites research.* Oak Ridge National Laboratory. doi:10.2172/777656

Das, S. K., & Sahoo, P. (2011). Tribological characteristics of electroless Ni–B coating and optimization of coating parameters using Taguchi based grey relational analysis. *Materials & Design, 32*(4), 2228–2238. doi:10.1016/j.matdes.2010.11.028

Davoodi, M. M., Sapuan, S. M., Ahmad, D., Ali, A., Khalina, A., & Jonoobi, M. (2010). Mechanical properties of hybrid kenaf/glass reinforced epoxy composite for passenger car bumper beam. *Materials & Design, 31*(10), 4927–4932. doi:10.1016/j.matdes.2010.05.021

Davoodi, M. M., Sapuan, S. M., Aidy, A., Osman, N. A. A., Oshkour, A. A., & Wan Abas, W. A. B. W. (2012). Development process of new bumper beam for passenger car: A review. *Materials & Design, 40*, 304–313. doi:10.1016/j.matdes.2012.03.060

Davoodi, M. M., Sapuan, S. M., & Yunus, R. (2008). Conceptual design of a polymer composite automotive bumper energy absorber. *J Mater Des., 29*(7), 1447–1452. doi:10.1016/j.matdes.2007.07.011

De Rosa, I. M., Santulli, C., Sarasini, F., & Valente, M. (2009). Effect of loading-unloading cycles on impact-damaged jute/glass hybrid laminates. *Polymer Composites, 30*(12), 1879–1887. doi:10.1002/pc.20789

DeArmitt, C. (2017). *23 - Functional Fillers for Plastics* (2nd ed.). Applied Plastics Engineering Handbook.

Decher, E. (1998). *Elemente şi structuri din materiale compozite pentru acoperişurile construcţiilor inginereşti*. Academic Press.

Deng, J. (1982a). Control problems of grey systems. *Systems & Control Letters*, (5): 288–294.

Deng, J. (1982b). Grey Control System. *Journal of Huazhong University of Science and Technology, 3*, 9–18.

Deng, J. (1989). Introduction to Grey System. *Journal of Grey System, 1*(1), 1–24.

Deng, J. (1990). *Theory of Grey System*. Wuhan: Huazhong University of Science and Technology Press.

Dennis, S. T., & Palazotto, A. N. (1990). Static response of a cylindrical composite panel with cutouts using a geometrically nonlinear theory. *AIAA Journal, 28*(6), 1082–1108. doi:10.2514/3.25170

Devi, L. G., Nagaraj, B., & Rajashekhar, K. E. (2012). Synergistic effect of Ag deposition and nitrogen doping in TiO_2 for the degradation of phenol under solar irradiation in presence of electron acceptor. *Chemical Engineering Journal, 181–182*, 259–266. doi:10.1016/j.cej.2011.11.076

Dey, A., Bandyopadhyay, J. N., & Sinha, P. K. (1992a). Finite element analysis of laminated composite conoidal shell structures. *Computers & Structures, 43*(3), 469–476. doi:10.1016/0045-7949(92)90281-4

Dey, A., Bandyopadhyay, J. N., & Sinha, P. K. (1992b). Finite element analysis of laminated composite paraboloid of revolution shells. *Computers & Structures*, *44*(3), 675–682. doi:10.1016/0045-7949(92)90400-T

Dey, A., Bandyopadhyay, J. N., & Sinha, P. K. (1994). Behaviour of paraboloid of revolution shell using cross-ply and antisymmetric angle-ply laminates. *Computers & Structures*, *52*(6), 1301–1308. doi:10.1016/0045-7949(94)90194-5

Dey, T. (2012). *Nanotechnology for water purification*. Universal-Publishers.

Dhakal, H. N., Zhang, Z. Y., Richardson, M. O. W., & Errajhi, O. A. Z. (2007). The low velocity impact response of non-woven hemp fibre reinforced unsaturated polyester composites. *Composite Structures*, *81*(4), 559–567. doi:10.1016/j.compstruct.2006.10.003

Dhoshi, N. P., Ingole, N. K., & Gulhane, U. D. (2011). Analysis and modification of leaf spring of tractor trailer using analytical and finite element method. *International Journal of Modern Engineering Research*, *1*(2), 719–722.

Diler, E. A., & Ipek, R. (2012). An experimental and statistical study of interaction effects of matrix particle size, reinforcement particle size and volume fraction on the flexural strength of Al-SiCp composites by P/M using central composite design. *Materials Science and Engineering A*, *548*, 43–55. doi:10.1016/j.msea.2012.03.066

Dolan, C. W. (1999). FRP prestressing in the USA. *Concrete International*, *21*(10), 21–24.

Dolat, D., Mozia, S., Wróbel, R. J., Moszyński, D., Ohtani, B., Guskos, N., & Morawski, A. W. (2015). Nitrogen-doped, metal-modified rutile titanium dioxide as photocatalysts for water remediation. *Applied Catalysis B: Environmental*, *162*, 310–318. doi:10.1016/j.apcatb.2014.07.001

Domone, P. L. (2006). Self-compacting concrete: An analysis of 11 years of case studies. *Cement and Concrete Composites*, *28*(2), 197–208. doi:10.1016/j.cemconcomp.2005.10.003

Domone, P. L. (2007). A review of the hardened mechanical properties of self-compacting concrete. *Cement and Concrete Composites*, *29*(1), 1–12. doi:10.1016/j.cemconcomp.2006.07.010

Dong, J., Locquet, A., Declercq, N. F., & Citrin, D. S. (2016). Polarization-resolved terahertz imaging of intra-and inter-laminar damages in hybrid fiber-reinforced composite laminate subject to low-velocity impact. *Composites. Part B, Engineering*, *92*, 167–174. doi:10.1016/j.compositesb.2016.02.016

Dong, S., Han, B., Ou, J., Li, Z., Han, L., & Yu, X. (2016). Electrically conductive behaviors and mechanisms of short-cut super-fine stainless wire reinforced reactive powder concrete. *Cement and Concrete Composites*, *72*, 48–65. doi:10.1016/j.cemconcomp.2016.05.022

Dowson, M., Pegg, I., Harrison, D., & Dehouche, Z. (2012). Predicted and in situ performance of a solar air collector incorporating a translucent granular aerogel cover. *Energy and Building*, *49*(2), 173–187. doi:10.1016/j.enbuild.2012.02.007

Droval, G., Feller, J. F., Salagnac, P., & Glouannec, P. (2006). Thermal conductivity enhancement of electrically insulating syndiotactic poly(styrene) matrix for diphasic conductive polymer composites. *Polymers for Advanced Technologies*, *17*(9-10), 732–745. doi:10.1002/pat.777

Du, T., Wang, W. H., Lin, H. L., Liu, Z. X., & Liu, J. (2010). Experimental study on interfacial strength of the high performance recycled aggregate concrete. In *Proceedings of the Workshop on Biennial International Conference on Engineering* (pp. 2821-2828). Academic Press. doi:10.1061/41096(366)266

Duan, S., Yang, X., Tao, Y., Mo, F., Xiao, Z., & Wei, K. (2017). Experimental and numerical investigation of Long Glass Fiber Reinforced Polypropylene composite and application in automobile components. *Transport*, *0*(0), 1–9. doi:10.3846/16484142.2017.1323231

Duari, S., Mukhopadhyay, A., Barman, T. K., & Sahoo, P. (2016a). Optimization of wear performance of electroless Ni-B coating under lubrication. *International Journal of Engineering and Technologies*, *7*, 94–103. doi:10.18052/www.scipress.com/IJET.7.94

Duari, S., Mukhopadhyay, A., Barman, T. K., & Sahoo, P. (2016b). Investigation of friction and wear properties of electroless Ni–P–Cu coating under dry condition. *Journal of Molecular and Engineering Materials*, *04*(04), 1640013. doi:10.1142/S225123731640013X

Duari, S., Mukhopadhyay, A., Barman, T. K., & Sahoo, P. (2017). Study of wear and friction of chemically deposited Ni-P-W coating under dry and lubricated condition. *Surfaces and Interfaces*, *6*, 177–189. doi:10.1016/j.surfin.2017.01.009

Dubey, J., Pandit, P. P., & Das, R. K. (2016). Weight Reduction Analysis of Leaf Spring using Epoxy Material. *International Journal of Engineering Technology Science and Research*, *3*(5), 87–91.

Du, F. P., Tang, H., & Huang, D. Y. (2013). Thermal Conductivity of Epoxy Resin Reinforced with Magnesium Oxide Coated Multiwalled Carbon Nanotube. *International Journal of Polymer Science*, *2013*, 1–5. doi:10.1155/2013/541823

Dugat, J., Roux, N., & Bernier, G. (1996). Mechanical properties of reactive powder concretes. *Materials and Structures*, *29*(4), 233–240. doi:10.1007/BF02485945

Dutta, S. (2014). A review on production, storage of hydrogen and its utilization as an energy resource. *Journal of Industrial and Engineering Chemistry*, *20*(4), 1148–1156. doi:10.1016/j.jiec.2013.07.037

Ebadi-Dehaghani, H., Reiszadeh, M., Chavoshi, A., Nazempour, M., & Vakili, M. H. (2014). A study on the effect of zinc oxide (ZnO) and calcium carbonate ($CaCO_3$) nanoparticles on the thermal conductivity (TC) of polypropylene. *Journal of Macromolecular Science, Part B: Physics*, *53*(1), 1364–1376. doi:10.1080/00222348.2013.810032

Edwards, K. (2004). Strategic substitution of new materials for old: Applications in automotive product development. *J Mater Des*, *25*(6), 529–533. doi:10.1016/j.matdes.2003.12.008

Edwin, B. D. (1992). Taguchi approach to design optimization for quality and cost: An overview. *Annual Conference of the International Society of Parametric Analysis.*

Elahinia, M. H., & Ahmadian, M. (2005). An enhanced SMA phenomenological model. Part II. The experimental study. *Journal of Smart Material and Structure*, *14*(1), 1309–1319. doi:10.1088/0964-1726/14/6/023

Elgafy, A., & Lafdi, K. (2005). Effect of carbon nanofiber additives on thermal behavior of phase change materials. *Carbon*, *43*(15), 3067–3074. doi:10.1016/j.carbon.2005.06.042

Es-Souni, M., Es-Souni, M., & Brandies, H. F. (2005). Assessing the Biocompatibility of NiTi shape memory alloys used for medical applications. *Analytical and Bioanalytical Chemistry*, *381*(3), 557–567. doi:10.1007/s00216-004-2888-3 PMID:15660223

Etaphop photo. (n.d.). *Shock absorber*. Retrieved August 4, 2017, from https://www.shutterstock.com/tr/image-photo/closeup-shock-absorber-suspension-pickup-truck-676655524

Fang, H., Bai, S. L., & Wong, C. P. (2017). Thermal, mechanical and dielectric properties of flexible BN foam and BN nanosheets reinforced polymer composites for electronic packaging application. *Composites. Part A, Applied Science and Manufacturing*, *100*(0), 71–80. doi:10.1016/j.compositesa.2017.04.018

Fan, X., Wan, J., Liu, E., Sun, L., Hu, Y., Li, H., ... Fan, J. (2015). High-efficiency photoelectrocatalytic hydrogen generation enabled by Ag deposited and Ce doped TiO_2 nanotube arrays. *Ceramics International*, *41*(3), 5107–5116. doi:10.1016/j.ceramint.2014.12.083

Felekoğlu, B., & Sarıkahya, H. (2008). Effect of chemical structure of polycarboxylate-based superplasticizers on workability retention of self-compacting concrete. *Construction & Building Materials*, *22*(9), 1972–1980. doi:10.1016/j.conbuildmat.2007.07.005

Felekoğlu, B., Türkel, S., & Baradan, B. (2007). Effect of water/cement ratio on the fresh and hardened properties of self-compacting concrete. *Building and Environment*, *42*(4), 1795–1802. doi:10.1016/j.buildenv.2006.01.012

Feng, P., Meng, X., Chen, J. F., & Ye, L. (2015). Mechanical properties of structures 3D printed with cementitious powders. *Construction & Building Materials*, *93*, 486–497. doi:10.1016/j.conbuildmat.2015.05.132

Feng, Y. C., Geng, L., Zheng, P. Q., Zheng, Z. Z., & Wang, G. S. (2008). Fabrication and characteristic of Al-based hybrid composite reinforced with tungsten oxide particle and aluminum borate whisker by squeeze casting. *Materials & Design*, *29*(10), 2023–2026. doi:10.1016/j.matdes.2008.04.006

Ferrara, L., Park, Y. D., & Shah, S. P. (2007). A method for mix-design of fiber-reinforced self-compacting concrete. *Cement and Concrete Research*, *37*(6), 957–971. doi:10.1016/j.cemconres.2007.03.014

Fieler, D. P. (2000). Hybrid polymer-inorganic nanocomposites. *Russian Chemical Reviews*, *69*(1), 53–80. doi:10.1070/RC2000v069n01ABEH000506

Fiore, V., Scalici, T., Calabrese, L., Valenza, A., & Proverbio, E. (2016). Effect of external basalt layers on durability behaviour of flax reinforced composites. *Composites. Part B, Engineering*, *84*, 258–265. doi:10.1016/j.compositesb.2015.08.087

Fisher, R. A. (1951). *Design of Experiments*. Edinburgh, UK: Oliver & Boyd.

Foulk, J. A., Chao, W. Y., Akin, D. E., Dodd, R. B., & Layton, P. A. (2006). Analysis of flax and cotton fiber fabric blends and recycled polyethylene composites. *Journal of Polymers and the Environment*, *14*(1), 15–25. doi:10.1007/s10924-005-8703-1

Franco, M., Sha, W., Aldic, G., Malinov, S., & Çimenoğlu, H. (2016). Effect of reinforcement and heat treatment on elevated temperature sliding of electroless Ni–P/SiC composite coatings. *Tribology International*, *97*, 265–271. doi:10.1016/j.triboint.2016.01.047

Frenzel, J., Zhang, Z., Somsen, C., Neukin, K., & Eggeler, G. (2007). Influence of carbon on martensitic phase transformations in NiTi shape memory alloys. *Acta Materialia*, *55*(4), 1331–1341. doi:10.1016/j.actamat.2006.10.006

Friedrich, K., & Almajid, A. A. (2013). Manufacturing Aspects of Advanced Polymer Composites for Automotive Applications. *Applied Composite Materials*, *20*(2), 107–128. doi:10.1007/s10443-012-9258-7

Friedrich, K., Zhang, Z., & Schlarb, A. K. (2005). Effects of various fillers on the sliding wear of polymer composites. *Composites Science and Technology*, *65*(15), 2329–2343. doi:10.1016/j.compscitech.2005.05.028

front-wheel drive. (2008). Retrieved July 24, 2017, from https://www.britannica.com/technology/front-wheel-drive

Fujishima, A., & Honda, K. (1972). Electrochemical photolysis of water at a semiconductor electrode. *Nature*, *238*(5358), 37–38. doi:10.1038/238037a0 PMID:12635268

Fukuyama, H. (1999). FRP composites in Japan. *Concrete International*, *21*(10), 29–32.

Furie, B., & Furie, B. C. (2008). Mechanisms of Thrombus Formation. *The New England Journal of Medicine*, *359*(9), 938–949. doi:10.1056/NEJMra0801082 PMID:18753650

Gadhari, P., & Sahoo, P. (2016). Electroless nickel-phosphorus composite coatings: A Review. *International Journal of Manufacturing, Materials, and Mechanical Engineering*, *6*(1), 14–50. doi:10.4018/IJMMME.2016010102

Gandhi, M. V., & Thompson, B. D. (1992). *Smart materials and structures*. Springer Science and Business Media.

Ganguli, S., Roy, A. K., & Anderson, D. P. (2008). Improved thermal conductivity for chemically functionalized exfoliated graphite/epoxy composites. *Carbon*, *46*(5), 806–817. doi:10.1016/j.carbon.2008.02.008

Gao, S. L., & Mäder, E. (2006). Jute/polypropylene composites I. Effect of matrix modification. *Composites Science and Technology*, *66*(7), 952–963.

Gawne, D. T., & Ma, U. (1987). Wear mechanisms in electroless nickel coatings. *Wear*, *120*(2), 125–149. doi:10.1016/0043-1648(87)90063-9

Genet, F., Schnitzler, A., Mathieu, S., Autret, K., Théfenne, L., Dizien, O., & Maldjian, A. (2010). Orthotic devices and gait in polio patients. *Annals of Physical and Rehabilitation Medicine*, *53*(1), 51–59. doi:10.1016/j.rehab.2009.11.005 PMID:20022835

George, J., Sreekala, M. S., & Thomas, S. (2001). A review on interface modification and characterization of natural fiber reinforced plastic composites. *Polymer Engineering and Science*, *41*(9), 1471–1485. doi:10.1002/pen.10846

Ger, M. D., Hou, K. H., Wang, L. M., & Hwang, B. J. (2002). The friction and wear of Ni–P–PTFE composite deposits under water lubrication. *Materials Chemistry and Physics*, *77*(3), 755–764. doi:10.1016/S0254-0584(02)00153-0

Ghaderi, M., Rezagholizadeh, M., Heidary, A., & Monirvaghefi, S. M. (2016b). The effect of Al_2O_3 nanoparticles on tribological and corrosion behavior of electroless Ni–B–Al_2O_3 composite coating. *Protection of Metals and Physical Chemistry of Surfaces*, *52*(5), 854–858. doi:10.1134/S2070205116050087

Ghaderi, M., Rezagholizadeh, M., Vaghefi, S. M., & Heidary, A. (2016a). Investigation of high temperature wear resistance of electroless nickel coating with different contents of phosphorous. *Protection of Metals and Physical Chemistry of Surfaces*, *52*(3), 538–542. doi:10.1134/S2070205116030126

Ghafoori, N., & Dutta, S. (1995). Development of no-fines concrete pavement applications. *Journal of Transportation Engineering*, *121*(3), 283–288. doi:10.1061/(ASCE)0733-947X(1995)121:3(283)

Ghilai, G., Green, A. K., Leibovich, H., & Miller, Z. (2012). Low-cost composite structures: how to avoid pitfalls. *JEC Composites*, (74), 41-45.

Ghorai, T. K., Chakraborty, M., & Pramanik, P. (2011). Photocatalytic performance of nano-photocatalyst from TiO_2 and Fe_2O_3 by mechanochemical synthesis. *Journal of Alloys and Compounds*, *509*(32), 8158–8164. doi:10.1016/j.jallcom.2011.05.069

Ghosh, A., & Chakravorty, D. (2014). Prediction of progressive failure behaviour of composite skewed hypar shells using finite element method. *Journal of Structures, ID, 147578*, 1–8.

Ghosh, A., & Chakravorty, D. (2017). Failure analysis of civil Engineering composite shell roofs. *Procedia Engineering*, *173*, 1642–1649. doi:10.1016/j.proeng.2016.12.258

Ghosh, B., & Bandyopadhyay, J. N. (1994). Bending analysis of conical shells with cutouts. *Computers & Structures*, *53*(1), 9–18. doi:10.1016/0045-7949(94)90124-4

Gibas, E. (2013). Plastics in the automotive industry can help reduce the weight of the car. *PlastNews*, (2), 29-31.

Gnanasekar, R. (2015). Influence of Additives on the Mechanical Behaviors of Natural Fiber-Polyester. *Journal of Multidisciplinary Engineering Science and Technology*, *2*(3), 289–297.

Gomes, L. F. A. M., & Lima, M. M. P. P. (1992). *Foundations of Computing and Decision Sciences*. Academic Press.

Goodier, C. I. (2003). Development of self-compacting concrete. *Structures & Buildings*, *156*(4), 405–414.

Gopalakrishnan, T., Raja, M., Prakash, V. M. J., & Gnanavel, C. (2017). Design and Fabrication of E-Glass /carbon/graphite epoxy hybrid composite leaf spring. *IOP Conference Series. Materials Science and Engineering*, *183*(1), 1–7. doi:10.1088/1757-899X/183/1/012004

Gopinatha, A., Kumar, S., & Elayaperumal, A. (2014). Experimental Investigations on Mechanical Properties of Jute Fiber Reinforced Composites with Polyester and Epoxy Resin Matrices. *Procedia Engineering*, *97*, 2052–2063. doi:10.1016/j.proeng.2014.12.448

Gosselin, C., Duballet, R., Roux, P., Gaudillière, N., Dirrenberger, J., & Morel, P. (2016). Large-scale 3D printing of ultra-high performance concrete-a new processing route for architects and builders. *Materials & Design*, *100*, 102–109. doi:10.1016/j.matdes.2016.03.097

Grealish, B., & Gray, N. (2013). World Rally Champion's super-car benefits from lightweight composite body panels. *JEC Composites Magazine, 82*, 36-37. Retrieved from http://www.jeccomposites.com/news/composites-news/hexmc-carbonfibreepoxy-moulded-parts-boeing-787-0

Gu, J., Lv, Z., Wu, Y., Guo, Y., Tian, L., Qiu, H., ... Zhang, Q. (2017). Dielectric thermally conductive boron nitride/polyimide composites with outstanding thermal stabilities via in-situ polymerization-electrospinning-hot press method. *Composites. Part A, Applied Science and Manufacturing*, *94*, 209–216. doi:10.1016/j.compositesa.2016.12.014

Gu, J., Meng, X., Tang, Y., Li, Y., Zhuang, Q., & Kong, J. (2017). Hexagonal boron nitride/polymethyl-vinyl siloxane rubber dielectric thermally conductive composites with ideal thermal stabilities. *Composites. Part A, Applied Science and Manufacturing*, *92*, 27–32. doi:10.1016/j.compositesa.2016.11.002

Gu, J., Zhang, Q., Dang, J., Zhang, J., & Yang, Z. (2009). Thermal conductivity and mechanical properties of aluminum nitride filled linear low-density polyethylene composites. *Polymer Engineering and Science*, *49*(5), 1030–1034. doi:10.1002/pen.21336

Güne, Å., & Uz, O. Ä. (2004). *A fracture-based approach to understanding debonding in FRP bonded structural members*. Academic Press.

Guo, Y. C., Wang, X., Sun, K. W., & Cheng, L. F. (2011). Experiment study on drying shrinkage deformation of recycled aggregate concrete. In *Proceedings of Electric Technology and Civil Engineering 2011 International Conference* (pp. 1219-1222). Academic Press. doi:10.1109/ICETCE.2011.5776036

Guo, H., Li, X., Li, B., Wang, J., & Wang, S. (2017). Thermal conductivity of graphene/poly(vinylidene fluoride) nanocomposite membrane. *Materials & Design*, *114*, 355–363. doi:10.1016/j.matdes.2016.11.010

Guo, W., & Kato, H. (2015). Submicron-porous NiTi and NiTiNb shape memory alloys with high damping capacity fabricated by a new top-down process. *Materials & Design*, *78*, 74–79. doi:10.1016/j.matdes.2015.04.028

Gupta, M. K. (2017). Effect of frequencies on dynamic mechanical properties of hybrid jute/sisal fibre reinforced epoxy composite. *Advances in Materials and Processing Technologies*, 1-14.

Gupta, R. D., Khanna, R., Kochar, G., & Sharma, N. (2016b). Experimental determination for optimal setting of parameters of EDM on Al/SiC Composite using Integrated Taguchi Grey Relational Analysis. *Journal of Machining and Forming Technologies*, *7*, 103–118.

Gupta, R.D., Khanna, R., Garg, M.P., & Sharma, N., & Preeti. (2016a). AWJ Circular Cutting and Advanced Parametric Optimization Using Taguchi, Grey Relational and Entropy Measurement Method. *Ciencia y Técnica (Buenos Aires)*, *31*, 57–86.

Gupta, S. M., & Tripathi, M. (2011). A review of TiO_2 nanoparticles. *Chinese Science Bulletin*, *56*(16), 1639–1657. doi:10.1007/s11434-011-4476-1

Gu, Y., Li, C., Bai, J., Wang, J., & Ma, T. (2016). One-step solvothermal synthesis of Au-TiO_2 loaded electrospun carbon fibers to enhance photocatalytic activity. *Vacuum*, *130*, 1–6. doi:10.1016/j.vacuum.2016.04.031

Güy, N., & Özacar, M. (2016). The influence of noble metals on photocatalytic activity of ZnO for Congo red degradation. *International Journal of Hydrogen Energy*, *41*(44), 20100–20112. doi:10.1016/j.ijhydene.2016.07.063

Hadipour, A., Bahrololoom, M. E., Monirvaghefi, S. M., & Bahramkia, A. R. (2016). Evaluation of wear properties for Ni–P electroless coatings. *Transactions of the Indian Institute of Metals*, *69*(9), 1733–1743. doi:10.1007/s12666-016-0833-8

Haggenmueller, R., Fischer, J. E., & Winey, K. I. (2007). Single Wall Carbon Nanotube/Polyethylene Nanocomposites: Thermal and Electrical Conductivity. *Macromolecules*, *40*(7), 2417–2421. doi:10.1021/ma0615046

Hamada, K., Kawano, F., & Asaoka, K. (2003). Shape recovery of shape memory alloy fiber embedded resin matrix smart composite after crack repair. *Dental Materials Journal*, *22*(2), 160–167. doi:10.4012/dmj.22.160 PMID:12873119

Hambali, A., Sapuan, S. M., Ismail, N., & Nukman, Y. (2010). Material selection of polymeric composite automotive bumper beam using analytical hierarchy process. *J. Cent. South Univ. Technol.*, *17*(2), 244–256. doi:10.1007/s11771-010-0038-y

Han, B., Ding, S., & Yu, X. (2015). Intrinsic self-sensing concrete and structures: A review. *Measurement*, *59*, 110–128. doi:10.1016/j.measurement.2014.09.048

Han, B., Dong, S., Zhang, C., Wang, Y., Yu, X., & Ou, J. (2016). Microstructure related mechanical behaviors of short-cut super-fine stainless wire reinforced reactive powder concrete. *Materials & Design*, *96*, 16–26. doi:10.1016/j.matdes.2016.02.004

Han, B., Li, Z., Zhang, L., Zeng, S. Z., Yu, X., Han, B., & Ou, J. (2017). Reactive powder concrete reinforced with nano SiO_2-coated TiO_2. *Construction & Building Materials*, *148*, 104–112. doi:10.1016/j.conbuildmat.2017.05.065

Han, B., Sun, S., Ding, S., Zhang, L., Yu, X., & Ou, J. (2015). Review of nanocarbon-engineered multifunctional cementitious composites. *Composites. Part A, Applied Science and Manufacturing*, *70*(70), 69–81. doi:10.1016/j.compositesa.2014.12.002

Han, B., Wang, Y., Dong, S., Zhang, L., Ding, S., Yu, X., & Ou, J. (2015). Smart concretes and structures: A review. *Journal of Intelligent Material Systems and Structures*, *26*(11), 1303–1345. doi:10.1177/1045389X15586452

Han, B., Zhang, L., & Ou, J. (2017). *Smart and multifunctional concrete toward sustainable infrastructures.* Springer. doi:10.1007/978-981-10-4349-9

Hancox, N. L. (1982). Developing a materials system for use in making orthopaedic calipers. *Composites*, *13*(2), 202–208. doi:10.1016/0010-4361(82)90058-1

Haneefa, A., Bindu, P., Aravind, I., & Thomas, S. (2008). Studies on tensile and flexural properties of short banana/glass hybrid fiber reinforced polystyrene composites. *Journal of Composite Materials*, *42*(15), 1471–1489. doi:10.1177/0021998308092194

Hansen, T. C. (1992). *Recycling of demolished concrete and masonry.* CRC Press.

Han, Z., & Fina, A. (2011). Thermal conductivity of carbon nanotubes and their polymer Nanocomposites: A review. *Progress in Polymer Science*, *36*(7), 914–944. doi:10.1016/j.progpolymsci.2010.11.004

Hapuarachchi, T. D., Ren, G., Fan, M., Hogg, P. J., & Peijs, T. (2007). Fire retardancy of natural fibre reinforced sheet moulding compound. *Applied Composite Materials*, *14*(4), 251–264. doi:10.1007/s10443-007-9044-0

Haq, A. N., Marimuthu, P., & Jeyapaul, R. (2008). Multiresponse optimization of machining parameters of drilling Al/SiC metal matrix composite using grey relational analysis in the Taguchi method. *International Journal of Advanced Manufacturing Technology*, *37*(3-4), 250–257. doi:10.1007/s00170-007-0981-4

Haque, M. M., & Sharif, A. (2001). Study on wear properties of aluminium–silicon piston alloy. *Journal of Materials Processing Technology*, *118*(1-3), 69–73. doi:10.1016/S0924-0136(01)00869-X

Hariharan, A. B. A., & Khalil, H. A. (2005). Lignocellulose-based hybrid bilayer laminate composite: Part I-studies on tensile and impact behavior of oil palm fiber-glass fiber-reinforced epoxy resin. *Journal of Composite Materials*, *39*(8), 663–684. doi:10.1177/0021998305047267

Harshit, Kumar, S., & Verma, A. (2016). Design And Simulation of Leaf Spring for TATA-ACE mini Loader Truck Using FEM. *Imperial Journal of Interdisciplinary Research*, *2*(10), 1421–1427.

Hayase, G., Kanamori, K., Hasegawa, G., Maeno, A., Kaji, H., & Nakanishi, K. (2013). A superamphiphobic macroporous silicone monolith with marshmallow-like flexibility. *Angewandte Chemie*, *52*(41), 10788–10791. doi:10.1002/anie.201304169 PMID:24009025

Hayase, G., Kanamori, K., & Nakanishi, K. (2012). Structure and properties of polymethylsilsesquioxane aerogels synthesized with surfactant n-hexadecyltrimethylammonium chloride. *Microporous and Mesoporous Materials*, *158*(4), 247–252. doi:10.1016/j.micromeso.2012.03.049

Hearle, J. W. (2001). *High-performance fibres*. Elsevier.

He, H., Fu, R., Han, Y., Shen, Y., & Song, X. (2007). Thermal conductivity of ceramic particle filled polymer composites and theoretical predictions. *Journal of Materials Science*, *42*(16), 6749–6754. doi:10.1007/s10853-006-1480-y

Himanshu, K., Mer, K. K. S., & Sandeep, K. (2014). A Review on Mechanical and Tribological Behaviors of Stir Cast Aluminum Matrix Composites. *Procedia Materials Science*, *6*, 1951–1960. doi:10.1016/j.mspro.2014.07.229

Hirano, K., Suzuki, E., Ishikawa, A., Moroi, T., Shiroishi, H., & Kaneko, M. (2000). Sensitization of TiO_2 particles by dyes to achieve H_2 evolution by visible light. *Journal of Photochemistry and Photobiology A Chemistry*, *136*(3), 157–161. doi:10.1016/S1010-6030(00)00342-7

Hirayama, Y., Jolly, S., & Batty, W. J. (1997). Investigation of thermal energy storage within building thermal mass in northern Japan through dynamic building and building services simulation. In *Proceedings of Seventh International Conference on Thermal Energy Storage* (pp. 355-360). Sapporo, Japan: Academic Press.

Horng, J. T., & Chiang, K. T. (2008). A grey and fuzzy algorithms integrated approach to the optimization of turning Hadfield steel with Al2O3/TiC mixed ceramic tool. *Journal of Materials Processing Technology*, *207*(1-3), 89–97. doi:10.1016/j.jmatprotec.2007.12.060

Hossain, M. F., Islam, M. K., & Islam, M. A. (2014). Effect of chemical treatment on the mechanical and physical properties of wood saw dust particles reinforced polymer matrix composites. *Procedia Engineering*, *90*, 39–45. doi:10.1016/j.proeng.2014.11.811

Hossain, M. F., Shuvo, S. N., & Islam, M. A. (2014). Effect of types of wood on the thermal conductivities of wood saw dust particle reinforced composites. *Procedia Engineering*, *90*, 46–51. doi:10.1016/j.proeng.2014.11.812

Hosseinzadeh, R., Shokrieh, M. M., & Lessard, L. B. (2005). Parametric study of automotive composite bumper beams subjected to low-velocity impacts. *Journal of Composite Structures.*, *68*(4), 419–427. doi:10.1016/j.compstruct.2004.04.008

Hota, S. S., & Padhi, P. (2007). Vibration of plates with arbitrary shapes of cutouts. *Journal of Sound and Vibration*, *302*(4-5), 1030–1036. doi:10.1016/j.jsv.2007.01.003

Hou, J. P., Cherruault, J. Y., Nairne, I., Jeronimidis, G., & Mayer, R. M. (2007). Evolution of the eye-end design of a composite leaf spring for heavy axle loads. *Composite Structures*, *78*(3), 351–358. doi:10.1016/j.compstruct.2005.10.008

Hou, J., & Jeronimidis, G. (2012). A novel bogie design made of glass fibre reinforced plastic. *Materials & Design*, *37*, 1–7. doi:10.1016/j.matdes.2011.12.026

Huang, C., Yao, W., Ali, T., & Muradov, N. (2011). Development of efficient photoreactors for solar hydrogen production. *Solar Energy*, *85*(1), 19–27. doi:10.1016/j.solener.2010.11.004

Huang, H., & Talreja, R. (2006). Numerical simulation of matrix micro-cracking in short fiber reinforced polymer composites: Initiation and propagation. *Composites Science and Technology*, *66*(15), 2743–2757. doi:10.1016/j.compscitech.2006.03.013

Huang, M., & Sakiyama, T. (1999). Free vibration analysis of rectangular plates with variously shaped holes. *Journal of Sound and Vibration*, *226*(4), 769–786. doi:10.1006/jsvi.1999.2313

Huang, X. Y., Jiang, P. K., & Tanaka, T. (2011). A review of dielectric polymer composites with high thermal conductivity. *IEEE Electrical Insulation Magazine*, *27*(4), 8–16. doi:10.1109/MEI.2011.5954064

Huang, X., Liu, C., Lv, X., Liu, G., & Li, F. (2011). Aluminum alloy pistons reinforced with SiC fabricated by centrifugal casting. *Journal of Materials Processing Technology*, *211*(9), 1540–1546. doi:10.1016/j.jmatprotec.2011.04.006

Huang, X., Ranade, R., Ni, W., & Li, V. C. (2013). On the use of recycled tire rubber to develop low e-modulus ECC for durable concrete repairs. *Construction & Building Materials*, *46*(8), 134–141. doi:10.1016/j.conbuildmat.2013.04.027

Hull, D. (1996). *An introduction to composite materials*. Cambridge University press. doi:10.1017/CBO9781139170130

Hu, R. H., Hu, K. H., & Yang, G. H. (2016). *Studies on jute fiber/polypropylene, jute fiber/Polylactide and glass fiber/Polypropylene composites for automotive application*. Presented at the *3rd International Conference on Mechatronics and Information Technology (ICMIT 2016) 3rd International Conference on Mechatronics and Information Technology (ICMIT 2016)*. doi:10.2991/icmit-16.2016.125

Hu, R. H., Ma, Z. G., Zheng, S., Li, Y. N., Yang, G. H., Kim, H. K., & Lim, J. K. (2012). A fabrication process of high volume fraction of jute fiber/polylactide composites for truck liner. *International Journal of Precision Engineering and Manufacturing*, *13*(7), 1243–1246. doi:10.1007/s12541-012-0165-5

Hwang, D. Y., & Foster, W. A. Jr. (1992). Analysis of axisymmetric free vibration of isotropic shallow spherical shells with a circular hole. *Journal of Sound and Vibration*, *157*(2), 331–343. doi:10.1016/0022-460X(92)90685-Q

Ichetaonye, S.I., Madufor, I.C., Yibowei, M.E., & Ichetaonye, D.N. (2015). Physico-Mechanical Properties of *Luffaaegyptiaca* Fiber Reinforced Polymer Matrix Composite. *Open Journal of Composite Materials, 5*(4), 110-117.

Idicula, M., Joseph, K., & Thomas, S. (2010). Mechanical performance of short banana/sisal hybrid fiber reinforced polyester composites. *Journal of Reinforced Plastics and Composites, 29*(1), 12–29. doi:10.1177/0731684408095033

Ishida, H., & Rimdusit, S. (1998). Very high thermal conductivity obtained by boron nitride filled polybenzoxazine. *ThermochimicaActa, 320*(1-2), 177–186. doi:10.1016/S0040-6031(98)00463-8

Islam, N., & Miyazaki, K. (2009). Nanotechnology innovation system: Understanding hidden dynamics of nanoscience fusion trajectories. *Technological Forecasting and Social Change, 76*(1), 128–140. doi:10.1016/j.techfore.2008.03.021

Ismail, A. A., & Bahnemann, D. W. (2014). Photochemical splitting of water for hydrogen production by photocatalysis: A review. *Solar Energy Materials and Solar Cells, 128*, 85–101. doi:10.1016/j.solmat.2014.04.037

Ivanova, T., & Harizanova, A. (2001). Characterization of TiO_2 and TiO_2–MnO oxides prepared by sol–gel method. *Solid State Ionics, 138*(3), 227–232. doi:10.1016/S0167-2738(00)00798-0

Iyer, R. (2013). New long-glass-fibre-reinforces polypropylene for automakers'. *JEC Composites Magazine, 82*, 33.

Jacob, M., Thomas, S., & Varughese, K. T. (2004). Mechanical properties of sisal/oil palm hybrid fiber reinforced natural rubber composites. *Composites Science and Technology, 64*(7), 955–965. doi:10.1016/S0266-3538(03)00261-6

Jacob, M., Varughese, K. T., & Thomas, S. (2006). Dielectric characteristics of sisal–oil palm hybrid biofibre reinforced natural rubber biocomposites. *Journal of Materials Science, 41*(17), 5538–5547. doi:10.1007/s10853-006-0298-y

Jagannatha, T. D., & Harish, G. (2015). Mechanical properties of carbon/glass fiber reinforced epoxy hybrid polymer composites. *International Journal of Mechanical Engineering and Robotics Research, 4*(2), 131 – 137.

Jagatheeshwaran, M. S., Elayaperumal, A., & Arulvel, S. (2015). Wear characteristics of electroless NiP/bio-composite coatings on En8 steel. *Journal of Manufacturing Processes, 20*, 206–214. doi:10.1016/j.jmapro.2015.08.002

Jagatheeshwaran, M. S., Elayaperumal, A., & Arulvel, S. (2016). The role of calcinated sea shell particles on friction-wear behavior of electroless NiP coating: Fabrication and characterization. *Surface and Coatings Technology, 304*, 492–501. doi:10.1016/j.surfcoat.2016.07.053

Jamshaid, H., Mishra, R., Militky, J., Pechociakova, M., & Noman, M. T. (2016). Mechanical, thermal and interfacial properties of green composites from basalt and hybrid woven fabrics. *Fibers and Polymers, 17*(10), 1675–1686. doi:10.1007/s12221-016-6563-z

Jang, I., Shin, K. H., Yang, I., Kim, H., Kim, J., Kim, W.-H., ... Kim, J.-P. (2017). Enhancement of thermal conductivity of BN/epoxy composite through surface modification with silane coupling agents. *Colloids and Surfaces. A, Physicochemical and Engineering Aspects*, *518*, 64–72. doi:10.1016/j.colsurfa.2017.01.011

Jangra, K.K., Sharma, N., & Khanna, R., & Deepak. (2016). An Experimental Investigation and Optimization of Friction Stir Welding process for AA6082 T6 (cryogenic treated and untreated) using an integrated approach of Taguchi, GRA and Entropy Method. *Proc. of IMeche Part L: Part L: Journal of Materials, Design and Applications*, *230*, 454–469.

Jarimopas, B., Singh, S. P., & Saengnil, W. (2005). Measurement and analysis of truck transport vibration levels and damage to packaged tangerines during transit. *Packaging Technology & Science*, *18*(4), 179–188. doi:10.1002/pts.687

Jarukumjorn, K., & Suppakarn, N. (2009). Effect of glass fiber hybridization on properties of sisal fiber–polypropylene composites. *Composites. Part B, Engineering*, *40*(7), 623–627. doi:10.1016/j.compositesb.2009.04.007

Jawaid, M. H. P. S., & Khalil, H. A. (2011). Cellulosic/synthetic fibre reinforced polymer hybrid composites: A review. *Carbohydrate Polymers*, *86*(1), 1–18. doi:10.1016/j.carbpol.2011.04.043

Jawaid, M., Alothman, O. Y., Paridah, M. T., & Khalil, H. S. A. (2014). Effect of oil palm and jute fiber treatment on mechanical performance of epoxy hybrid composites. *International Journal of Polymer Analysis and Characterization*, *19*(1), 62–69. doi:10.1080/1023666X.2014.858429

Jawaid, M., Khalil, H. A., Bakar, A. A., & Khanam, P. N. (2011). Chemical resistance, void content and tensile properties of oil palm/jute fibre reinforced polymer hybrid composites. *Materials & Design*, *32*(2), 1014–1019. doi:10.1016/j.matdes.2010.07.033

Jeff, L. (2015). *3D Printing Concrete: A 2,500-Square-Foot House in 20 Hours and an Eye on a Moon Shot*. Retrieved July 30, 2017, from https://redshift.autodesk.com/3d-printing-concrete/

Jeon, K. W., Shin, K. B., & Kim, J. S. (2011). A study on fatiguelife and strength of a GFRP composite bogie frame for urban subway trains. *Procedia Engineering*, *10*, 2405–2410. doi:10.1016/j.proeng.2011.04.396

Jha, A. K., Dan, T. K., Prasad, S. V., & Rohatgi, K. (1986). Aluminum alloy solid lubricant talc particle composites. *Journal of Materials Science*, *21*(10), 3681–3685. doi:10.1007/BF02403021

Jiang, Y., Sun, R., Zhang, H. B., Min, P., Yang, D., & Yu, Z. Z. (2017). Graphene-coated ZnO tetrapod whiskers for thermally and electrically conductive epoxy composites. *Composites. Part A, Applied Science and Manufacturing*, *94*, 104–112. doi:10.1016/j.compositesa.2016.12.009

Jiang, Z. W., Sun, Z. P., & Wang, P. M. (2005). Effects of some factors on properties of porous pervious concrete. *Journal of Building Materials*, *8*(5), 513–519.

Jing, D., Guo, L., Zhao, L., Zhang, X., Liu, H., Li, M., ... Zhang, K. (2010). Efficient solar hydrogen production by photocatalytic water splitting: From fundamental study to pilot demonstration. *International Journal of Hydrogen Energy*, *35*(13), 7087–7097. doi:10.1016/j.ijhydene.2010.01.030

Jin, Z., Zhang, X., Li, Y., Li, S., & Lu, G. (2007). 5.1% Apparent quantum efficiency for stable hydrogen generation over eosin-sensitized CuO/TiO_2 photocatalyst under visible light irradiation. *Catalysis Communications*, *8*(8), 1267–1273. doi:10.1016/j.catcom.2006.11.019

John, K., & Naidu, S. V. (2007). Chemical resistance of sisal/glass reinforced unsaturated polyester hybrid composites. *Journal of Reinforced Plastics and Composites*, *26*(4), 373–376. doi:10.1177/0731684406072524

John, M. J., & Anandjiwala, R. D. (2008). Recent developments in chemical modification and characterization of natural fiber-reinforced composites. *Polymer Composites*, *29*(2), 187–207. doi:10.1002/pc.20461

Joseph, K., Thomas, S., & Pavithran, C. (1996). Effect of chemical treatment on the tensile properties of short sisal fibre-reinforced polyethylene composites. *Polymer*, *37*(23), 5139–5149. doi:10.1016/0032-3861(96)00144-9

Kabir, M. M., Wang, H., Lau, K. T., & Cardona, F. (2012). Chemical treatments on plant-based natural fibre reinforced polymer composites: An overview. *Composites. Part B, Engineering*, *43*(7), 2883–2892. doi:10.1016/j.compositesb.2012.04.053

Kaewtatip, K., & Thongmee, J. (2012). Studies on the structure and properties of thermoplastic starch/luffa fiber composites. *Materials & Design*, *40*, 314–318. doi:10.1016/j.matdes.2012.03.053

Kaieda, Y. (2003). Fabrication of composition controlled TiNi shape memory wire using combustion synthesis process and the influence of Ni content on phase transformation behaviour. *Science and Technology of Advanced Materials*, *4*(3), 239–246. doi:10.1016/S1468-6996(03)00031-7

Kalaprasad, G., Francis, B., Thomas, S., Kumar, C. R., Pavithran, C., Groeninckx, G., & Thomas, S. (2004). Effect of fibre length and chemical modifications on the tensile properties of intimately mixed short sisal/glass hybrid fibre reinforced low density polyethylene composites. *Polymer International*, *53*(11), 1624–1638. doi:10.1002/pi.1453

Kalia, S., Kaith, B. S., & Kaur, I. (2009). Pretreatments of natural fibers and their application as reinforcing material in polymer composites—a review. *Polymer Engineering and Science*, *49*(7), 1253–1272. doi:10.1002/pen.21328

Kamada, T., & Li, V. C. (2000). The effects of surface preparation on the fracture behavior of ECC/concrete repair system. *Cement and Concrete Composites*, *22*(6), 423–431. doi:10.1016/S0958-9465(00)00042-1

Kamath, M. G., Bhat, G. S., Parikh, D. V., & Mueller, D. (2005). Cotton fiber nonwovens for automotive composites. *International Nonwovens Journal*, *14*(1), 34–40.

Kamila, S. (2013). Introduction, classification and applications of smart materials: An overview. *American Journal of Applied Sciences*, *10*(8), 876–880. doi:10.3844/ajassp.2013.876.880

Kananu, S. K. (2003). Dielectric properties of matrix composite materials with high volume concentrations of inclusions (effective field approach). *International Journal of Engineering Science*, *41*(12), 1287–1312. doi:10.1016/S0020-7225(03)00042-9

Kango, S., Kalia, S., Celli, A., Njuguna, J., Habibi, Y., & Kumar, R. (2013). Surface modification of inorganic nanoparticles for development of organic–inorganic nanocomposites—a review. *Progress in Polymer Science, 38*(8), 1232–1261. doi:10.1016/j.progpolymsci.2013.02.003

Kan, L. L., & Shi, H. S. (2012). Investigation of self-healing behavior of engineered cementitious composites (ECC) materials. *Construction & Building Materials, 29*(4), 348–356. doi:10.1016/j.conbuildmat.2011.10.051

Karácsonyi, É., Baia, L., Dombi, A., Danciu, V., Mogyorósi, K., Pop, L. C., ... Pap, Z. (2013). The photocatalytic activity of TiO_2/WO_3/noble metal (Au or Pt) nanoarchitectures obtained by selective photodeposition. *Catalysis Today, 208*, 19–27. doi:10.1016/j.cattod.2012.09.038

Karaipekli, A., Biçer, A., Sarıc, A., & Tyagie, V. V. (2017). Thermal characteristics of expanded perlite/paraffin composite phase change material with enhanced thermal conductivity using carbon nanotubes. *Energy Conversion and Management, 134*(15), 373–381. doi:10.1016/j.enconman.2016.12.053

Kassie, A. A., Kumar, R. R., Rao, A., & ... (2014). Design of single composite leaf spring for light weight vehicle. *International Journal of Mechanical Engineering and Robotics Research, 3*(1), 191–197.

Katake, K. A., Mankar, S. H., Kale, S. A., Dabeer, P. S., & Deshmukh, S. J. (2016). Numerical and Experimental Stress Analysis of a Composite Leaf Spring. *IACSIT International Journal of Engineering and Technology, 8*(5), 2098–2104. doi:10.21817/ijet/2016/v8i5/160805428

Keener, T. J., Stuart, R. K., & Brown, T. K. (2004). Maleated coupling agents for natural fibre composites. *Composites. Part A, Applied Science and Manufacturing, 35*(3), 357–362. doi:10.1016/j.compositesa.2003.09.014

Kelman, J., & Nelson, G. V. (1998). Composite motor vehicle bumper beam. Academic Press.

Kevern, J., Wang, K., Suleiman, M. T., & Schaefer, V. R. (2006). Mix design development for pervious concrete in cold weather climates. *Binder Content.* In *Proceedings of the 2005 Mid-Continent Transportation Research Symposium.* Ames, IA: Academic Press.

Khanna, R., Kumar, A., Garg, M. P., Singh, A., & Sharma, N. (2015). Multiple Performance Characteristics Optimization for Al 7075 on Electric Discharge Drilling by Taguchi Grey Relational Theory. *Journal of Industrial Engineering International, 11*(4), 459–472. doi:10.1007/s40092-015-0112-z

Kielb, R. E., Leissa, A. W., & Macbain, J. C. (1985). Vibrations of twisted cantilever plates—a comparison of theoretical results. *International Journal for Numerical Methods in Engineering, 21*(8), 1365–1380. doi:10.1002/nme.1620210802

Kim, K., Ryu, S., & Kim, J. (2017). Melt-processable aggregated boron nitride particle via polysilazane coating for thermal conductive composite. *Ceramics International, 43*(2-1), 2441–2447.

Kim, D. H., Choi, D. H., & Kim, H. S. (2014). Design optimization of a carbon fiber reinforced composite automotive lower arm. *Composites. Part B, Engineering*, *58*, 400–407. doi:10.1016/j.compositesb.2013.10.067

Kim, D. H., Kim, H. G., & Kim, H. K. (2015). Design optimization and manufacture of hybrid glass/carbon fiber reinforced composite bumper beam for automobile vehicle. *Composite Structures*, *131*, 742–752. doi:10.1016/j.compstruct.2015.06.028

Kim, G. S., & Hyun, S. H. (2003a). Effect of mixing on thermal and mechanical properties of aerogel-PVB composites. *Journal of Materials Science*, *38*(9), 1961–1966. doi:10.1023/A:1023560601911

Kim, G. S., & Hyun, S. H. (2003b). Synthesis of window glazing coated with silica aerogel films via ambient drying. *Journal of Non-Crystalline Solids*, *320*(1), 125–132. doi:10.1016/S0022-3093(03)00027-9

Kim, H. S., Kim, J. H., Yang, C. M., & Kim, S. Y. (2017). Synergistic enhancement of thermal conductivity in composites filled with expanded graphite and multi-walled carbon nanotube fillers via melt-compounding based on polymerizable low-viscosity oligomer matrix. *Journal of Alloys and Compounds*, *690*, 274–280. doi:10.1016/j.jallcom.2016.08.141

Kim, J. K., Kim, J. S., Ha, G. J., & Yun, Y. K. (2007). Tensile and fiber dispersion performance of ECC (engineered cementitious composites) produced with ground granulated blast furnace slag. *Cement and Concrete Research*, *37*(7), 1096–1105. doi:10.1016/j.cemconres.2007.04.006

Kim, J. S., Lee, S. J., & Shin, K. B. (2007). Manufacturing and structural safety evaluation of a composite train carbody. *Composite Structures*, *78*(4), 468–476. doi:10.1016/j.compstruct.2005.11.006

Kim, J. S., & Yoon, H. J. (2011). Structural behaviors of a GFRP composite bogie frame for urban subway trains under critical load conditions. *Procedia Engineering*, *10*, 2375–2380. doi:10.1016/j.proeng.2011.04.391

Kim, J.-S., & Jeong, J.-C. (2006). Natural frequency evaluation of a composite train carbody with length of 23 m. *Composites Science and Technology*, *66*(13), 2272–2283. doi:10.1016/j.compscitech.2005.11.036

Kim, S., Moon, G. H., Kim, G., Kang, U., Park, H., & Choi, W. (2017). TiO_2 complexed with dopamine-derived polymers and the visible light photocatalytic activities for water pollutants. *Journal of Catalysis*, *346*, 92–100. doi:10.1016/j.jcat.2016.11.027

Kim, Y. Y., & Fischer, G. (2009). Performance of bridge deck link slabs designed with ductile engineered cementitious composite. *ACI Structural Journal*, *101*(6), 792–801.

Kirkby, E. L., Rule, J. D., Michaud, V. L., Sottos, N. R., White, S. R., & Manson, J. A. (2008). Embedded shape-memory alloy wires for improved performance of self-healing polymers. *Advanced Functional Materials*, *18*(15), 2253–2260. doi:10.1002/adfm.200701208

Kistler, S. S. (1931). Coherent expanded aerogels and jellies. *Nature*, *127*(3211), 741. doi:10.1038/127741a0

Koebel, M., Rigacci, A., & Achard, P. (2012). Aerogel-based thermal super insulation: An overview. *Journal of Sol-Gel Science and Technology, 63*(3), 315–339. doi:10.1007/s10971-012-2792-9

Kohnke, P. C., & Schnobrich, W. C. (1972). Analysis of eccentrically stiffened cylindrical shells. *Journal of the Structural Division, 98*(ST7), 1493–1510.

Kollmannsberger, A., Brand, M., & Drechsler, K. (2012). Braiding a convertible windscreen frame with concave sections. *JEC Composites*, (74), 24-25.

Koncherry, V., Potluri, P., & Fernando, A. (2017). Multifunctional Carbon Fibre Tapes for Automotive Composites. *Applied Composite Materials, 24*(2), 477–493. doi:10.1007/s10443-016-9550-z

Kou, S. C., Chi, S. P., & Chan, D. (2007). Influence of fly ash as cement replacement on the properties of recycled aggregate concrete. *Journal of Materials in Civil Engineering, 19*(9), 709–717. doi:10.1061/(ASCE)0899-1561(2007)19:9(709)

Kou, S. C., & Poon, C. S. (2009). Properties of self-compacting concrete prepared with coarse and fine recycled concrete aggregates. *Cement and Concrete Composites, 31*(9), 622–627. doi:10.1016/j.cemconcomp.2009.06.005

Kowalski, L., Duszczyk, J., & Katgerman, L. (1999). Thermal conductivity of metal powder polymer feedstock for powder injection moulding. *Journal of Materials Science, 34*(1), 1–5. doi:10.1023/A:1004424401427 PMID:15347988

Kranthi, G., & Satapathy, A. (2010). Evaluation and prediction of wear response of pine wood dust filled epoxy composites using neural computation. *Computational Materials Science, 49*(3), 609–614. doi:10.1016/j.commatsci.2010.06.001

Kretsis, G. (1987). A review of the tensile, compressive, flexural and shear properties of hybrid fibre-reinforced plastics. *Composites, 18*(1), 13–23. doi:10.1016/0010-4361(87)90003-6

Krishna, B. V., Bose, S., & Bandyopadhyay, A. (2007). Laser processing of net-shape NiTi shape memory alloy. *Metallurgical and Materials Transactions. A, Physical Metallurgy and Materials Science, 38*(5), 1096–1103. doi:10.1007/s11661-007-9127-4

Krishnaveni, K., Narayanan, T. S., & Seshadri, S. K. (2005). Electroless Ni–B coatings: Preparation and evaluation of hardness and wear resistance. *Surface and Coatings Technology, 190*(1), 115–121. doi:10.1016/j.surfcoat.2004.01.038

Ku, H., Wang, H., Pattarachaiyakoop, N., & Trada, M. (2011). A review on the tensile properties of natural fiber reinforced polymer composites. *Composites. Part B, Engineering, 42*(4), 856–873. doi:10.1016/j.compositesb.2011.01.010

Kumar A, Johnston C (1998). Comparative performance of a compression: Molded I-section bumper beam with integrated mounting stays versus other GMT bumper design. *SAE Technical Papers,* 980111.

Kumar, A., Jain, R., & Patil, P. P. (2016). Dynamic Analysis of Heavy Vehicle Medium Duty Drive Shaft Using Conventional and Composite Material. *IOP Conference Series. Materials Science and Engineering*, *149*(1), 012156. doi:10.1088/1757-899X/149/1/012156

Kumar, A., Singh, D., & Kaur, D. (2012). Variation in phase transformation paths of NiTi films as a function of film thickness. *Sensors and Actuators. A, Physical*, *178*, 57–63. doi:10.1016/j.sna.2012.02.014

Kumar, N. M., Reddy, G. V., Naidu, S. V., Rani, T. S., & Subha, M. C. S. (2009). Mechanical properties of coir/glass fiber phenolic resin based composites. *Journal of Reinforced Plastics and Composites*, *28*(21), 2605–2613. doi:10.1177/0731684408093092

Kumar, R. S., Ariraman, M., & Alagar, M. (2015). Studies on dielectric properties of GO reinforced bisphenol-Z polybenzoxazine hybrids. *RSC Advances*, *5*(30), 23787–23797. doi:10.1039/C5RA00472A

Kumar, S. G., & Rao, K. K. (2017). Comparison of modification strategies towards enhanced charge carrier separation and photocatalytic degradation activity of metal oxide semiconductors (TiO_2, WO_3 and ZnO). *Applied Surface Science*, *391*, 124–148. doi:10.1016/j.apsusc.2016.07.081

Kumar, S., & Balasubramanian, V. (2010). Effect of reinforcement size and volume fraction on abrasive wear behavior of AA7075 Al/SiCp P/M composites-A statistical analysis. *Tribology International*, *43*(1-2), 414–422. doi:10.1016/j.triboint.2009.07.003

Kum, J. M., Yoo, S. H., Ali, G., & Cho, S. O. (2013). Photocatalytic hydrogen production over CuO and TiO_2 nanoparticles mixture. *International Journal of Hydrogen Energy*, *38*(31), 13541–13546. doi:10.1016/j.ijhydene.2013.08.004

Kumlutas, D., Tavman, I. H., & Coban, M. T. (2003). Thermal conductivity of particle filled polyethylene composite materials. *Composites Science and Technology*, *63*(1), 113–117. doi:10.1016/S0266-3538(02)00194-X

Kundu, S., Das, S. K., & Sahoo, P. (2016). Tribological behaviour of electroless Ni-P deposits under elevated temperature. Silicon. doi:10.1007/s12633-016-9450-8

Kuo, Y., Yang, T., & Huang, G. W. (2008). The use of grey relational analysis in solving multiple attribute decision-making problems. *Computers & Industrial Engineering*, *55*(1), 80–93. doi:10.1016/j.cie.2007.12.002

Kuznik, F., Virgone, J., & Noel, J. (2008). Optimization of a phase change material wallboard for building use. *Applied Thermal Engineering*, *28*(11), 1291–1298. doi:10.1016/j.applthermaleng.2007.10.012

Lakshmi, B. V., & Satyanarayana, I. (2012). Static and dynamic analysis on composite leaf spring in heavy vehicle. *International Journal of Advanced Engineering Research and Studies*, *2*(1), 80–84.

Lau, K. T., & Cheung, H. Y. (2009). Preface Natural fiber composites. *Composites. Part B, Engineering*, *40*(7), 591–593. doi:10.1016/j.compositesb.2009.04.002

Lee, C. K. (2008). Corrosion and wear-corrosion resistance properties of electroless Ni–P coatings on GFRP composite in wind turbine blades. *Surface and Coatings Technology*, *202*(19), 4868–4874. doi:10.1016/j.surfcoat.2008.04.079

Lee, C. K. (2009). Structure, electrochemical and wear-corrosion properties of electroless nickel–phosphorus deposition on CFRP composites. *Materials Chemistry and Physics*, *114*(1), 125–133. doi:10.1016/j.matchemphys.2008.08.088

Lee, D. G., Lee, C. S., Lee, H. G., Hwang, H. Y., & Kim, J. W. (2004). Novel applications of composite structures to robots, machine tools and automobiles. *Composite Structures*, *66*(1–4), 17–39. doi:10.1016/j.compstruct.2004.04.044

Lee, E. S., Lee, S. M., Shanefield, D. J., & Cannon, W. R. (2008). Enhanced thermal conductivity of polymer matrix composite via high solids loading of aluminum nitride in epoxy resin. *Journal of the American Ceramic Society*, *91*(4), 1169–1174. doi:10.1111/j.1551-2916.2008.02247.x

Lee, K. S., Yang, D. Y., Park, S. H., & Kim, R. H. (2006). Recent Developments in the Use of Two-photon Polymerization in Precise 2D and 3D Microfabrication. *Polymers for Advanced Technologies*, *17*(2), 72–82. doi:10.1002/pat.664

Lee, M. G., Wang, Y. C., & Chiu, C. T. (2007). A preliminary study of reactive powder concrete as a new repair material. *Construction & Building Materials*, *21*(1), 182–189. doi:10.1016/j.conbuildmat.2005.06.024

Lee, S. S., Bai, H., Liu, Z., & Sun, D. D. (2013). Optimization and an insightful properties—Activity study of electrospun TiO_2/CuO composite nanofibers for efficient photocatalytic H_2 generation. *Applied Catalysis B: Environmental*, *140*, 68–81. doi:10.1016/j.apcatb.2013.03.033

Lee, S. Y., & Park, S. J. (2013). TiO_2 photocatalyst for water treatment applications. *Journal of Industrial and Engineering Chemistry*, *19*(6), 1761–1769. doi:10.1016/j.jiec.2013.07.012

León, O. A., Staia, M. H., & Hintermann, H. E. (2003). High temperature wear of an electroless Ni–P–BN (h) composite coating. *Surface and Coatings Technology*, *163*, 578–584. doi:10.1016/S0257-8972(02)00663-1

Lepech, M. D., & Li, V. C. (2009). Application of ECC for bridge deck link slabs. *Materials and Structures*, *42*(9), 1185–1195. doi:10.1617/s11527-009-9544-5

Lepore, G. P., Persaud, L., & Langford, C. H. (1996). Supporting titanium dioxide photocatalysts on silica gel and hydrophobically modified silica gel. *Journal of Photochemistry and Photobiology A Chemistry*, *98*(1-2), 103–111. doi:10.1016/1010-6030(95)04242-3

Le, T. T., Austin, S. A., Lim, S., Buswell, R. A., Law, R., Gibb, A. G. F., & Thorpe, T. (2012). Hardened properties of high-performance printing concrete. *Cement and Concrete Research*, *42*(3), 558–566. doi:10.1016/j.cemconres.2011.12.003

Li, K.S., & Wang, Q. (2002). Advances in thermal conductive polymeric materials. *Journal of Functional Materials of China*, *33*(1), 201-232.

Li, V. C. (2008). Engineered cementitious composites (ECC)-material, structural, and durability performance. In Nawy (Ed.), Concrete construction engineering handbook. CRC Press.

Lian, C., & Zhuge, Y. (2010). Optimum mix design of enhanced permeable concrete-an experimental investigation. *Construction & Building Materials*, *24*(12), 2664–2671. doi:10.1016/j.conbuildmat.2010.04.057

Liao, C. H., Huang, C. W., & Wu, J. (2012). Hydrogen production from semiconductor-based photocatalysis via water splitting. *Catalysts*, *2*(4), 490–516. doi:10.3390/catal2040490

Liao, S., Donggen, H., Yu, D., Su, Y., & Yuan, G. (2004). Preparation and characterization of ZnO/TiO_2, $SO_4^{2-}/ZnO/TiO_2$ photocatalyst and their photocatalysis. *Journal of Photochemistry and Photobiology A Chemistry*, *168*(1), 7–13. doi:10.1016/j.jphotochem.2004.05.010

Li, B., Li, R., & Xie, Y. (2017). Properties and effect of preparation method of thermally conductive polypropylene/aluminum oxide composite. *Journal of Materials Science*, *52*(5), 2524–2533. doi:10.1007/s10853-016-0546-8

Li, C., Li, Q., Cheng, L., Li, T., Lu, H., Tang, L., & Zhang, K., E, S., Jun Zhang, J., Li, Z., & Yao, Y. (2017). Conductivity enhancement of polymer composites using high-temperature short-time treated silver fillers. *Composites. Part A, Applied Science and Manufacturing*, *100*(0), 64–70.

Li, J., & Zhang, J. Z. (2009). Optical properties and applications of hybrid semiconductor nanomaterials. *Coordination Chemistry Reviews*, *253*(23), 3015–3041. doi:10.1016/j.ccr.2009.07.017

Li, L., & Chung, D. D. L. (1994). Thermally conducting polymer–matrix composites containing both AlN particles and SiC whiskers. *Journal of Electronic Materials*, *23*(6), 557–564. doi:10.1007/BF02670659

Lim, S., Buswell, R. A., Le, T. T., Austin, S. A., Gibb, A. G. F., & Thorpe, T. (2012). Developments in construction-scale additive manufacturing processes. *Automation in Construction*, *21*(1), 262–268. doi:10.1016/j.autcon.2011.06.010

Ling, W., Gu, A., Liang, G., & Yuan, L. (2010). New composites with thermal conductivity and low dielectric constant for microelectronic packaging. *Polymer Composites*, *31*(2), 307–313.

Liu, C. H., Huang, Y. C., Chiu, C. H., Lai, Y. C., & Pai, T. Y. (2016). Design and Analysis of Automotive Bumper Covers in Transient Loading Conditions. *Key Engineering Materials*, *715*, 174–179. doi:10.4028/www.scientific.net/KEM.715.174

Liu, H., Guo, R. X., & Liu, Z. (2012). Characteristics of microstructure and performance of laser-treated electroless Ni–P/Ni–W–P duplex coatings. *Transactions of Nonferrous Metals Society of China*, *22*(12), 3012–3020. doi:10.1016/S1003-6326(11)61564-1

Liu, H., Lv, Y. Y., Liu, Z., & Thompson, G. E. (2016). Dry sliding wear behaviour and structural characteristics of laser-annealed electroless Ni–P/Ni–Mo–P duplex coatings. *Tribology International*, *103*, 343–351. doi:10.1016/j.triboint.2016.07.014

Liu, J., Wang, X., Tian, Z., Yuan, M., & Ma, X. (2015). Effect of copper content on the properties of electroless Ni–Cu–P coatings prepared on magnesium alloys. *Applied Surface Science*, *356*, 289–293. doi:10.1016/j.apsusc.2015.08.072

Liu, Q., Lin, Y., Zong, Z., Sun, G., & Li, Q. (2013). Lightweight design of carbon twill weave fabric composite body structure for electric vehicle. *Composite Structures*, *97*, 231–238. doi:10.1016/j.compstruct.2012.09.052

Liu, R., Peng, Y., Cao, J., & Chen, Y. (2014). Comparison on properties of lignocellulosic flour/polymer composites by using wood, cellulose, and lignin flours as fillers. *Composites Science and Technology*, *103*, 1–7. doi:10.1016/j.compscitech.2014.08.005

Liu, X., Zhu, G., Wang, X., Yuan, X., Lin, T., & Huang, F. (2016). Progress in Black Titania: A New Material for Advanced Photocatalysis. *Advanced Energy Materials*, *6*(17), 1600452. doi:10.1002/aenm.201600452

Liu, Y., Shi, Y., Liu, X., & Li, H. (2017). A facile solvothermal approach of novel Bi_2S_3/TiO_2/RGO composites with excellent visible light degradation activity for methylene blue. *Applied Surface Science*, *396*, 58–66. doi:10.1016/j.apsusc.2016.11.028

Liu, Y., Wang, Z., Fan, W., Geng, Z., & Feng, L. (2014). Enhancement of the photocatalytic performance of Ni-loaded TiO_2 photocatalyst under sunlight. *Ceramics International*, *40*(3), 3887–3893. doi:10.1016/j.ceramint.2013.08.030

Liu, Z., Lu, J., & Zhu, P. (2016). Lightweight design of automotive composite bumper system using modified particle swarm optimizer. *Composite Structures*, *140*, 630–643. doi:10.1016/j.compstruct.2015.12.031

Li, V. C. (2002). Reflections on the research and development of engineered cementitious composites (ECC). *Proceedings of the JCI International Workshop on Ductile Fiber Reinforced Cementitious Composites (DDFRCC)-Application and Evaluation.*

Li, V. C. (2003). Hygral behavior of engineered cementitious composites (ECC) / vergleich der hygrischen eigenschaften von ECC mit beton: Restoration of buildings and monuments. *Restoration of Buildings and Monuments*, *9*(5), 513–534.

Li, V. C. (2004). High performance fiber reinforced cementitious composites as durable material for concrete structure repair. *International Journal for Restoration*, *10*, 163–180.

Li, V. C. (2007). Progress and application of engineered cementitious composites. *Journal of the Chinese Ceramic Society*, *35*, 531–536.

Li, V. C., & Leung, C. Y. (1992). Theory of steady state and multiple cracking of random discontinunous fiber reinforced brittle matrix composites. *Journal of Engineering Mechanics*, *118*(11), 2246–2264. doi:10.1061/(ASCE)0733-9399(1992)118:11(2246)

Li, X., Tabil, L. G., & Panigrahi, S. (2007). Chemical treatments of natural fiber for use in natural fiber-reinforced composites: A review. *Journal of Polymers and the Environment*, *15*(1), 25–33. doi:10.1007/s10924-006-0042-3

Li, X., Zhang, J., Kang, S., Li, G., & Mu, J. (2014). Visible light photocatalytic activity of CuO/Cr_2O_3 co-loaded multiwalled carbon nanotubes sensitized with eosin Y for hydrogen evolution from water. *Ceramics International*, *40*(7), 10171–10176. doi:10.1016/j.ceramint.2014.02.055

Li, Y., Lin, Z., Jiang, A., & Chen, G. (2003). Use of high strength steel sheet for lightweight and crashworthy car body. *Materials & Design*, *24*(3), 177–182. doi:10.1016/S0261-3069(03)00021-9

Li, Z., Wang, J., Lu, J., & Meng, J. (2013). Tribological characteristics of electroless Ni–P–MoS_2 composite coatings at elevated temperatures. *Applied Surface Science*, *264*, 516–521. doi:10.1016/j.apsusc.2012.10.055

Llano, B., Restrepo, G., Marín, J. M., Navío, J. A., & Hidalgo, M. C. (2010). Characterisation and photocatalytic properties of titania–silica mixed oxides doped with Ag and Pt. *Applied Catalysis A, General*, *387*(1), 135–140. doi:10.1016/j.apcata.2010.08.021

Lo, C. C., Huang, C. W., Liao, C. H., & Wu, J. C. (2010). Novel twin reactor for separate evolution of hydrogen and oxygen in photocatalytic water splitting. *International Journal of Hydrogen Energy*, *35*(4), 1523–1529. doi:10.1016/j.ijhydene.2009.12.032

López, S. M., Hidalgo, M. C., Navío, J. A., & Colón, G. (2011). Novel Bi_2WO_6–TiO_2 heterostructures for Rhodamine B degradation under sunlike irradiation. *Journal of Hazardous Materials*, *185*(2), 1425–1434. doi:10.1016/j.jhazmat.2010.10.065 PMID:21074938

Loto, C. A. (2016). Electroless nickel plating–A review. *Silicon*, *8*(2), 177–186. doi:10.1007/s12633-015-9367-7

Lu, Y. C., Pilla, S., & Society of Automotive Engineers (Eds.). (2014). *Design of automotive composites*. Warrendale, PA: Society of Automotive Engineers.

Madah, F., Amadeh, A. A., & Dehghanian, C. (2016). Investigation on the phase transformation of electroless Ni–B coating after dry sliding against alumina ball. *Journal of Alloys and Compounds*, *658*, 272–279. doi:10.1016/j.jallcom.2015.10.156

Madah, F., Dehghanian, C., & Amadeh, A. A. (2015). Investigations on the wear mechanisms of electroless Ni–B coating during dry sliding and endurance life of the worn surfaces. *Surface and Coatings Technology*, *282*, 6–15. doi:10.1016/j.surfcoat.2015.09.003

Madenci, E., & Barut, A. (1994). Pre and post-buckling response of curved thin composite panels with cutouts under compression. *International Journal for Numerical Methods in Engineering*, *37*(9), 1499–1510. doi:10.1002/nme.1620370906

Maeda, K. (2011). Photocatalytic water splitting using semiconductor particles: History and recent developments. *Journal of Photochemistry and Photobiology C, Photochemistry Reviews*, *12*(4), 237–268. doi:10.1016/j.jphotochemrev.2011.07.001

Mahapatra, S. S., & Patnaik, A. (2007). Optimization of wire electrical discharge machining (WEDM) process parameters using Taguchi method. *International Journal of Advanced Manufacturing Technology*, *34*(9-10), 911–925. doi:10.1007/s00170-006-0672-6

Maheta, V. V., Patel, A. B., Maheta, V. V., & Patel, A. B. (2015). Design, Analysis and optimization in Automobile Drive shaft. *International Journal (Toronto, Ont.)*, *1*, 432–439.

Maicu, M., Hidalgo, M. C., Colón, G., & Navío, J. A. (2011). Comparative study of the photodeposition of Pt, Au and Pd on pre-sulphated TiO_2 for the photocatalytic decomposition of phenol. *Journal of Photochemistry and Photobiology A Chemistry*, *217*(2), 275–283. doi:10.1016/j.jphotochem.2010.10.020

Malaeb, Z., Hachem, H., Tourbah, A., Maalouf, T., Zarwi, N. E., & Hamzeh, F. (2016). 3D concrete printing: Machine and mix design. *International Journal of Civil Engineering and Technology*, *6*(6), 14–22.

Malhotra, S. K., Ganesan, N., & Veluswami, M. A. (1989). Vibration of composite plates with cut-outs. *Journal of Aeronautical Society of India*, *41*, 61–64.

Mallick, P. K. (2008). *Fiber-reinforced composites: Materials, manufacturing, and design*. Boca Raton, FL: CRC Press.

Mamunya, Y. P., Davydenko, V. V., Pissis, P., & Lebedev, E. V. (2002). Electrical and thermal conductivity of polymers filled with metal powders. *European Polymer Journal*, *38*(9), 1887–1897. doi:10.1016/S0014-3057(02)00064-2

Manjunath, V., & Udupa, N. G. S. (2016). A Study on Hybrid Composite using Areca and Eucalyptus Fiber by using Epoxy Resin. *Journal of Industrial Mechanics*, *1*(1-2).

Mao, S. S., Shen, S., & Guo, L. (2012). Nanomaterials for renewable hydrogen production, storage and utilization. *Progress in Natural Science: Materials International*, *22*(6), 522–534. doi:10.1016/j.pnsc.2012.12.003

Marconnet, A. M., Yamamoto, N., Panzer, M. A., Wardle, B. L., & Goodson, K. E. (2011). Thermal Conduction in Aligned Carbon Nanotube–Polymer Nanocomposites with High Packing Density. *ACS Nano*, *5*(6), 4818–4825. doi:10.1021/nn200847u PMID:21598962

Marí, A., Etxeberria, M., & Vázquez, E. (2006). Microstructure analysis of hardened recycled aggregate concrete. *Magazine of Concrete Research*, *58*(10), 683–690. doi:10.1680/macr.2006.58.10.683

Masoumi, F., Ghasemi, H. R., Ziaei, A. A., & Shahriari, D. (2012). Tribological characterization of electroless Ni–10% P coatings at elevated test temperature under dry conditions. *The International Journal of Advanced Manufacturing Technology*, *62*(9-12), 1063–1070. doi:10.1007/s00170-011-3832-2

Mayer, C., Wang, X., & Neitzel, M. (1998). Macro-and micro-impregnation phenomena in continuous manufacturing of fabric reinforced thermoplastic composites. *Composites. Part A, Applied Science and Manufacturing*, *29*(7), 783–793. doi:10.1016/S1359-835X(98)00056-6

Mazaheripour, H., Ghanbarpour, S., Mirmoradi, S. H., & Hosseinpour, I. (2011). The effect of polypropylene fibers on the properties of fresh and hardened lightweight self-compacting concrete. *Construction & Building Materials*, *25*(1), 351–358. doi:10.1016/j.conbuildmat.2010.06.018

Mcneil, K., & Kang, H. K. (2013). Recycled concrete aggregates: A review. *International Journal of Concrete Structures and Materials*, *7*(1), 61–69. doi:10.1007/s40069-013-0032-5

Mehndiratta, A., Singh, N. K., & Singh, K. K. (2015). Analysis of GFRP Leaf Spring. *International Journal Of Modern Engineering Research*, *5*(5), 22–27.

Mehul, S., Shah, D. B., & Bhojawala, V. (2012). Analysis of composite leaf spring using FEA for light vehicle mini truck. *Journal of Information. Knowledge and Research in Mechanical Engineering*, *2*(2), 424–428.

Melián, E. P., López, C. R., Méndez, A. O., Díaz, O. G., Suárez, M. N., Rodríguez, J. D., ... Hevia, D. F. (2013). Hydrogen production using Pt-loaded TiO_2 photocatalysts. *International Journal of Hydrogen Energy*, *38*(27), 11737–11748. doi:10.1016/j.ijhydene.2013.07.006

Meng, F., Huang, F., Guo, Y., Chen, J., Chen, X., Hui, D., ... Zhou, Z. (2017). In situ intercalation polymerization approach to polyamide-6/graphite nanoflakes for enhanced thermal conductivity. *Composites. Part B, Engineering*, *117*, 165–173. doi:10.1016/j.compositesb.2017.02.043

Merkisz, J., & Bajerlein, M. (2011). Composite materials used in the aerospace industry. *Logistyka*, 2829-2837

Meschut, G., Janzen, V., & Olfermann, T. (2014). Innovative and Highly Productive Joining Technologies for Multi-Material Lightweight Car Body Structures. *Journal of Materials Engineering and Performance*, *23*(5), 1515–1523. doi:10.1007/s11665-014-0962-3

Meyer, B. (2012). Lighter than water. *JEC Composites Magazine*, *74*, 40.

Micelli, F., Mazzotta, R., Leone, M., & Aiello, M. A. (2015). Review study on the durability of FRP-confined concrete. *Journal of Composites for Construction*, *19*(3), 4014056. doi:10.1061/(ASCE)CC.1943-5614.0000520

Michaud. (2010). *Simulation-based Design Optimization and Control of Thick Composite Laminates Manufactured by Resin Transfer Molding* (Doctoral dissertation). Department of Chemical Engineering, University of Delaware.

Mishra, D., Raj, P. M., Tishler, J., Sun, T., Shipton, E., & Tummala, R. (2016). Multilayered Ferromagnetic Polymer Composite Structures for High-Density Power Inductors. *IEEE Transactions on Magnetics*, *52*(11), 1–5. doi:10.1109/TMAG.2016.2589925

Mishra, S., Misra, M., Tripathy, S. S., Nayak, S. K., & Mohanty, A. K. (2001). Graft copolymerization of acrylonitrile on chemically modified sisal fibers. *Macromolecular Materials and Engineering*, *286*(2), 107–113. doi:10.1002/1439-2054(20010201)286:2<107::AID-MAME107>3.0.CO;2-0

Mishra, S., Mohanty, A. K., Drzal, L. T., Misra, M., Parija, S., Nayak, S. K., & Tripathy, S. S. (2003). Studies on mechanical performance of biofibre/glass reinforced polyester hybrid composites. *Composites Science and Technology*, *63*(10), 1377–1385. doi:10.1016/S0266-3538(03)00084-8

Misnon, M. I., Islam, M. M., Epaarachchi, J. A., & Lau, K. T. (2013). Textile material forms for reinforcement materials - a review. In *Third Malaysian Postgraduate Conference (MPC) 2013* (p. 105). MPC.

Misra, S., Misra, M., Tripathy, S. S., Nayak, S. K., & Mohanty, A. K. (2002). The influence of chemical surface modification on the performance of sisal-polyester biocomposites. *Polymer Composites*, *23*(2), 164–170. doi:10.1002/pc.10422

Miwa, T., Kaneco, S., Katsumata, H., Suzuki, T., Ohta, K., Verma, S. C., & Sugihara, K. (2010). Photocatalytic hydrogen production from aqueous methanol solution with $CuO/Al_2O_3/TiO_2$ nanocomposite. *International Journal of Hydrogen Energy*, *35*(13), 6554–6560. doi:10.1016/j.ijhydene.2010.03.128

Mkaddem, A., Demirci, I., & El Mansori, M. (2008). A micro–macro combined approach using FEM for modelling of machining of FRP composites: Cutting forces analysis. *Composites Science and Technology*, *68*(15), 3123–3127. doi:10.1016/j.compscitech.2008.07.009

Modi, O. P., Yadav, R. P., Mondal, D. P., Dasgupta, R., Das, S., & Yegneswaran, A. H. (2001). Abrasive wear behaviour of zinc-aluminium alloy 10% alumina composite through factorial design of experiment. *Journal of Materials Science*, *36*(7), 1601–1607. doi:10.1023/A:1017523214073

Mohammed, H. A., & Uttandaraman, S. (2009). A review of vapor grown carbon nanofiber/polymer conductive composites. *Carbon*, *47*(1), 2–22. doi:10.1016/j.carbon.2008.09.039

Mohan, R. (1987). Composites for automotive structural applications. *Proceedings of the Third Annual Conference on Advanced Composites,* 57–62.

Mohanty, A. K., Misra, M., & Drzal, L. T. (2001). Surface modifications of natural fibers and performance of the resulting biocomposites: An overview. *Composite Interfaces*, *8*(5), 313–343. doi:10.1163/156855401753255422

Mohanty, A. K., Misra, M., & Hinrichsen, G. (2000). Biofibers, biodegradable polymers and biocomposites: An overview. *Macromolecular Materials and Engineering*, *276*(1), 1–24. doi:10.1002/(SICI)1439-2054(20000301)276:1<1::AID-MAME1>3.0.CO;2-W

Mohapatra, R.C., Mishra, A., & Choudhury, B. B. (2013). Investigations on tensile and flexural strength of wood dust and glass fibre filled epoxy hybrid composites. *International Journal of Mechanical Engineering and Technology, 4*(4), 180-187.

Molinari, R., Marino, T., & Argurio, P. (2014). Photocatalytic membrane reactors for hydrogen production from water. *International Journal of Hydrogen Energy*, *39*(14), 7247–7261. doi:10.1016/j.ijhydene.2014.02.174

Moniruzzaman, M. (2012). PEI/carbon as metal substitute in aircraft food tray arms. *JEC Composites*, (70), 48-53.

Montgomery, D. C. (2001). *Design and analysis of experiments*. New York: John Wiley.

Moore, A. L., & Shi, L. (2014). Emerging challenges and materials for thermal management of electronics. *Materials Today*, *17*(4), 163–174. doi:10.1016/j.mattod.2014.04.003

Morishit, T., Katagiri, Y., Matsunag, T., Muraok, Y., & Fukumori, K. (2017). Design and fabrication of morphologically controlled carbon nanotube/polyamide-6-based composites as electrically insulating materials having enhanced thermal conductivity and elastic modulus. *Composites Science and Technology*, *142*, 41–49. doi:10.1016/j.compscitech.2017.01.022

Muhammad Nasiruddin, S., Hambali, A., Rosidah, J., Widodo, W. S., & Ahmad, M. N. (2017). A Review of Energy Absorption of Automotive Bumper Beam. *International Journal of Applied Engineering Research*, *12*(2), 238–245.

Mukhopadhyay, A., Barman, T. K., & Sahoo, P. (2017c). Effect of heat treatment on tribological behavior of electroless Ni-B-Mo coatings at different operating temperatures. Silicon. doi:10.1007/s12633-017-9594-1

Mukhopadhyay, A., Duari, S., Barman, T. K., & Sahoo, P. (2016d). Tribological performance optimization of electroless Ni–B coating under lubricated condition using hybrid grey fuzzy logic. *Journal of the Institution of Engineers (India): Series D, 97*(2), 215-231.

Mukhopadhyay, A., Barman, T. K., & Sahoo, P. (2016g). Effect of operating temperature on tribological behavior of as-plated Ni-B coating deposited by electroless method. *Tribology Transactions*, 1–12. doi:10.1080/10402004.2016.1271929

Mukhopadhyay, A., Barman, T. K., & Sahoo, P. (2017b). Tribological behavior of sodium borohydride reduced electroless nickel alloy coatings at room and elevated temperatures. *Surface and Coatings Technology*, *321*, 464–476. doi:10.1016/j.surfcoat.2017.05.015

Mukhopadhyay, A., Barman, T. K., & Sahoo, P. (2017d). Effects of heat treatment on tribological behavior of electroless Ni–B coating at elevated temperatures. *Surface Review and Letters*, 24(Supp01), 1850014. doi:10.1142/S0218625X18500142

Mukhopadhyay, A., Duari, S., Barman, T. K., & Sahoo, P. (2016a). Application of grey fuzzy logic for the optimization of tribological performance of electroless Ni-B coating. *Journal of Manufacturing Technology Research*, *7*(1/2), 1–24.

Mukhopadhyay, A., Duari, S., Barman, T. K., & Sahoo, P. (2016b). Wear analysis of electroless Ni-P coating under lubricated condition using fuzzy logic. *Portugaliae Electrochimica Acta*, *34*(1), 63–82. doi:10.4152/pea.201601063

Mukhopadhyay, A., Duari, S., Barman, T. K., & Sahoo, P. (2016c). Optimization of friction and wear properties of electroless Ni–P coatings under lubrication using grey fuzzy logic. *Journal of the Institution of Engineers*, *98*(2), 255–268. doi:10.1007/s40033-016-0133-9

Mukhopadhyay, A., Duari, S., Barman, T. K., & Sahoo, P. (2016e). Investigation of Wear Behavior of Electroless Ni-P-W Coating under Dry and Lubricated Conditions Using RSM and Fuzzy Logic. *Portugaliae Electrochimica Acta*, *34*(4), 231–255. doi:10.4152/pea.201604231

Mukhopadhyay, A., Duari, S., Barman, T. K., & Sahoo, P. (2016f). Optimization of wear behavior of electroless Ni-P-W coating under dry and lubricated conditions using genetic algorithm (GA). *Jurnal Tribologi*, *11*, 24–48.

Mukhopadhyay, A., Duari, S., Barman, T. K., & Sahoo, P. (2017a). Tribology of electroless Ni-P coating under lubricated condition: An RSM and GA approach. *International Journal of Surface Engineering and Interdisciplinary Materials Science*, *5*(1), 37–57. doi:10.4018/IJSEIMS.2017010103

Mulinari, D. R., Baptista, C. A. R. P., Souza, J. V. C., & Voorwald, H. J. C. (2011). Mechanical Properties of Coconut Fibers Reinforced Polyester Composites. *Procedia Engineering*, *10*, 2074–2079. doi:10.1016/j.proeng.2011.04.343

Munshi, K. (2012). A modular composite houseboat for tourism in Kerala. *JEC Composites*, (75), 36-37.

Mu, Q., & Feng, S. (2007). Thermal conductivity of graphite/silicone rubber prepared by solution intercalation. *Thermochimica Acta*, *462*(1-2), 70–75. doi:10.1016/j.tca.2007.06.006

Mu, Q., Feng, S., & Diao, G. (2007). Thermal conductivity of silicone rubber filled with ZnO. *Polymer Composites*, *28*(2), 125–130. doi:10.1002/pc.20276

Murugan, R., Ramesh, R., Padmanabhan, K., Jeyaraam, R., & Krishna, S. (2014). Experimental investigation on static mechanical properties of glass/carbon hybrid woven fabric composite laminates. *Advanced Materials Research*, *903*, 96–101. doi:10.4028/www.scientific.net/AMR.903.96

Mustakangas, M. (2012). Milestone celebration of Finnish aerospace composites expertise. *JEC Composites*, (72), 49-51.

Naeem, K., & Ouyang, F. (2013). Influence of supports on photocatalytic degradation of phenol and 4-chlorophenol in aqueous suspensions of titanium dioxide. *Journal of Environmental Sciences (China)*, *25*(2), 399–404. doi:10.1016/S1001-0742(12)60055-2 PMID:23596962

Nagarajan, V., & Subramanian, K. (2008). Studies on the influence of water cement ratio on the properties of pervious concrete made with natural and recycled aggregate. *Journal on Future Engineering and Technology*, *3*(3), 62–67.

Naghmouchi, I., Espinach, F. X., Mutjé, P., & Boufi, S. (2015). Polypropylene composites based on lignocellulosic fillers: How the filler morphology affects the composite properties. *Materials & Design*, *65*, 454–461. doi:10.1016/j.matdes.2014.09.047

Naik, B., Moon, S. Y., Kim, S. H., & Park, J. Y. (2015). Enhanced photocatalytic generation of hydrogen by Pt-deposited nitrogen-doped TiO_2 hierarchical nanostructures. *Applied Surface Science*, *354*, 347–352. doi:10.1016/j.apsusc.2015.05.102

Namratha, K., & Byrappa, K. (2012). Novel solution routes of synthesis of metal oxide and hybrid metal oxide nanocrystals. *Progress in Crystal Growth and Characterization of Materials*, *58*(1), 14–42. doi:10.1016/j.pcrysgrow.2011.10.005

Nanni, A. (1993). *Fiber-reinforced-plastic for concrete structures: Properties and applications.* Amsterdam: Elsevier Science.

Narita, Y., & Leissa, A. W. (1984). Vibrations of corner point supported shallow shells of rectangular planform. *Earthquake Engineering & Structural Dynamics, 12*(5), 651–661. doi:10.1002/eqe.4290120506

Nasiruddin, S. M., Hambali, A., & Rosidah, J. (2017). Energy absorption analysis of different cross section bumper beam using finite element analysis. *Proceedings of Innovative Research and Industrial Dialogue, 2016*(1), 97–98.

Nayak, S. K., & Mohanty, S. (2010). Sisal glass fiber reinforced PP hybrid composites: Effect of MAPP on the dynamic mechanical and thermal properties. *Journal of Reinforced Plastics and Composites, 29*(10), 1551–1568. doi:10.1177/0731684409337632

Neithalath, N., Bentz, D. P., & Sumanasooriya, M. S. (2010). Predicting the permeability of pervious concrete. *Concrete International, 32*, 35–40.

Neogi, S. D., Karmakar, A., & Chakravorty, D. (2011). Impact response of simply supported skewed hypar shell roofs by finite element. *Journal of Reinforced Plastics and Composites, 30*(21), 1795–1805. doi:10.1177/0731684411418865

Neogi, S. D., Karmakar, A., & Chakravorty, D. (2013). Study of dynamic behavior of multilayered clamped composite skewed hypar shell roofs under impact load. *Journal of Engineering, ID, 192176*, 1–10.

Neogi, S. D., Karmakar, A., & Chakravorty, D. (2017). Finite element analysis of laminated composite skewed hypar shell roofs under oblique impact with friction. *Procedia Engineering, 173*, 314–322. doi:10.1016/j.proeng.2016.12.023

Ni, X., Wang, B., Shen, J., Zhang, Z., Zhou, B., Wu, G., & Wang, H. (2010) Insulator properties of SiO2 nano aerogel in energy-saving structural wallboard. In *Proceedings of 7th China Functional Materials and Applications of Academic Meeting* (pp. 350-352). Changsha, China: Academic Press.

Nicolettoa, G., Rivaa, E., & Di Filippo, A. (2014). High Temperature Fatigue Behavior of Eutectic Al-Si-Alloys used for Piston Production. *Procedia Engineering, 74*, 157–160. doi:10.1016/j.proeng.2014.06.241

Ni, M., Leung, M. K., Leung, D. Y., & Sumathy, K. (2007). A review and recent developments in photocatalytic water-splitting using TiO_2 for hydrogen production. *Renewable & Sustainable Energy Reviews, 11*(3), 401–425. doi:10.1016/j.rser.2005.01.009

Ning, H., Janowski, G. M., Vaidya, U. K., & Husman, G. (2007). Thermoplastic sandwich structure design and manufacturing for the body panel of mass transit vehicle. *Composite Structures, 80*(1), 82–91. doi:10.1016/j.compstruct.2006.04.090

Ning, H., Pillay, S., & Vaidya, U. K. (2009). Design and development of thermoplastic composite roof door for mass transit bus. *Materials & Design, 30*(4), 983–991. doi:10.1016/j.matdes.2008.06.066

Ning, H., Vaidya, U., Janowski, G. M., & Husman, G. (2007). Design, manufacture and analysis of a thermoplastic composite frame structure for mass transit. *Composite Structures*, *80*(1), 105–116. doi:10.1016/j.compstruct.2006.04.036

Niyomwas, S. (2011). Synthesis and Characterization of TiC and TiC-Al2O3 Composite from Wood Dust by Self-Propagating High Temperature Synthesis. *Energy Procedia*, *9*, 522–531. doi:10.1016/j.egypro.2011.09.060

Noor, A. K., Starnes, J. H. Jr, & Peters, J. M. (1996). Nonlinear and postbuckling responses of curved composite panels with cutouts. *Composite Structures*, *34*(2), 213–240. doi:10.1016/0263-8223(95)00147-6

Novak, M., Vojtěch, D., Novak, P., & Vítů, T. (2011). Tribological properties of heat-treated electroless Ni–P coatings on AZ91 alloy. *Applied Surface Science*, *257*(23), 9982–9985. doi:10.1016/j.apsusc.2011.06.119

Nuengmatcha, P., Chanthai, S., Mahachai, R., & Oh, W. C. (2016). Visible light-driven photocatalytic degradation of rhodamine B and industrial dyes (texbrite BAC-L and texbrite NFW-L) by ZnO-graphene-TiO_2 composite. *Journal of Environmental Chemical Engineering*, *4*(2), 2170–2177. doi:10.1016/j.jece.2016.03.045

Nunes dos Santos, W. (2007). Thermal properties of polymers by non-steady-state techniques. *Polymer Testing*, *26*(4), 556–566. doi:10.1016/j.polymertesting.2007.02.005

Obele, C., & Ishidi, E. (2015). Mechanical Properties of Coir Fiber Reinforced Epoxy Resin Composites for Helmet Shell. *Industrial Engineering Letters*, *5*(7), 67–74.

Ohashi, M., Kawakami, S., Yokogawa, Y., & Lai, G.-C. (2005). Spherical aluminum nitride fillers for heat conducting plastic packages. *Journal of the American Ceramic Society*, *88*(9), 2615–2618. doi:10.1111/j.1551-2916.2005.00456.x

Ohno, M., & Li, V. C. (2014). A feasibility study of strain hardening fiber reinforced fly ash-based geopolymer composites. *Construction & Building Materials*, *57*(5), 163–168. doi:10.1016/j.conbuildmat.2014.02.005

Oprisan, G., Taranu, N., Munteanu, V., & Entuc, I. (2010). Application of modern polymeric composite materials in industrial construction. Buletinul Institutului Politehnic din lasi. Sectia Constructii. *Arhitectura*, *56*(3), 121.

Otsuka, K., & Ren, X. (2005). Physical metallurgy of Ti–Ni-based shape memory alloys. *Progress in Materials Science*, *50*(5), 511–678. doi:10.1016/j.pmatsci.2004.10.001

Otto, G. P., Moisés, M. P., Carvalho, G., Rinaldi, A. W., Garcia, J. C., Radovanovic, E., & Fávaro, S. L. (2017). Mechanical properties of a polyurethane hybrid composite with natural lignocellulosic fibers. *Composites. Part B, Engineering*, *110*, 459–465. doi:10.1016/j.compositesb.2016.11.035

Ozbulut, O. E., Hurlebaus, S., & Desroches, R. (2011). Seismic Response Control Using Shape Memory Alloys: A Review. *Journal of Intelligent Material Systems and Structures*, *22*(14), 1531–1549. doi:10.1177/1045389X11411220

Padey, D. (2012). PEEK thermoplastic powers state-of-the-art aerospace engineering for metal replacement. *JEC Composites*, (72), 34-37.

Palaniappa, M., & Seshadri, S. K. (2008). Friction and wear behavior of electroless Ni–P and Ni–W–P alloy coatings. *Wear*, *265*(5), 735–740. doi:10.1016/j.wear.2008.01.002

Palmas, S., Da Pozzo, A., Mascia, M., Vacca, A., & Ricci, P. C. (2012). Sensitization of TiO_2 nanostructures with coumarin 343. *Chemical Engineering Journal*, *211*, 285–292. doi:10.1016/j.cej.2012.09.093

Pal, U., Sandoval, A., Madrid, S. I. U., Corro, G., Sharma, V., & Mohanty, P. (2016). Mixed titanium, silicon, and aluminum oxide nanostructures as novel adsorbent for removal of rhodamine 6G and methylene blue as cationic dyes from aqueous solution. *Chemosphere*, *163*, 142–152. doi:10.1016/j.chemosphere.2016.08.020 PMID:27529381

Pan, C., Kou, K., Jia, Q., Zhang, Y., Wu, G., & Ji, T. (2017). Improved thermal conductivity and dielectric properties of h-BN/PTFE composites via surface treatment by silane coupling agent. *Composites. Part B, Engineering*, *111*, 83–90. doi:10.1016/j.compositesb.2016.11.050

Pandya, K. S., Veerraju, C., & Naik, N. K. (2011). Hybrid composites made of carbon and glass woven fabrics under quasi-static loading. *Materials & Design*, *32*(7), 4094–4099. doi:10.1016/j.matdes.2011.03.003

Panja, B., Das, S. K., & Sahoo, P. (2015). Tribological behaviour of electroless Ni–P coatings in brine solution and optimization of coating parameters using Taguchi based grey relation analysis. *Journal of the Institution of Engineers (India): Series C, 96*(3), 299-309.

Panja, B., Das, S. K., & Sahoo, P. (2016). Tribological behavior of electroless Ni–P coatings in various corrosive environments. *Surface Review and Letters, 23*(5) 1650040. doi: 10.1142/S0218625X16500402

Panja, B., & Sahoo, P. (2014). Taguchi-grey based optimization of friction and wear of electroless Ni-P coatings in acidic environment. *International Journal of Surface Engineering and Interdisciplinary Materials Science*, *2*(2), 53–69. doi:10.4018/IJSEIMS.2014070104

Panja, B., & Sahoo, P. (2015). Tribological behavior of electroless Ni-P coatings in alkaline environment and optimization of coating parameters using Taguchi based grey relational analysis. *Indian Journal of Engineering and Materials Sciences*, *22*(5), 503–512.

Pan, L., Zou, J. J., Wang, S., Huang, Z. F., Zhang, X., & Wang, L. (2013). Enhancement of visible-light-induced photodegradation over hierarchical porous TiO_2 by nonmetal doping and water-mediated dye sensitization. *Applied Surface Science*, *268*, 252–258. doi:10.1016/j.apsusc.2012.12.074

Panneerdhass, R., Gnanavelbabu, A., & Rajkumar, K. (2014). Mechanical Properties of Luffa Fiber and Ground nut Reinforced Epoxy Polymer Hybrid Composites. *Procedia Engineering*, *97*, 2042–2051. doi:10.1016/j.proeng.2014.12.447

Panthapulakkal, S., & Sain, M. (2007). Injection-molded short hemp fiber/glass fiber-reinforced polypropylene hybrid composites—Mechanical, water absorption and thermal properties. *Journal of Applied Polymer Science*, *103*(4), 2432–2441. doi:10.1002/app.25486

Parkinson, R. (1997). Properties and applications of electroless nickel. *Nickel Development Institute Technical Series*, (10081), 1-37.

Park, J. M., Quang, S. T., Hwang, B. S., & DeVries, K. L. (2006). Interfacial evaluation of modified Jute and Hemp fibers/polypropylene (PP)-maleic anhydride polypropylene copolymers (PP-MAPP) composites using micromechanical technique and nondestructive acoustic emission. *Composites Science and Technology*, *66*(15), 2686–2699. doi:10.1016/j.compscitech.2006.03.014

Park, J. O., Kim, K. J., Kang, D. Y., Lee, Y. S., & Kim, Y. H. (2001). An experimental study on the optimization of powder forging process parameter for an aluminium alloy piston. *Journal of Materials Processing Technology*, *113*(1-3), 486–492. doi:10.1016/S0924-0136(01)00663-X

Parra, C., Valcuende, M., & Gómez, F. (2011). Splitting tensile strength and modulus of elasticity of self-compacting concrete. *Construction & Building Materials*, *25*(1), 201–207. doi:10.1016/j.conbuildmat.2010.06.037

Patel, V. A., Vasoya, P. J., Bhuva, B. D., & Parsania, P. H. (2008). Preparation and physicochemical study of hybrid glass-jute (treated and untreated) bisphenol-C-based mixed epoxy phenolic resin composites. *Polymer-Plastics Technology and Engineering*, *47*(8), 842–846. doi:10.1080/03602550802188839

Patil, D. D., Shinde, R. M., & Sawant, S. M. (2016). Design of Carbon/Epoxy Composite Drive Shaft for Low Passenger Vehicles. *International Journal of Current Engineering and Technology*, *6*(5), 1532–1536.

Paul, S. A., Joseph, K., Mathew, G. G., Pothen, L. A., & Thomas, S. (2010). Influence of polarity parameters on the mechanical properties of composites from polypropylene fiber and short banana fiber. *Composites. Part A, Applied Science and Manufacturing*, *41*(10), 1380–1387. doi:10.1016/j.compositesa.2010.04.015

Peng, Y., Xu, Y., Geng, J., Dearn, K. D., & Hu, X. (2017). Tribological assessment of coated piston ring-cylinder liner contacts under bio-oil lubricated conditions. *Tribology International*, *107*, 283–293. doi:10.1016/j.triboint.2016.12.004

Pérez-Larios, A., Lopez, R., Hernandez-Gordillo, A., Tzompantzi, F., Gómez, R., & Torres-Guerra, L. M. (2012). Improved hydrogen production from water splitting using TiO_2–ZnO mixed oxides photocatalysts. *Fuel*, *100*, 139–143. doi:10.1016/j.fuel.2012.02.026

Perraudin, M., & Hummer, E. (2012). BioMobile. ch: incorporating green reinforcements into an automotive body frame. *JEC Composites*, (74).

Peters, S. T. (1998). *Introduction, composite basics and road map*. Springer. doi:10.1007/978-1-4615-6389-1_1

Pickering, K. L., Efendy, M. A., & Le, T. M. (2016). A review of recent developments in natural fibre composites and their mechanical performance. *Composites. Part A, Applied Science and Manufacturing*, *83*, 98–112. doi:10.1016/j.compositesa.2015.08.038

Plengudomkit, R., Okhawilai, M., & Rimdusit, S. (2016). Highly filled graphene-benzoxazine composites as bipolar plates in fuel cell applications. *Polymer Composites*, *37*(6), 1715–1727. doi:10.1002/pc.23344

Pleşa, I., Noţingher, P. V., Schlögl, S., Sumereder, C., & Muhr, M. (2016). Properties of Polymer Composites Used in High-Voltage Applications. *Polymers*, *8*(5), 173–236. doi:10.3390/polym8050173

Popoola, A. P. I., Loto, C. A., Osifuye, C. O., Aigbodion, V. S., & Popoola, O. M. (2016). Corrosion and wear properties of Ni-Sn-P ternary deposits on mild steel via electroless method. *Alexandria Engineering Journal*, *55*(3), 2901–2908. doi:10.1016/j.aej.2016.06.018

Prabu, L., Vinoth, A., & Senthilkumar, D. M. (2016). Experimental investigation on mechanical properties of hybrid and rice husk fiber reinforced polyester based composite for truck body application. *International Journal of Innovative Technology and Research*, *4*(2), 2857–2863.

Pradyumna, S., & Bandyopadhyay, J. N. (2010). Buckling of laminated composite hypars and conoids using a higher-order theory. *The IES Journal Part A: Civil & Structural Engineering*, *3*(2), 85–95.

Preethi, V., & Kanmani, S. (2013). Photocatalytic hydrogen production. *Materials Science in Semiconductor Processing*, *16*(3), 561–575. doi:10.1016/j.mssp.2013.02.001

Premalal, H. G., Ismail, H., & Baharin, A. (2002). Comparison of the mechanical properties of rice husk powder filled polypropylene composites with talc filled polypropylene composites. *Polymer Testing*, *21*(7), 833–839. doi:10.1016/S0142-9418(02)00018-1

Priya, R., & Kanmani, S. (2009). Batch slurry photocatalytic reactors for the generation of hydrogen from sulfide and sulfite waste streams under solar irradiation. *Solar Energy*, *83*(10), 1802–1805. doi:10.1016/j.solener.2009.06.012

Progelhof, R. C., Throne, J. L., & Ruetsch, R. R. (1976). Methods of Predicting the Thermal Conductivity of Composite Systems. *Polymer Engineering and Science*, *16*(9), 615–625. doi:10.1002/pen.760160905

Prusty, B. G., & Satsangi, S. K. (2001). Analysis of stiffened shell for ships and ocean structures by finite element method. *Ocean Engineering*, *28*(6), 621–638. doi:10.1016/S0029-8018(00)00021-4

Puglia, D., Biagiotti, J., & Kenny, J. M. (2005). A review on natural fibre-based composites—Part II: Application of natural reinforcements in composite materials for automotive industry. *Journal of Natural Fibers*, *1*(3), 23–65. doi:10.1300/J395v01n03_03

Pujari, S., Ramakrishna, A., & Kumar, M. S. (2014). Comparison of Jute and Banana Fiber Composites: A Review. *International Journal of Current Engineering and Technology*, (2), 121-126.

Puvanasvaran, A. P., Hisham, S., & Kamil Sued, M. (2011). Investigation on The Mechanical Characteristics Of Sawdust And Chipwood Filled Epoxy. *Journal of Mechanical Engineering and Technology.*, *3*(1), 71–78.

Qatu, M. S., & Leissa, A. W. (1991). Vibration studies for laminated composite twisted cantilever plates. *International Journal of Mechanical Sciences*, *33*(11), 927–940. doi:10.1016/0020-7403(91)90012-R

Qatu, M. S., & Leissa, A. W. (1993). Vibration of shallow shells with two adjacent edges clamped and others free. *Mechanics of Structures and Machines*, *21*(3), 285–301. doi:10.1080/08905459308905190

Rady, M. H. (2016). Computational Stress Analysis of a Composite Multi Leaf Spring. *International Journal of Current Engineering and Technology*, *6*(6), 2239–2242.

Raghavendra Rao, H., Varada Rajulu, A., Ramachandra Reddy, G., & Hemachandra Reddy, K. (2010). Flexural and compressive properties of bamboo and glass fiber-reinforced epoxy hybrid composites. *Journal of Reinforced Plastics and Composites*, *29*(10), 1446–1450. doi:10.1177/0731684409105077

Raghu Raja Pandiyan, K., Kundu, G., Neogi, S., & Patel, J. (2010). Development of Manufacturing Technology for Cab Front using Resin Transfer Molding Process. *Journal of Composite Materials*, *44*(18), 2217–2231. doi:10.1177/0021998310364168

Raghu Raja Pandiyan, K., & Neogi, S. (2014). Development of Resin Transfer Molding Process Using Scaling Down Strategy. *Polymer Composites*, *35*(9), 1683–1639. doi:10.1002/pc.22821

Raghu, K., Noorunnisa Khanam, P., & Venkata Naidu, S. (2010). Chemical resistance studies of silk/sisal fiber-reinforced unsaturated polyester-based hybrid composites. *Journal of Reinforced Plastics and Composites*, *29*(3), 343–345. doi:10.1177/0731684408097770

Rahal, K. (2007). Mechanical properties of concrete with recycled coarse aggregate. *Building and Environment*, *42*(1), 407–415. doi:10.1016/j.buildenv.2005.07.033

Rahman, M. M., Mallik, A. K., & Khan, M. A. (2007). Influences of various surface pretreatments on the mechanical and degradable properties of photografted oil palm fibers. *Journal of Applied Polymer Science*, *105*(5), 3077–3086. doi:10.1002/app.26481

Rahman, R. A., Tamin, M. N., & Kurdi, O. (2008). Stress analysis of heavy duty truck chassis as a preliminary data for its fatigue life prediction using FEM. *Jurnal Mekanikal*, *26*, 76–85.

Rahul, K., Kaushik, K., & Sumit, B. (2014). Optimisation of mechanical properties of epoxy based wood dust reinforced green composite using Taguchi method. *Procedia Material Science*, *5*, 688-696.

Ramalho, A., & Miranda, J. C. (2007). Tribological characterization of electroless NiP coatings lubricated with biolubricants. *Wear*, *263*(1), 592–597. doi:10.1016/j.wear.2006.10.027

Ramdani, N., Derradji, M., Feng, T., Tong, Z., Wang, J., Mokhnache, E., & Liu, W. (2015). Preparation and characterization of thermally-conductive silane-treated silicon nitride filled polybenzoxazine nanocomposites. *Materials Letters*, *155*, 34–37. doi:10.1016/j.matlet.2015.04.097

Ramdani, N., Derradji, M., Wang, J., Mokhnache, E., Liu, W., Liu, Y., & Dong, W. (2016). Experimental and modeling of thermal and dielectric properties of aluminum nitride-reinforce polybenzoxazine hybrids. *Journal of Thermal Analysis and Calorimetry*, *126*(2), 561–570. doi:10.1007/s10973-016-5492-9

Ranganatha, S., Venkatesha, T. V., & Vathsala, K. (2012). Development of high performance electroless Ni–P–HNT composite coatings. *Applied Surface Science*, *263*, 149–156. doi:10.1016/j.apsusc.2012.09.020

Ranjan, R., Bajpai, P. K., & Tyagi, R. K. (2013). Mechanical Characterization of Banana/Sisal Fibre Reinforced PLA Hybrid Composites for Structural Application. *Engineering International*, *1*(1), 39–48. doi:10.18034/ei.v1i1.118

Rao, D. L. V., Venkatappaiah, B., & Shanmugam, I. S. (1985). Two Dimensional Cinematographic Analysis-Handicapped Human Gait. *Biomedical Engineering*, *4*, 288–291.

Rao, K. P. (1978). A rectangular laminated anisotropic shallow thin shell finite element. *Computer Methods in Applied Mechanics and Engineering*, *15*(1), 13–33. doi:10.1016/0045-7825(78)90003-8

Rathnakar, G., & Shivanand, H. K. (2015). A Review of Polymer Based Laminated Composites for Flexural and Shear Properties. *International Journal of Innovative Research in Science Engineering and Technology*, *4*(2), 40–43.

Ravindra, S. R., Rajesh, P., Anil, S., & Saraswati, R. (2014). Optimization of Wear Performance of Aa 5083/10 Wt. % Sic Composites Using Taguchi Method. *Procedia Materials Science*, *6*, 503–511. doi:10.1016/j.mspro.2014.07.064

Ray, D., Sarkar, B. K., Rana, A. K., & Bose, N. R. (2001). Effect of alkali treated jute fibres on composite properties. *Bulletin of Materials Science*, *24*(2), 129–135. doi:10.1007/BF02710089

Rebba, B., & Ramanaiah, N. (2014). Evaluation of mechanical properties of aluminium alloy (Al-2024) reinforced with molybdenum disulphide (MOS2) metal matrix composites. *Procedia Materials Science*, *6*, 1161–1169. doi:10.1016/j.mspro.2014.07.189

Reddy, A. C., & Zitoun, E. (2010). Matrix Al-alloys for silicon carbide particle reinforced metal matrix composites. *Indian Journal of Science and Technology*, *3*(12), 1184–1187.

Reddy, J. N. (1982). Large amplitude flexural vibration of layered composite plates with cutouts. *Journal of Sound and Vibration*, *83*(1), 1–10. doi:10.1016/S0022-460X(82)80071-0

Reddy, J. N., & Chadrashekhara, K. (1985). Geometrically nonlinear transient analysis of laminated doubly curved shells. *International Journal of Non-linear Mechanics*, *20*(2), 79–90. doi:10.1016/0020-7462(85)90002-2

Redman, L., Friman, M., Gärling, T., & Hartig, T. (2013). Quality attributes of public transport that attract car users: A research review. *Transport Policy*, *25*, 119–127. doi:10.1016/j.tranpol.2012.11.005

Regin, A. F., Solanki, S. C., & Saini, J. S. (2008). Heat transfer characteristics of thermal energy storage system using PCM capsules: A review. *Renewable & Sustainable Energy Reviews*, *12*(9), 2438–2458. doi:10.1016/j.rser.2007.06.009

Richard, P., & Cheyrezy, M. (1995). Composition of reactive powder concretes. *Cement and Concrete Research*, *25*(7), 1501–1511. doi:10.1016/0008-8846(95)00144-2

Riddle, Y. W., & Bailer, T. O. (2005). Friction and wear reduction via an Ni-B electroless bath coating for metal alloys. *JOM*, *57*(4), 40–45. doi:10.1007/s11837-005-0080-7

Rissi, G. O., Singh, S. P., Burgess, G., & Singh, J. (2008). Measurement and analysis of truck transport environment in Brazil. *Packaging Technology & Science*, *21*(4), 231–246. doi:10.1002/pts.797

ROHACO. (2015). *TU eindhoven helps builders pioneer recyclable concrete constructions with giant 3D printer.* Retrieved July 30, 2017, from https://www.designboom.com/technology/tu-eindhoven-concrete-3d-printer-06-28-2016/

Rohatgi, P. (1991). Cast aluminum-matrix composites for automotive applications. *Journal of Materials*, *43*(4), 10–15.

Roslan, S. A. H., Hassan, M. Z., Rasid, Z. A., Zaki, S. A., Daud, Y., Aziz, S., ... Ismail, Z. (2015). Mechanical properties of bamboo reinforced epoxy sandwich structure composites. *International Journal of Automotive and Mechanical Engineering*, *12*, 2882–2892. doi:10.15282/ijame.12.2015.7.0242

Rossi, R.E. (1999). Transverse vibrations of thin orthotropic rectangular plates with rectangular cutouts with fixed boundaries. *Journal of Sound and Vibration, 221*(4), 733-736.

Rothon, R., & DeArmitt, C. (2017). *Fillers (Including Fiber Reinforcements)* (8th ed.). Brydson's Plastics Materials.

Rout, J., Misra, M., Tripathy, S. S., Nayak, S. K., & Mohanty, A. K. (2001). The influence of fibre treatment on the performance of coir-polyester composites. *Composites Science and Technology*, *61*(9), 1303–1310. doi:10.1016/S0266-3538(01)00021-5

Rowell, R. M., Caulfield, D. F., Chen, G., Ellis, W. D., Jacobson, R. E., Lange, S. E., . . . Sain, M. (1998). Recent advances in agro-fiberhhermoplastic composites. Proc. Second International Symposium on Natural Polymers and Composites.

Rowell, R. M. (1998). *Economic opportunities in natural fiber-thermoplastic composites.* In P. N. Prasad, J. E. Mark, S. H. Kandil, & Z. H. Kafafi (Eds.), *Science and technology of polymers and advanced materials* (pp. 869–872). New York, NY: Plenum Publishing Corp. doi:10.1007/978-1-4899-0112-5_76

Roy, R. K. (1990). A primer on the Taguchi method. Society of Manufacturing Engineers.

Roy, S., Davim, J. P., Bhowmik, S., & Kumar, K. (2016). *Estimation of mechanical and tribological properties of epoxy based green composites*. Academic Press.

Rozière, E., Granger, S., Turcry, P., & Loukili, A. (2007). Influence of paste volume on shrinkage cracking and fracture properties of self-compacting concrete. *Cement and Concrete Composites*, *29*(8), 626–636. doi:10.1016/j.cemconcomp.2007.03.010

Rozman, H. D., Tay, G. S., Kumar, R. N., Abusamah, A., Ismail, H., & Ishak, Z. M. (2001). Polypropylene–oil palm empty fruit bunch–glass fibre hybrid composites: A preliminary study on the flexural and tensile properties. *European Polymer Journal*, *37*(6), 1283–1291. doi:10.1016/S0014-3057(00)00243-3

Ruban, P., & Sellappa, K. (2014). Development and performance of bench-scale reactor for the photocatalytic generation of hydrogen. *Energy*, *73*, 926–932. doi:10.1016/j.energy.2014.06.107

Safri, S. N. A., Sultan, M. T. H., Jawaid, M., & Jayakrishna, K. (2017). Impact behaviour of hybrid composites for structural applications: A review. *Composites. Part B, Engineering*.

Sagoe-Crentsil, K. K., Brown, T., & Taylor, A. H. (2001). Performance of concrete made with commercially produced coarse recycled concrete aggregate. *Cement and Concrete Research*, *31*(5), 707–712. doi:10.1016/S0008-8846(00)00476-2

Sahin, Y. (2005). The prediction of wear resistance model for the metal matrix composites. *Wear*, *258*(11-12), 1717–1722. doi:10.1016/j.wear.2004.11.024

Şahmaran, M., & Li, V. C. (2009). Durability properties of micro-cracked ECC containing high volumes fly ash. *Cement and Concrete Research*, *39*(11), 1033–1043. doi:10.1016/j.cemconres.2009.07.009

Sahmaran, M., & Yaman, I. O. (2007). Hybrid fiber reinforced self-compacting concrete with a high-volume coarse fly ash. *Construction & Building Materials*, *21*(1), 150–156. doi:10.1016/j.conbuildmat.2005.06.032

Sahoo, P. (2009). Wear behaviour of electroless Ni–P coatings and optimization of process parameters using Taguchi method. *Materials & Design*, *30*(4), 1341–1349. doi:10.1016/j.matdes.2008.06.031

Sahoo, P., & Das, S. K. (2011). Tribology of electroless nickel coatings–a review. *Materials & Design*, *32*(4), 1760–1775. doi:10.1016/j.matdes.2010.11.013

Sahoo, P., & Roy, S. (2017). Tribological behavior of electroless Ni-P, Ni-P-W and Ni-P-Cu coatings: A comparison. *International Journal of Surface Engineering and Interdisciplinary Materials Science*, *5*(1), 1–15. doi:10.4018/IJSEIMS.2017010101

Sahoo, S. (2011). Free vibration of laminated composite hypar shell roofs with cutouts. *Advances in Acoustics and Vibration, ID*, *403684*, 1–13.

Sahoo, S. (2012). Behaviour and optimization aids of composite stiffened hypar shell roofs with cutout under free vibration. *ISRN Civil Engineering, ID, 989785*, 1–14.

Sahoo, S., & Chakravorty, D. (2004). Finite element bending behaviour of composite hyperbolic paraboloidal shells with various edge conditions. *Journal of Strain Analysis for Engineering Design, 39*(5), 499–513. doi:10.1243/0309324041896434

Sahoo, S., & Chakravorty, D. (2005). Finite element vibration characteristics of composite hypar shallow shells with various edge supports. *Journal of Vibration and Control, 11*(10), 1291–1309.

Sahoo, S., & Chakravorty, D. (2006). Stiffened composite hypar shell roofs under free vibration: Behaviour and optimization aids. *Journal of Sound and Vibration, 295*(1-2), 362–377. doi:10.1016/j.jsv.2006.01.012

Sahoo, S., & Chakravorty, D. (2008). Bending of composite stiffened hypar shell roofs under point load. *Journal of Engineering Mechanics, 134*(6), 441–454. doi:10.1061/(ASCE)0733-9399(2008)134:6(441)

Sai Ram, K. S., & Babu, T. S. (2001a). Study of bending of laminated composite shells, Part I: Shells without a Cutout. *Composite Structures, 51*(1), 103–116. doi:10.1016/S0263-8223(00)00129-X

Sai Ram, K. S., & Babu, T. S. (2001b). Study of bending of laminated composite shells, Part II: Shells with a Cutout. *Composite Structures, 51*(1), 117–126. doi:10.1016/S0263-8223(00)00130-6

Sakhtivel, M., Vijayakumar, S., & Ramesh, S. (2014). Production and Characterization of Luffa/coir Reinforced polyproplenecomposite. *Procedia Materials Science, 5*, 739–745. doi:10.1016/j.mspro.2014.07.323

Salerno, M., Landoni, P., & Verganti, R. (2008). Designing foresight studies for Nanoscience and Nanotechnology (NST) future developments. *Technological Forecasting and Social Change, 75*(8), 1202–1223. doi:10.1016/j.techfore.2007.11.011

Samal, S. K., Mohanty, S., & Nayak, S. K. (2009). Banana/Glass Fiber-Reinforced Polypropylene Hybrid Composites: Fabrication and Performance Evaluation. *Polymer-Plastics Technology and Engineering, 48*(4), 397–414. doi:10.1080/03602550902725407

Samsudin, E. M., Hamid, S. B. A., Juan, J. C., Basirun, W. J., & Kandjani, A. E. (2015). Surface modification of mixed-phase hydrogenated TiO_2 and corresponding photocatalytic response. *Applied Surface Science, 359*, 883–896. doi:10.1016/j.apsusc.2015.10.194

Sanada, K., Tada, Y., & Shindo, Y. (2009). Thermal conductivity of polymer composites with close-packed structure of nano and micro fillers. *Composites. Part A, Applied Science and Manufacturing, 40*(6–7), 724–730. doi:10.1016/j.compositesa.2009.02.024

Sanders, J. L. Jr. (1970). Cutouts in Shallow shells. *Journal of Applied Mechanics, 37*(2), 374–383. doi:10.1115/1.3408516

Sankara Narayanan, T. S. N., Krishnaveni, K., & Seshadri, S. K. (2003). Electroless Ni–P/Ni–B duplex coatings: Preparation and evaluation of microhardness, wear and corrosion resistance. *Materials Chemistry and Physics*, *82*(3), 771–779. doi:10.1016/S0254-0584(03)00390-0

Sannino, A. P., & Rack, H. J. (1995). Dry sliding wear of discontinuously reinforced aluminium composites: Review and discussion. *Wear*, *189*(1-2), 1–19. doi:10.1016/0043-1648(95)06657-8

Sapuan, S., Maleque, M., Hameedullah, M., Suddin, M., & Ismail, N. (2005). A note on the conceptual design of polymeric composite automotive bumper system. *Journal of Materials Processing Technology*, *159*(2), 145–151. doi:10.1016/j.jmatprotec.2004.01.063

Sarasini, F., Tirillò, J., D'Altilia, S., Valente, T., Santulli, C., Touchard, F., ... Gaudenzi, P. (2016). Damage tolerance of carbon/flax hybrid composites subjected to low velocity impact. *Composites. Part B, Engineering*, *91*, 144–153. doi:10.1016/j.compositesb.2016.01.050

Sari, M., Prat, E., & Labastire, J. F. (1999). High strength self-compacting concrete original solutions associating organic and inorganic admixtures. *Cement and Concrete Research*, *29*(6), 813–818. doi:10.1016/S0008-8846(99)00037-X

Sasi Kumar, R., Ariraman, M., & Alagar, M. (2015). Studies on dielectric properties of GO reinforced bisphenol-Z polybenzoxazine hybrids. *RSC Advances, 5*, 23787-23797.

Sasikala, R., & Bharadwaj, S. R. (2012). Photocatalytic hydrogen generation from water using solar radiation. *Chemistry Division. BARC Newsletter*, (325), 10-15.

Savi, M. A., De Paula, A. S., & Lagoudas, D. C. (2011). Numerical investigation of an adaptive vibration absorber using shape memory alloys. *Journal of Intelligent Material Systems and Structures*, *22*(1), 67–80. doi:10.1177/1045389X10392612

Saw, S. K., & Datta, C. (2009). Thermo mechanical properties of jute/bagasse hybrid fibre reinforced epoxy thermoset composites. *BioResources*, *4*(4), 1455–1475.

Saw, S. K., Sarkhel, G., & Choudhury, A. (2012). Preparation and characterization of chemically modified Jute–Coir hybrid fiber reinforced epoxy novolac composites. *Journal of Applied Polymer Science*, *125*(4), 3038–3049. doi:10.1002/app.36610

Scholz, M., & Grabowiecki, P. (2007). Review of permeable pavement systems. *Building and Environment*, *42*(11), 3830–3836. doi:10.1016/j.buildenv.2006.11.016

Schwarte, J. (1994). Vibrations of corner point supported rhombic hypar-shells. *Journal of Sound and Vibration*, *175*(1), 105–114. doi:10.1006/jsvi.1994.1314

Sedlacek, F., Bernardin, P., & Lasova, V. (2016). *Design of a composite leaf spring for railway vehicles*. In *Proceedings of the 27th DAAAM International Symposium on Intelligent Manufacturing and Automation*. Vienna, Austria: DAAAM International.

Sen, F., & Kahraman, M. V. (2004). Thermal conductivity and properties of cyanate Ester/nanodiamond composites. *Polymers for Advanced Technologies*, *25*(9), 1020–1026. doi:10.1002/pat.3346

Şerbescu, A., Kypros, P., & Ţăranu, N. (2006). The Efficiency of Basalt Fibres in Strengthening the Reinforced Concrete Beams. The Bulletin of the Polytechnic Institute of Jassy, Construction. *Architecture Section*, *52*(3-4), 47–58.

Serin, I. G., & Göksenli, A. (2015). Effect of annealing temperature on hardness and wear resistance of electroless Ni–B–Mo coatings. *Surface Review and Letters, 22*(5) 1550058. doi: 10.1142/S0218625X15500584

Seshu, P., & Ramamurti, V. (1989). Vibrations of twisted composite plates. *Journal of Aeronautical Society of India*, *41*, 65–70.

Sethi, P. K. (1990). Designing Aids for Physically Handicapped in Developing Countries. *Indian Journal of Physical Medicine and Rehabilitation*, *3*, 1–10.

Sezgin, H., & Berkalp, O. B. (2016). The effect of hybridization on significant characteristics of jute/glass and jute/carbon-reinforced composites. *Journal of Industrial Textiles*. doi:10.1177/1528083716644290

Shahin, K., Zou, G. P., & Taheri, F. (2005). Shape memory alloy wire reinforced composites for structural damage repairs. *Mechanics of Advanced Materials and Structures*, *12*(6), 425–435. doi:10.1080/15376490500259251

Shalin, R. E. (2012). *Polymer Matrix Composites*. Springer Science & Business Media.

Shanmugam, D., & Thiruchitrambalam, M. (2013). Static and dynamic mechanical properties of alkali treated unidirectional continuous palmyra palm leaf stalk fiber/jute fiber reinforced hybrid polyester composites. *Materials & Design*, *50*, 533–542. doi:10.1016/j.matdes.2013.03.048

Sharma, N., Jangra, K.K., & Raj, T. (2015). Fabrication of NiTi Alloy: A Review. *Proc. IMechE Part L: Journal of Materials: Design and Applications*. doi: 10.1177/1464420715622494

Sharma, N., Kumar, K., Raj, T., & Kumar, V. (2016). Porosity exploration of SMA by Taguchi, Regression Analysis and Genetic Programming. *Journal of Intelligent Manufacturing*. doi:10.1007/s10845-016-1236-8

Sharma, N., Raj, T., & Jangra, K. K. (2015). Applications of NiTi: A Review. *Journal of Engineering Technology*, *5*(1), 1–7. doi:10.4103/0976-8580.149472

Sharma, N., Raj, T., & Jangra, K. K. (2016). Micro-structural Evaluation of NiTi powder, Steatite and Steel balls after different Milling Conditions. *Materials and Manufacturing Processes*, *31*(5), 628–632. doi:10.1080/10426914.2015.1004710

Sharma, N., Raj, T., & Jangra, K. K. (2017). Physical and Tribological Characterization of NiTi alloy Fabricated by P/M. *Particulate Science and Technology: An International Journal.*, *35*, 541–546. doi:10.1080/02726351.2016.1171814

Shaw, A., Sriramula, S., Gosling, P. D., & Chryssanthopoulos, M. K. (2010). A critical reliability evaluation of fibre reinforced composite materials based on probabilistic micro and macro-mechanical analysis. *Composites. Part B, Engineering*, *41*(6), 446–453. doi:10.1016/j.compositesb.2010.05.005

Shifu, C., Lei, C., Shen, G., & Gengyu, C. (2005). The preparation of coupled WO_3/TiO_2 photocatalyst by ball milling. *Powder Technology*, *160*(3), 198–202. doi:10.1016/j.powtec.2005.08.012

Shifu, C., Lei, C., Shen, G., & Gengyu, C. (2006). The preparation of coupled SnO_2/TiO_2 photocatalyst by ball milling. *Materials Chemistry and Physics*, *98*(1), 116–120. doi:10.1016/j.matchemphys.2005.08.073

Shifu, C., Sujuan, Z., Wei, L., & Wei, Z. (2008). Preparation and activity evaluation of p–n junction photocatalyst NiO/TiO_2. *Journal of Hazardous Materials*, *155*(1), 320–326. doi:10.1016/j.jhazmat.2007.11.063 PMID:18166268

Shimizu, K. I., Kaneko, T., Fujishima, T., Kodama, T., Yoshida, H., & Kitayama, Y. (2002). Selective oxidation of liquid hydrocarbons over photoirradiated TiO_2 pillared clays. *Applied Catalysis A, General*, *225*(1), 185–191. doi:10.1016/S0926-860X(01)00863-8

Shimoto, T., Baba, K., Matsui, K., Tsukano, J., Maeda, T., & Oyachi, K. (2005). Ultra-thin high-density LSI packaging substrate for advanced CSPs and SiPs. *Microelectronics and Reliability*, *45*(3-4), 567–574. doi:10.1016/j.microrel.2004.08.006

Shokrieh, M. M., & Rezaei, D. (2003). Analysis and optimization of a composite leaf spring. *Composite Structures*, *60*(3), 317–325. doi:10.1016/S0263-8223(02)00349-5

Sim, J., & Park, C. (2011). Compressive strength and resistance to chloride ion penetration and carbonation of recycled aggregate concrete with varying amount of fly ash and fine recycled aggregate. *Waste Management (New York, N.Y.)*, *31*(11), 2352–2360. doi:10.1016/j.wasman.2011.06.014 PMID:21784626

Simsek, E. B. (2017). Solvothermal synthesized boron doped TiO_2 catalysts: Photocatalytic degradation of endocrine disrupting compounds and pharmaceuticals under visible light irradiation. *Applied Catalysis B: Environmental*, *200*, 309–322. doi:10.1016/j.apcatb.2016.07.016

Singh, J., Singh, S. P., & Joneson, E. (2006). Measurement and analysis of US truck vibration for leaf spring and air ride suspensions, and development of tests to simulate these conditions. *Packaging Technology & Science*, *19*(6), 309–323. doi:10.1002/pts.732

Singh, R., Kumari, P., Chavan, P., Datta, S., & Dutta, S. (2017). Synthesis of solvothermal derived TiO_2 nanocrystals supported on ground nano egg shell waste and its utilization for the photocatalytic dye degradation. *Optical Materials*, *73*, 377–383. doi:10.1016/j.optmat.2017.08.040

Sinha, G., & Mukhopadhyay, M. (1995). Static and dynamic analysis of stiffened shells- A review. *Proceedings of Indian National Science Academy*, *61A*(3&4), 195–219.

Sinha, G., Sheikh, A. H., & Mukhopadhyay, M. (1992). A new finite element model for the analysis of arbitrary stiffened shells. *Finite Elements in Analysis and Design*, *12*(3-4), 241–271. doi:10.1016/0168-874X(92)90036-C

Sivakandhan, C., & Prabhu, P. S. (2014). Investigation of Hybrid Composite Drive Shaft under Tensile, Bending and Torsion Testing. *International Journal of Vehicle Structures & Systems*, *6*(4), 110–114.

Sivakumar, K., Iyengar, N. G. R., & Deb, K. (1999). Free vibration of laminated composite plates with cutout. *Journal of Sound and Vibration*, *221*(3), 443–470. doi:10.1006/jsvi.1998.2034

Sivasubramonian, B., Kulkarni, A. M., Rao, G. V., & Krishnan, A. (1997). Free vibration of curved panels with cutouts. *Journal of Sound and Vibration*, *200*(2), 227–234. doi:10.1006/jsvi.1996.0637

Smith, G. (2012). Flax fabric used to create an eco-friendly surfboard. *JEC Composites*, (75).

Solano, J. (2012). OAK Aerocomposit chooses innovative solutions to build the MS-21 composite wing. *JEC Composites*, (72), 44-46.

Sonawane, P. M., & Karnik, M. G. (2017). Experimental Investigation of Optimum level of Reinforcements with in 6061 Al Alloy for better Tribological and Mechanical Properties. *Materials Today: Proceedings*, *4*(2), 3290–3299. doi:10.1016/j.matpr.2017.02.215

Soner, B., & Serdar, O. Y. (2007). Relationship between thermal and sliding wear behavior, of Al6061/Al2O3 metal matrix composites. *Journal of Springer Science*, *42*, 4485–4493.

Song, S. H., Park, K. H., Kim, B. H., Choi, Y. W., Jun, G. H., Lee, D. J., ... Jeon, S. (2013). Enhanced thermal conductivity of epoxy-graphene composites by using non-oxidized graphene flakes with non-covalent functionalization. *Advanced Materials*, *25*(5), 732–735. doi:10.1002/adma.201202736 PMID:23161437

Song, Y. H., Kim, S. G., Lee, S. B., Rhee, K. J., & Kim, T. S. (2003). A study of considering the reliability issues on ASIC/memory integration by SIP (system-in-package) technology. *Microelectronics and Reliability*, *43*(9–11), 1405–1410. doi:10.1016/S0026-2714(03)00251-8

Soutis, C. (2005). Fibre reinforced composites in aircraft construction. *Progress in Aerospace Sciences*, *41*(2), 143–151. doi:10.1016/j.paerosci.2005.02.004

Spoelstra, M. R., & Monti, G. (1999). FRP-confined concrete model. *Journal of Composites for Construction*, *3*(3), 143–150. doi:10.1061/(ASCE)1090-0268(1999)3:3(143)

Sreekala, M. S., Kumaran, M. G., Joseph, S., Jacob, M., & Thomas, S. (2000). Oil palm fibre reinforced phenol formaldehyde composites: Influence of fibre surface modifications on the mechanical performance. *Applied Composite Materials*, *7*(5), 295–329. doi:10.1023/A:1026534006291

Srinivasan, C., Sathish, S., & Vignesh, K. (2014). Mechanical Properties of Chemically Treated LuffaAegyptiaca Fiber Reinforced Epoxy Matrix Composites. *International Journal of Scientific Research and Management*, *2*(10), 1515-1524.

Srivathsan, S. (2014). Comparative study of aluminium alloy pistons manufactured by casting and powder forging. *International Journal of Scientific & Engineering Research*, *5*(6), 662–664.

Staia, M. H., Castillo, E. J., Puchi, E. S., Lewis, B., & Hintermann, H. E. (1996). Wear performance and mechanism of electroless Ni-P coating. *Surface and Coatings Technology*, *86*, 598–602. doi:10.1016/S0257-8972(96)03086-1

Staia, M. H., Conzono, A., Cruz, M. R., Roman, A., Lesage, J., Chicot, D., & Mesmacque, G. (2002). Wear behaviour of silicon carbide/electroless nickel composite coatings at high temperature. *Surface Engineering*, *18*(4), 265–269. doi:10.1179/026708401225005359

Statista. (2016). *Worldwide car production through 2016*. Retrieved November 6, 2017, from https://www.statista.com/statistics/262747/worldwide-automobile-production-since-2000/

Statista. (2017). *Number of vehicles in use worldwide 2015*. Retrieved November 8, 2017, from https://www.statista.com/statistics/281134/number-of-vehicles-in-use-worldwide/

Subbaya, K. M., Suresha, B., Rajendra, N., & Varadarajan, Y. S. (2012). Grey-based Taguchi approach for wear assessment of SiC filled carbon–epoxy Composites. *Materials & Design*, *41*, 124–130. doi:10.1016/j.matdes.2012.04.051

Sudagar, J., Lian, J., & Sha, W. (2013). Electroless nickel, alloy, composite and nano coatings–A critical review. *Journal of Alloys and Compounds*, *571*, 183–204. doi:10.1016/j.jallcom.2013.03.107

Suddin, M. N., Harun, M. R., Hamzah, A. T., & Anuar, S. (2007). A prototype of KBS for material selection in bumper beam design. *Suranaree Journal Science Technology.*, *14*, 215–222.

Sudeepan, J., Kumar, K., Barman, T. K., & Sahoo, P. (2014). Study of friction and wear properties of ABS/Kaolin polymer composites using grey relational technique. *Procedia Technology*, *14*, 196–203. doi:10.1016/j.protcy.2014.08.026

Sulima, I., Jaworska, L., Wyzga, P., & Perek-Nowak, M. (2010). The influence of reinforcing particles on mechanical and tribological properties and microstructure of the steel-TiB2 composites. *Journals of Achievements in Materials and Manufacturing Engineering*, *48*, 52–57.

Summerscales, J., & Short, D. (1978). Carbon fibre and glass fibre hybrid reinforced plastics. *Composites*, *9*(3), 157–166. doi:10.1016/0010-4361(78)90341-5

Suprith, N., Annamalai, K., Naiju, C. D., & Mahadevan, A. (2013). Design and Analysis of Automotive Multi-Leaf Springs Using Composite Materials. *Applied Mechanics and Materials*, *372*, 533–537. . doi:10.4028/www.scientific.net/AMM.372.533

Surappa, M. K. (2003). Aluminium matrix composites: Challenges and opportunitie. *Sadhana*, *28*(1), 319–334. doi:10.1007/BF02717141

Surappa, M. K., Prasad, S. V., & Rohatgi, P. K. (1982). Wear and abrasion of cast Al-alumina particles composites. *Wear*, *77*(3), 295–302. doi:10.1016/0043-1648(82)90055-2

Sutter, Kenner, Pelham, Miller, Polis, Nailadi, … Fikes. (2010). *Comparison of Autoclave and Out-of-Autoclave Composites*. NASA Report. Retrieved from ntrs.nasa.gov/archive/nasa/casi.ntrs.nasa.gov/20110016095.pdf

Swolfs, Y., Gorbatikh, L., & Verpoest, I. (2014). Fibre hybridisation in polymer composites: A review. *Composites. Part A, Applied Science and Manufacturing*, *67*, 181–200. doi:10.1016/j.compositesa.2014.08.027

Taerwe, L. R., & Matthys, S. (1999). FRP for concrete construction: Activities in Europe. *Concrete International*, *21*(10), 33–36.

Taguchi, G. (1990). *Introduction to quality engineering*. Asian Productivity Organization.

Taguchi, G. (1993). *Taguchi on Robust Technology Development: Bringing Quality Engineering Upstream*. New York, NY: ASME press. doi:10.1115/1.800288

Taheri, R., Oguocha, I. N. A., & Yannacopoulos, S. (2001). The tribological characteristics of electroless NiP coatings. *Wear*, *249*(5), 389–396. doi:10.1016/S0043-1648(01)00539-7

Tam, V. W. Y., Gao, X. F., & Tam, C. M. (2005). Microstructural analysis of recycled aggregate concrete produced from two-stage mixing approach. *Cement and Concrete Research*, *35*(6), 1195–1203. doi:10.1016/j.cemconres.2004.10.025

Tan, C., Fung, B. M., Newman, J. K., & Vu, C. (2001). Organic aerogels with very high impact strength. *Advanced Materials*, *13*(9), 644–646. doi:10.1002/1521-4095(200105)13:9<644::AID-ADMA644>3.0.CO;2-#

Tang, B., Wei, H., Zhao, D., & Zhang, S. (2017). Light-heat conversion and thermal conductivity enhancement of PEG/SiO_2 composite PCM by in situ Ti_4O_7 doping. *Solar Energy Materials and Solar Cells*, *161*, 183–189. doi:10.1016/j.solmat.2016.12.003

Taranu, N. (2009). *Composite materials-course notes*. The University of Sheffield Printing Office.

Tataroğlu, S., Ok, E., Kaya, S., Şen, A., Doğan, V. Z., Mecitoğlu, Z., ... Türkmen, H. S. (2016). Design and Manufacturing of a Composite Drive Shaft. *Key Engineering Materials*, *706*, 73–77. doi:10.4028/www.scientific.net/KEM.706.73

Tavman, I. H. (1996). Thermal and mechanical properties of aluminum powder-filled high density polyethylene composites. *Journal of Applied Polymer Science*, *62*(12), 2161–2167. doi:10.1002/(SICI)1097-4628(19961219)62:12<2161::AID-APP19>3.0.CO;2-8

Tay, T. E., Tan, V. B., & Liu, G. (2006). A new integrated micro–macro approach to damage and fracture of composites. *Materials Science and Engineering B*, *132*(1), 138–142. doi:10.1016/j.mseb.2006.02.023

Tee, D. I., Mariatti, M., Azizan, A., See, C. H., & Chong, K. F. (2007). Effect of Silane based coupling agent on the properties of silver nanoparticles filled epoxy composites. *Composites Science and Technology*, *67*(11-12), 2584–2591. doi:10.1016/j.compscitech.2006.12.007

Tekce, H. S., Kumlutas, D., & Tavman, I. H. (2007). Effect of particle shape on thermal conductivity of copper reinforced polymer composites. *Journal of Reinforced Plastics and Composites*, *26*(1), 113–121. doi:10.1177/0731684407072522

Teng, C. C., Ma, C. C. M., Chiou, K. C., Lee, T. M., & Shih, Y. F. (2011). Synergetic effect of hybrid boron nitride and multi-walled carbon nanotubes on the thermal conductivity of epoxy composites. *Materials Chemistry and Physics*, *126*(3), 722–728. doi:10.1016/j.matchemphys.2010.12.053

Tessem, A., Zhao, D., Moll, J., Xu, S., Yang, R., & Li, C. (2017). Effect of filler loading, geometry, dispersion and temperature on thermal conductivity of polymer Nanocomposites. *Polymer Testing*, *57*, 101–106. doi:10.1016/j.polymertesting.2016.11.015

Thierry, B., Merhi, Y., Bilodeau, L., Trépanier, C., & Tabrizian, M. (2002). Nitinol versus stainless steel stents: Acute thrombogenicity study in an ex vivo porcine model. *Biomaterials*, *23*(14), 2997–3005. doi:10.1016/S0142-9612(02)00030-3 PMID:12069342

Thomason, J. L. (1995). The interface region in glass fibre-reinforced epoxy resin composites: 1. Sample preparation, void content and interfacial strength. *Composites*, *26*(7), 467–175. doi:10.1016/0010-4361(95)96804-F

Tian, J., Leng, Y., Cui, H., & Liu, H. (2015). Hydrogenated TiO_2 nanobelts as highly efficient photocatalytic organic dye degradation and hydrogen evolution photocatalyst. *Journal of Hazardous Materials*, *299*, 165–173. doi:10.1016/j.jhazmat.2015.06.019 PMID:26118828

Tibbetts, G. G., Lake, M. L., Strong, K. L., & Rice, B. P. (2007). A review of the fabrication and properties of vapor-grown carbon nanofiber/polymer composites. *Composites Science and Technology*, *67*(7-8), 1709–1718. doi:10.1016/j.compscitech.2006.06.015

Tjong, S. C., & Tam, K. F. (2006). Mechanical and thermal expansion behavior of hipped aluminum-TiB2 composites. *Materials Chemistry and Physics*, *97*(1), 91–97. doi:10.1016/j.matchemphys.2005.07.075

Tomar, A., & Singh, D. (2016). Modelling and Analysis of a Chassis Frame by Using Carbon Fiber and E-Glass Epoxy as Composite Material: A Comparative Study'. *International Research Journal of Engineering and Technology*, *3*(04), 2395–0072.

Torimoto, T., Okawa, Y., Takeda, N., & Yoneyama, H. (1997). Effect of activated carbon content in TiO_2-loaded activated carbon on photodegradation behaviors of dichloromethane. *Journal of Photochemistry and Photobiology A Chemistry*, *103*(1-2), 153–157. doi:10.1016/S1010-6030(96)04503-0

Torres, F. G., & Cubillas, M. L. (2005). Study of the interfacial properties of natural fibre reinforced polyethylene. *Polymer Testing*, *24*(6), 694–698. doi:10.1016/j.polymertesting.2005.05.004

Tosun, N. (2006). Determination of Optimum Parameters for Multi-Performance Characteristics in Drilling by Using Grey Relational Analysis. *International Journal of Advanced Manufacturing Technology*, *28*(5-6), 450–455. doi:10.1007/s00170-004-2386-y

TPI Composites Inc - Home. (2006). Retrieved July 22, 2017, from http://www.tpicomposites.com/English/home/default.aspx

Tsui, K. L. (2007). An overview of Taguchi method and newly developed statistical methods for Robust Design. *IIE Transactions*, *24*(5), 44–56. doi:10.1080/07408179208964244

Tucker, N., & Lindsey, K. (2002). *An introduction to automotive composites*. iSmithers Rapra Publishing.

Tu, H., & Ye, L. (2009). Thermal conductive PS/graphite composites. *Polymers for Advanced Technologies*, *20*(1), 21–27. doi:10.1002/pat.1236

Turkish Composites Manufacturers Association. (2014, September). Kompozitler Dünyası ve Otomotiv. *Composites Turkey*, (8), 28–37.

Uddin, N. (2013). *Developments in Fiber-Reinforced Polymer (FRP) Composites for Civil Engineering*. Elsevier. doi:10.1533/9780857098955

Upadhya, A. R., Dayananda, G. N., Kamalakannan, G. M., Ramaswamy Setty, J., & Christopher Daniel, J. (2011). Autoclaves for Aerospace Applications: Issues and Challenges. International Journal of Aerospace Engineering, Article ID 985871.

Uskoković, V. (2007). Nanotechnologies: What we do not know. *Technology in Society*, *29*(1), 43–61. doi:10.1016/j.techsoc.2006.10.005

Uthayakumar, M., Manikandan, V., Rajini, N., & Jeyaraj, P. (2014). Influence of redmud on the mechanical, damping and chemical resistance properties of banana/polyester hybrid composites. *Materials & Design*, *64*, 270–279. doi:10.1016/j.matdes.2014.07.020

Vaidya, U. K., Samalot, F., Pillay, S., Janowski, G. M., Husman, G., & Gleich, K. (2004). Design and Manufacture of Woven Reinforced Glass/Polypropylene Composites for Mass Transit Floor Structure. *Journal of Composite Materials*, *38*(21), 1949–1971. doi:10.1177/0021998304048418

Vakili, M. H., Ebadi-Dehaghani, H., & Haghshenas-Fard, M. (2011). Crystallization and Thermal Conductivity of $CaCO_3$ Nanoparticle Filled Polypropylene. *Journal of Macromolecular Science, Part B: Physics*, *50*(8), 1637–1645. doi:10.1080/00222348.2010.543033

Van de Weyenberg, I., Ivens, J., De Coster, A., Kino, B., Baetens, E., & Verpoest, I. (2003). Influence of processing and chemical treatment of flax fibres on their composites. *Composites Science and Technology*, *63*(9), 1241–1246. doi:10.1016/S0266-3538(03)00093-9

Varada Rajulu, A., & Devi, R. R. (2008). Flexural properties of ridge gourd/phenolic composites and glass/ridge gourd/phenolic hybrid composites. *Journal of Composite Materials*, *42*(6), 593–601. doi:10.1177/0021998307086197

Vargas Mendoza, L., Barba, A., Bolarin, A., & Sanchez, F. (2006). Age hardening of Ni–P–Mo electroless deposit. *Surface Engineering*, *22*(1), 58–62. doi:10.1179/174329406X84976

Venkata Subba Reddy, E., Varada Rajulu, A., Hemachandra Reddy, K., & Ramachandra Reddy, G. (2010). Chemical resistance and tensile properties of glass and bamboo fibers reinforced polyester hybrid composites. *Journal of Reinforced Plastics and Composites*, *29*(14), 2119–2123. doi:10.1177/0731684409349520

Venkatesh, R. P., Ramanathan, K., & Raman, V. S. (2016). Tensile, Flexual, Impact and Water Absorption Properties of Natural Fibre Reinforced Polyester Hybrid Composites. *Fibres & Textiles in Eastern Europe*, *24*(3), 90–94. doi:10.5604/12303666.1196617

Venkatprasad, S., & Subramanian, R. (2011). Influence of parameters on the dry sliding wear behavior of aluminium/flyash/graphite hybrid metal matrix composite. *European Journal of Scientific Research*, *53*(2), 280–290.

Verma, , Dinesh, Singh, Gaddikeri, Srinivasa, Kumar, & Sundaram. (2013). Development of Vacuum Enhanced Resin Infusion Technology (VERITy) Process for Manufacturing of Primary Aircraft Structures. *Journal of the Indian Institute of Science*, *93*(4), 621–633.

Verma, C. S., Chariar, V. M., & Purohit, R. (2012). Tensile Strength Analysis of bamboo and Layered Laminate Bamboo composites. *International Journal of Engineering Research and Applications.*, *2*(2), 1253–1264.

Verma, D., Jain, S., Zhang, X., Vaidya, A. A., Gaugler, M., & Smith, D. A. (2016). Green route to modification of wood waste, cellulose and hemicellulose using reactive extrusion. [PubMed]. *Carbohydrate Polymers*, *136*, 1238–1250.

Vignesh, P., & Sengottaiyan, M. (2014). A Study on Mechanical Behaviour of Hybrid Reinforced Composite. *IOSR Journal of Mechanical and Civil Engineering*, *11*(6), 68–70. doi:10.9790/1684-11626870

Vijaya Kumar, , Safiulla, & Ahmed. (2013). Analysis of Vacuum Failures during Curing of CFRP Composites. *International Journal of Scientific & Technology Research*, *2*(5), 220–225.

Villa, K., Black, A., Domènech, X., & Peral, J. (2012). Nitrogen doped TiO_2 for hydrogen production under visible light irradiation. *Solar Energy*, *86*(1), 558–566. doi:10.1016/j.solener.2011.10.029

Vishal, S., & Sanjeev, D. (2007). *Synthesis and Interfacial Characterization of Al-4.5wt%Cu/ Zircon Sand/ Silicon Carbide Hybrid Composite* (Master's thesis). Department of Physics and Materials Sciences, Thapar University, Patiala, India.

Vitry, V., & Bonin, L. (2017a). Formation and characterization of multilayers borohydride and hypophosphite reduced electroless nickel deposits. *Electrochimica Acta*, *243*, 7–17. doi:10.1016/j.electacta.2017.04.152

Vitry, V., & Bonin, L. (2017b). Increase of boron content in electroless nickel-boron coating by modification of plating conditions. *Surface and Coatings Technology*, *311*, 164–171. doi:10.1016/j.surfcoat.2017.01.009

Vitry, V., Kanta, A. F., & Delaunois, F. (2010). Initiation and formation of electroless nickel–boron coatings on mild steel: Effect of substrate roughness. *Materials Science and Engineering B, 175*(3), 266–273. doi:10.1016/j.mseb.2010.08.003

Vitry, V., Kanta, A. F., & Delaunois, F. (2011). Mechanical and wear characterization of electroless nickel-boron coatings. *Surface and Coatings Technology, 206*(7), 1879–1885. doi:10.1016/j.surfcoat.2011.08.008

Vitry, V., Kanta, A. F., Dille, J., & Delaunois, F. (2012a). Structural state of electroless nickel–boron deposits (5wt.% B): Characterization by XRD and TEM. *Surface and Coatings Technology, 206*(16), 3444–3449. doi:10.1016/j.surfcoat.2012.02.003

Vitry, V., Sens, A., Kanta, A. F., & Delaunois, F. (2012b). Experimental study on the formation and growth of electroless nickel–boron coatings from borohydride-reduced bath on mild steel. *Applied Surface Science, 263*, 640–647. doi:10.1016/j.apsusc.2012.09.126

Vitry, V., Sens, A., Kanta, A. F., & Delaunois, F. (2012c). Wear and corrosion resistance of heat treated and as-plated Duplex NiP/NiB coatings on 2024 aluminum alloys. *Surface and Coatings Technology, 206*(16), 3421–3427. doi:10.1016/j.surfcoat.2012.01.049

Vivekanandan, R., & Prakash, M. (2014). Selection of alternative material for car roof panel. In *Second International Conference on Current Trends In Engineering and Technology - ICCTET 2014* (pp. 32–40). Academic Press. doi:10.1109/ICCTET.2014.6966259

Vo, H. T., Todd, M., Shi, F. G., Shapiro, A. A., & Edwards, M. (2001). Towards model based engineering of underfill materials: CTE modeling. *Microelectronics Journal, 32*(4), 331–338. doi:10.1016/S0026-2692(00)00152-X

Walczyk, D., & Kuppers, J. (2012). Thermal Press Curing of advanced Thermoset Composite Laminate Parts. *Composites. Part A, Applied Science and Manufacturing, 43*(4), 635–646. doi:10.1016/j.compositesa.2011.12.008

Wan Busu, W. N., Anuar, H., Ahmad, S. H., Rasid, R., & Jamal, N. A. (2010). The mechanical and physical properties of thermoplastic natural rubber hybrid composites reinforced with Hibiscus cannabinus, L and short glass fiber. *Polymer-Plastics Technology and Engineering, 49*(13), 1315–1322. doi:10.1080/03602559.2010.496408

Wang, B., Panigrahi, S., Tabil, L., & Crerar, W. (2007). Pre-treatment of flax fibers for use in rotationally molded biocomposites. *Journal of Reinforced Plastics and Composites, 26*(5), 447–463. doi:10.1177/0731684406072526

Wang, C., Hu, Q., Huang, J., Wu, L., Deng, Z., Liu, Z., & Cao, Y. (2013). Efficient hydrogen production by photocatalytic water splitting using N-doped TiO_2 film. *Applied Surface Science, 283*, 188–192. doi:10.1016/j.apsusc.2013.06.080

Wang, J. J., & Yi, X. S. (2004). Effects of interfacial thermal barrier resistance and particle shape and size on the thermal conductivity of AlN/PI composites. *Composites Science and Technology, 64*(10-11), 1623–1628. doi:10.1016/j.compscitech.2003.11.007

Wang, J., & Mak, C. (2014). Adaptive-passive vibration isolation between non rigid machines and non rigid foundations using a dual-beam periodic structure with shape memory alloy transverse connection. *Journal of Sound and Vibration*, *333*(23), 6005–6023. doi:10.1016/j.jsv.2014.07.010

Wang, J., Yan, F., & Xue, Q. (2009). Friction and wear behavior of ultra-high molecular weight polyethylene sliding against GCr15 steel and electroless Ni–P alloy coating under the lubrication of seawater. *Tribology Letters*, *35*(2), 85–95. doi:10.1007/s11249-009-9435-5

Wang, Q., Jin, R., Zhang, M., & Gao, S. (2017). Solvothermal preparation of Fe-doped TiO_2 nanotube arrays for enhancement in visible light induced photoelectrochemical performance. *Journal of Alloys and Compounds*, *690*, 139–144. doi:10.1016/j.jallcom.2016.07.281

Wang, Q., Xia, H. S., & Zhang, C. H. (2000). Preparation of polymer/inorganic nanoparticles composites through ultrasonic inadiation. *Journal of Applied Polymer Science*, *80*(7-10), 1478–1488.

Wang, S. L., Cheng, Y., Wang, R. R., Sun, J., & Gao, L. (2014). Highly thermal conductive copper nanowire composites with ultralow loading: Toward applications as thermal interface materials. *ACS Applied Materials & Interfaces*, *6*(9), 6481–6486. doi:10.1021/am500009p PMID:24716483

Wang, X., Dong, H., Hu, Z., Qi, Z., & Li, L. (2017). Fabrication of a $Cu_2O/Au/TiO_2$ composite film for efficient photocatalytic hydrogen production from aqueous solution of methanol and glucose. *Materials Science and Engineering B*, *219*, 10–19. doi:10.1016/j.mseb.2017.02.011

Wang, Y., He, Y., Lai, Q., & Fan, M. (2014). Review of the progress in preparing nano TiO_2: An important environmental engineering material. *Journal of Environmental Sciences (China)*, *26*(11), 2139–2177. doi:10.1016/j.jes.2014.09.023 PMID:25458670

Wang, Z., Yang, C., Lin, T., Yin, H., Chen, P., Wan, D., ... Jiang, M. (2013). H-Doped Black Titania with Very High Solar Absorption and Excellent Photocatalysis Enhanced by Localized Surface Plasmon Resonance. *Advanced Functional Materials*, *23*(43), 5444–5450. doi:10.1002/adfm.201300486

Wan, Y., Yu, Y., Cao, L., Zhang, M., Gao, J., & Qi, C. (2016). Corrosion and tribological performance of PTFE-coated electroless nickel boron coatings. *Surface and Coatings Technology*, *307*, 316–323. doi:10.1016/j.surfcoat.2016.09.001

Weller, S. (2012). Out-of-autoclave CF monocoque sports car cockpit chassis. *JEC Composites*, (76), 81-82.

Wen, J., Li, X., Liu, W., Fang, Y., Xie, J., & Xu, Y. (2015). Photocatalysis fundamentals and surface modification of TiO_2 nanomaterials. *Chinese Journal of Catalysis*, *36*(12), 2049–2070. doi:10.1016/S1872-2067(15)60999-8

Werkheiser, M. J., Fiske, M., Edmunson, J., & Khoshnevis, B. (2015). On the development of additive construction technologies for application to development of lunar/martian surface structures using in-situ materials. In *Proceedings of the AIAA SPACE 2015 Conference and Exposition*. AIAA. doi:10.2514/6.2015-4451

Wheels and bogie. (n.d.). Retrieved August 4, 2017, from https://www.shutterstock.com/tr/image-vector/wheels-bogie-part-wagonvector-illustration-eps-237431683

Whitcher, F. D. (1997). Simulation of in vivo loading conditions of nitinol vascular stent structures. *Computers & Structures*, *64*(5-6), 1005–1011. doi:10.1016/S0045-7949(97)00014-X

Williams, E., Shaw, G., & Elahinia, M. (2010). Control of an automotive shape memory alloy mirror actuator. *Mechatronics*, *20*(5), 527–534. doi:10.1016/j.mechatronics.2010.04.002

Wong, Y. W., Lo, K. L., & Shin, F. G. (2001). Electrical and thermal properties of composite of liquid crystalline polymer filled with carbon black. *Journal of Applied Polymer Science*, *82*(6), 1549–1555. doi:10.1002/app.1993

Wooster, T. J., Abrol, S., Hey, J. M., & MacFarlane, D. R. (2004). Thermal, mechanical, and conductivity properties of cyanate ester composites. *Composites. Part A, Applied Science and Manufacturing*, *35*(1), 75–82. doi:10.1016/j.compositesa.2003.09.002

Wu, H., & Kessler, M. R. (2015). Multifunctional cyanate ester nanocomposites reinforced by hexagonal boron nitride after noncovalent biomimetic functionalization. *ACS Applied Materials & Interfaces*, *7*(10), 5915–5926. doi:10.1021/acsami.5b00147 PMID:25726956

Wu, X., Yang, Z., Kuang, W., Tang, Z., & Guo, B. (2017). Coating polyrhodanine onto boron nitride nanosheets for thermally conductive elastomer composites. *Composites. Part A, Applied Science and Manufacturing*, *94*, 77–85. doi:10.1016/j.compositesa.2016.12.015

Wu, X., Yin, S., Dong, Q., Guo, C., Li, H., Kimura, T., & Sato, T. (2013). Synthesis of high visible light active carbon doped TiO_2 photocatalyst by a facile calcination assisted solvothermal method. *Applied Catalysis B: Environmental*, *142*, 450–457. doi:10.1016/j.apcatb.2013.05.052

Xiao, J., Li, J., & Zhang, C. (2005). Mechanical properties of recycled aggregate concrete under uniaxial loading. *Cement and Concrete Research*, *35*(6), 1187–1194. doi:10.1016/j.cemconres.2004.09.020

Xiao, Z., Fang, J., Sun, G., & Li, Q. (2015). Crashworthiness design for functionally graded foam-filled bumper beam. *Advances in Engineering Software*, *85*, 81–95. doi:10.1016/j.advengsoft.2015.03.005

Xiong, J. Y., Li, Y. C., Wang, X. J., Hodgson, P. D., & Wen, C. E. (2008). Titanium- nickel shape memory alloy foams for bone tissue engineering. *Journal of the Mechanical Behavior of Biomedical Materials*, *1*(3), 269–273. doi:10.1016/j.jmbbm.2007.09.003 PMID:19627791

Xu, S., & Sun, D. D. (2009). Significant improvement of photocatalytic hydrogen generation rate over TiO_2 with deposited CuO. *International Journal of Hydrogen Energy*, *34*(15), 6096–6104. doi:10.1016/j.ijhydene.2009.05.119

Xu, W., Zhu, S., Li, Z., Cui, Z., & Yang, X. (2013). Synthesis and catalytic properties of Pd nanoparticles loaded nanoporous TiO_2 material. *Electrochimica Acta*, *114*, 35–41. doi:10.1016/j.electacta.2013.09.084

Xu, Y., & Langford, C. H. (1997). Photoactivity of titanium dioxide supported on MCM41, zeolite X, and zeolite Y. *The Journal of Physical Chemistry B*, *101*(16), 3115–3121. doi:10.1021/jp9624941

Xu, Y., Zheng, W., & Liu, W. (1999). Enhanced photocatalytic activity of supported TiO_2: Dispersing effect of SiO_2. *Journal of Photochemistry and Photobiology A Chemistry*, *122*(1), 57–60. doi:10.1016/S1010-6030(98)00470-5

Xu, Y., Zheng, X., Hu, X., Yin, Y., & Lei, T. (2014). Preparation of the electroless Ni–P and Ni–Cu–P coatings on engine cylinder and their tribological behaviors under bio-oil lubricated conditions. *Surface and Coatings Technology*, *258*, 790–796. doi:10.1016/j.surfcoat.2014.07.079

Yahaya, R., Sapuan, S. M., Jawaid, M., Leman, Z., & Zainudin, E. S. (2015). Effect of layering sequence and chemical treatment on the mechanical properties of woven kenaf–aramid hybrid laminated composites. *Materials & Design*, *67*, 173–179. doi:10.1016/j.matdes.2014.11.024

Yang, J., & Jiang, G. (2003). Experimental study on properties of pervious concrete pavement materials. *Cement and Concrete Research*, *33*(3), 381–386. doi:10.1016/S0008-8846(02)00966-3

Yang, S. Y., Ma, C. C. M., Teng, C. C., Huang, Y. W., Liao, S. H., Huang, Y.-L., ... Chiou, K.-C. (2010). Effect of functionalized carbon nanotubes on the thermal conductivity of epoxy composites. *Carbon*, *48*(3), 592–603. doi:10.1016/j.carbon.2009.08.047

Yang, S., Lin, W., Huang, Y., Tien, H.-W., Wang, J.-Y., Ma, C.-C. M., ... Wang, Y.-S. (2011). Synergetic effects of graphene platelets and carbon nanotubes on the mechanical and thermal properties of epoxy composites. *Carbon*, *49*(3), 793–803. doi:10.1016/j.carbon.2010.10.014

Yan, M., Ying, H. G., & Ma, T. Y. (2008). Improved microhardness and wear resistance of the as-deposited electroless Ni–P coating. *Surface and Coatings Technology*, *202*(24), 5909–5913. doi:10.1016/j.surfcoat.2008.06.180

Yazıcı, H., Yardımcı, M. Y., Aydın, S., & Karabulut, A. Ş. (2009). Mechanical properties of reactive powder concrete containing mineral admixtures under different curing regimes. *Construction & Building Materials*, *23*(3), 1223–1231. doi:10.1016/j.conbuildmat.2008.08.003

Yazıcı, H., Yiğiter, H., Karabulut, A. Ş., & Baradan, B. (2008). Utilization of fly ash and ground granulated blast furnace slag as an alternative silica source in reactive powder concrete. *Fuel*, *87*(12), 2401–2407. doi:10.1016/j.fuel.2008.03.005

Ye, C. M., Shentu, B. Q., & Weng, Z. X. (2006). Thermal conductivity of high density polyethylene filled with graphite. *Journal of Applied Polymer Science*, *101*(6), 3806–3810. doi:10.1002/app.24044

Yeo, H., Islam, A. M., You, N. H., Ahn, S., Goh, M., Hahn, J. R., & Jang, S. G. (2017). Characteristic correlation between liquid crystalline epoxy and alumina filler on thermal conducting properties. *Composites Science and Technology*, *141*, 99–105. doi:10.1016/j.compscitech.2017.01.016

Yoneyama, H., & Torimoto, T. (2000). Titanium dioxide/adsorbent hybrid photocatalysts for photodestruction of organic substances of dilute concentrations. *Catalysis Today*, *58*(2), 133–140. doi:10.1016/S0920-5861(00)00248-0

Yoong, L. S., Chong, F. K., & Dutta, B. K. (2009). Development of copper-doped TiO_2 photocatalyst for hydrogen production under visible light. *Energy*, *34*(10), 1652–1661. doi:10.1016/j.energy.2009.07.024

Yu, A., Bekyarova, E., Zhao, B., & Haddon, R. C. (2007). Graphite nanoplatelet–epoxy composite thermal interface materials. *Journal of Physics and Chemistry Part C*, *111*(21), 7565–7569. doi:10.1021/jp071761s

Yu, A., Ramesh, P., Sun, X., Bekyarova, E., Itkis, M. E., & Haddon, R. C. (2008). Enhanced thermal conductivity in a hybrid graphite nanoplatelet-carbon nanotube filler for epoxy composites. *Advanced Materials*, *20*(24), 4740–4744. doi:10.1002/adma.200800401

Yu, H. S., Luo, S. F., & Wang, Y. R. (2001). A comparative study on the crystallization behavior of electroless Ni-P and Ni-Cu-P deposits. *Surface and Coatings Technology*, *148*(2), 143–148. doi:10.1016/S0257-8972(01)01345-7

Yung, K. C., & Liem, H. (2007). Enhanced thermal conductivity of boron nitride epoxy-matrix composite through multi-modal particle size mixing. *Applied Polymer Science*, *106*(6), 3587–3591. doi:10.1002/app.27027

Yu, S., Hing, P., & Hu, X. (2002). Thermal conductivity of polystyrene–aluminum nitride composite. *Composites. Part A, Applied Science and Manufacturing*, *33*(2), 289–292. doi:10.1016/S1359-835X(01)00107-5

Yusoff, R. B., Takagi, H., & Nakagaito, A. N. (2016). Tensile and flexural properties of polylactic acid-based hybrid green composites reinforced by kenaf, bamboo and coir fibers. *Industrial Crops and Products*, *94*, 562–573. doi:10.1016/j.indcrop.2016.09.017

Zäch, M., Hägglund, C., Chakarov, D., & Kasemo, B. (2006). Nanoscience and nanotechnology for advanced energy systems. *Current Opinion in Solid State and Materials Science*, *10*(3), 132–143. doi:10.1016/j.cossms.2007.04.004

Zahi, S., & Daud, A. R. (2010). Fly ash characterization and application in Al–based Mg alloys. *Journal of Elsevier-. Materials & Design*, *32*(3), 1–10.

Zeng, F., Xie, H., Liu, Q., Li, F., & Tan, W. (2016). Design and optimization of a new composite bumper beam in high-speed frontal crashes. *Structural and Multidisciplinary Optimization*, *53*(1), 115–122. doi:10.1007/s00158-015-1312-2

Zhang, B., Ye, F., Gao, Y., Liu, S. C., Liu, Q., & Ding, D. H. (2016). Fabrication and dielectric properties of BADCy/$Ni_{0.5}Ti_{0.5}NbO_4$ composites for ultra-low-loss printed circuit board application. *Ceramics International*, *42*(1), 234–241. doi:10.1016/j.ceramint.2015.08.099

Zhang, J., & Khoshnevis, B. (2013). Optimal machine operation planning for construction by contour crafting. *Automation in Construction*, *29*(1), 50–67. doi:10.1016/j.autcon.2012.08.006

Zhang, J., Li, Q., Du, J., & Li, X. (2009). Experimental study on mineral admixture and recycled aggregates affecting the rapid chloride permeability of high-performance recycled concrete. *Concrete (London)*, *232*, 94–96.

Zhang, S., Wang, H., Yeung, M., Fang, Y., Yu, H., & Peng, F. (2013). $Cu(OH)_2$-modified TiO_2 nanotube arrays for efficient photocatalytic hydrogen production. *International Journal of Hydrogen Energy*, *38*(18), 7241–7245. doi:10.1016/j.ijhydene.2013.04.027

Zhang, Z., Zhang, L., & Mai, Y. (1995). Wear of ceramic particle-reinforced metal-matrix composites'. *Journal of Materials Science*, *30*(8), 1967–1971. doi:10.1007/BF00353019

Zhao, C., Xu, S., Qin, Y., Chen, H., Zhao, W., Sun, F., & Zhu, X. (2014). Thermally conductive cyanate ester nanocomposites filled with graphene nanosheets and multiwalled carbon nanotubes. *Polymers for Advanced Technologies*, *25*(12), 1546–1552. doi:10.1002/pat.3399

Zhi, C., Bando, Y., Terao, T., Tang, C., Kuwahara, H., & Golberg, D. (2009). Towards thermoconductive, electrically insulating polymeric composites with boron nitride nanotubes as fillers. *Advanced Functional Materials*, *19*(12), 1857–1862. doi:10.1002/adfm.200801435

Zhong, J. S., Wang, Q. Y., & Yu, Y. F. (2015). Solvothermal preparation of Ag nanoparticles sensitized TiO_2 nanotube arrays with enhanced photoelectrochemical performance. *Journal of Alloys and Compounds*, *620*, 168–171. doi:10.1016/j.jallcom.2014.08.205

Zhong, K. (2006). Review of the application of phase change energy-storing material on building energy saving. *Guangzhou Architecture*, *2*, 37–40.

Zhou, H., Zhang, S., & Yang, M. (2007). The effect of heat-transfer passages on the effective thermal conductivity of high filler loading composite materials. *Composites Science and Technology*, *67*(6), 1035–1040. doi:10.1016/j.compscitech.2006.06.004

Zhou, J., Qian, S., Ye, G., Copuroglu, O., Breugel, K. V., & Li, V. C. (2012). Improved fiber distribution and mechanical properties of engineered cementitious composites by adjusting the mixing sequence. *Cement and Concrete Composites*, *34*(3), 342–348. doi:10.1016/j.cemconcomp.2011.11.019

Zhou, T., Wang, X., Liu, X., & Xiong, D. (2017). Improved thermal conductivity of epoxy composites using a hybrid multi-walled carbon nanotube/micro-SiC filler. *Carbon*, *48*(4), 1171–1176. doi:10.1016/j.carbon.2009.11.040

Zhou, W., Qi, S., An, Q., Zhao, H., & Liu, N. (2007). Thermal conductivity of boron nitride reinforced polyethylene composites. *Materials Research Bulletin*, *42*(10), 1863–1873. doi:10.1016/j.materresbull.2006.11.047

Zhou, Y., Liu, F., & Wang, H. (2017). Novel organic–inorganic composites with high thermal conductivity for electronic packaging applications: A key issue review. *Polymer Composites*, *38*(4), 803–813. doi:10.1002/pc.23641

Zhuang, X. J., & Yang, L. (2006). Investigation on the tensile behavior of recycled aggregate concrete. *Journal of Building Materials*, *9*(2), 154–158.

Zhu, G., Yin, H., Yang, C., Cui, H., Wang, Z., Xu, J., ... Huang, F. (2015). Black Titania for Superior Photocatalytic Hydrogen Production and Photoelectrochemical Water Splitting. *ChemCatChem*, *7*(17), 2614–2619. doi:10.1002/cctc.201500488

Zhu, J., & Zäch, M. (2009). Nanostructured materials for photocatalytic hydrogen production. *Current Opinion in Colloid & Interface Science*, *14*(4), 260–269. doi:10.1016/j.cocis.2009.05.003

Zhu, W., & Bartos, P. J. M. (2003). Permeation properties of self-compacting concrete. *Cement and Concrete Research*, *33*(6), 921–926. doi:10.1016/S0008-8846(02)01090-6

Zindani, D., Maity, S. R., Bhowmik, S., & Chakraborty, S. (2017). A material selection approach using the TODIM (TOmada de Decisao Interativa Multicriterio) method and its analysis. *International Journal of Materials Research*, *108*(5), 345–354. doi:10.3139/146.111489

Zou, C. Y., Wang, Y., & Hu, Q. (2009). Experimental study and model predictive of recycled aggregate concrete creep. *Journal of Wuhan University of Technology*, *12*, 94–98.

Zou, Y., Kang, S. Z., Li, X., Qin, L., & Mu, J. (2014). TiO_2 nanosheets loaded with Cu: A low-cost efficient photocatalytic system for hydrogen evolution from water. *International Journal of Hydrogen Energy*, *39*(28), 15403–15410. doi:10.1016/j.ijhydene.2014.07.174

About the Contributors

K. Kumar, B.Tech (Mechanical Engineering, REC [Now NIT], Warangal), MBA (Marketing, IGNOU), and Ph.D. (Engineering, Jadavpur University), is presently an Associate Professor in the Department of Mechanical Engineering, Birla Institute of Technology, Mesra, Ranchi, India. He has 14 years of Teaching & Research and over 11 years of industrial experience in a manufacturing unit of Global repute. His areas of teaching and research interest are Quality Management Systems, Optimization, Non-conventional machining, CAD / CAM, Rapid Prototyping and Composites. He has 9 Patents,12 Book, 3 Edited Book 22 Book Chapters, 113 international Journal publications, 18 International and 8 National Conference publications to his credit. He is on the editorial board and review panel of 7 International and 1 National Journals of repute. He has been felicitated with many awards and honours.

J. Paulo Davim received the Ph.D. degree in Mechanical Engineering in 1997, the M.Sc. degree in Mechanical Engineering (materials and manufacturing processes) in 1991, the Mechanical Engineer degree (MEng-5 years) in 1986, from the University of Porto (FEUP), the Aggregate title (Full Habilitation) from the University of Coimbra in 2005 and the D.Sc. from London Metropolitan University in 2013. He is Eur Ing by FEANI-Brussels and Senior Chartered Engineer by the Portuguese Institution of Engineers with a MBA and Specialist title in Engineering and Industrial Management. Currently, he is Professor at the Department of Mechanical Engineering of the University of Aveiro, Portugal. He has more than 30 years of teaching and research experience in Manufacturing, Materials and Mechanical Engineering with special emphasis in Machining & Tribology. He has also interest in Management & Industrial Engineering and Higher Education for Sustainability & Engineering Education. He has guided large numbers of postdoc, Ph.D. and masters students. He has received several scientific awards. He has worked as evaluator of projects for international research agencies as well as examiner of Ph.D. thesis for many universities. He is the Editor in Chief of several international journals, Guest Editor of journals, books Editor, book Series Editor and Scientific Advisory for many international journals and conferences. Presently, he is an Editorial Board member of 25 international journals and acts as reviewer for more than 80 prestigious Web of Science journals. In addition, he has also published as editor (and co-editor) more than 100 books and as author (and co-author) more than 10 books, 70 book chapters and 400 articles in journals and conferences (more than 200 articles in journals indexed in Web of Science core collection/h-index 43+/5500+ citations and SCOPUS/h-index 52+/7500+ citations).

* * *

Mehdi Azibi graduated from university of science & technology houari Boumedienne, master in process engineering, environmental engineering, in recent years, he worked on the materials specific for water treatment, and composite materials for detection and waste treatment, now, he works at research center as researcher in the field of detection and waste treatment with different materials.

Tapan Kr. Barman received his BME, MME and PhD from Jadavpur University, Kolkata, India. He is presently working as Associate Professor in Mechanical Engineering at Jadavpur University, Kolkata. His areas of interests are characterization of machined surface topography and coatings tribology. He has 67 journals and 40 conference publications to his credit.

Ömer Berk Berkalp, born 1973, is working at ITU since 1995. He has studied on performance characteristics of woven carpets for his M.Sc. degree and then studied on objective evaluation of hydroentangled nonwovens for his Ph.D. Degree. He had been in the United States and worked for NCRC (Nonwovens Cooperative Research Center; North Carolina State University) during his PhD studies for 18 months. He is a full-time professor at Istanbul Technical University, Department of Textile Engineering since 2003.He has worked in various positions at the Department and at the Faculty since 1995. He is the Dean of the Faculty and Department Head of Textile Engineering Program at this time. He is teaching various courses at the undergraduate and graduate level. He is teaching woven fabric design, weaving technology II, nonwovens, statistics and specific fabric structures (coating-laminating techniques, composites etc.). He has research activities on woven and nonwoven fabrics and also in textile recycling and textile composites. He has many publications on these subjects.

Sumit Bhowmik is an Assistant Professor in Department of Mechanical Engineering, National Institute of Technology Silchar, Assam. He received his B. E. in Mechanical Engineering from Tripura Engineering College (Tripura University), India, Master in Mechanical Engineering and PhD from Jadavpur University, Kolkata. He has over 15 years of teaching and research experience. He has more than 20 international journals and international Conference publications to his credit. His areas of interests are Fracture Mechanics, Materials properties, Composites, and Optimization, etc.

Anuj Dixit received his B.Tech in Mechanical Engineering from U.P. Technical University, Lucknow, India, and Master's in Engineering from Department of Mechanical Engineering, Birla Institute of Technology, Mesra, Ranchi, India. He has over 4 years of academic, research and industrial experience. Currently, he is Research Scholar at the Department of Mechanical Engineering, Birla Institute of Technology and Science, Pilani, India. His areas of interests are optimisation, supply chain and composites. He has 7 international journals, 3 international and 1national conference publications to his credit.

Suman Dutta is an Assistant Professor in the Department of Chemical Engineering, Indian Institute of Technology (ISM), Dhanbad, Jharkhand.

Baoguo Han received his PhD in the field of smart materials and structures from the Harbin Institute of Technology, China, in 2005. He is currently a professor of civil engineering in the Dalian University of Technology, China. His main research interests include multifunctional/smart materials and structures, high performance concrete and structures, nanotechnology in civil engineering, and structural health monitoring and traffic detection. He is a member of the editorial board of three international journals and has published 2 book, 9 book chapters and more than 120 technical papers. He was invited to the University of Minnesota and has worked as a visiting research scholar there for 3 years. He was also awarded the New Century Excellent Talents in University by the Ministry of Education of China and the First Prize of Natural Science by the Ministry of Education of China.

Kamal Kumar is an Assistant Professor at Department of Mechanical Engineering, Punjab Engineering College, Chandigarh since 2013. He has completed his PhD in 2012 in the area of non-conventional machining process. He has 20 publications in SCI Journals. His area of research is Advance Manufacturing Processes, Biodegradable Metallic Materials, process optimization etc.

Nisha Kumari, B.E. (Mechanical Engineering, Institute of Technical Education and Research, Bhubaneswar) and M.E. (Design of Mechanical Equipment, BIT Mesra). She has over 1 year of Industrial experience. Her areas of interests are Bio-mechanics, Product and Process Design, CAD/CAM/CAE and Rapid Prototyping. She has 2 Edited Books, 2 Book, 4 Book Chapters, 4 International Journal and 1 National Conference publications to her credit. She has been felicitated with awards and honours.

Saikat Ranjan Maity is an Assistant Professor in Mechanical Engineering Department of National Institute of Technology Silchar. He obtained his Bachelors in Production Engineering from Maulana Abul Kalam Azad University of Technology (Previously known as West Bengal University of Technology), Masters and Ph.D (Engg.) from Jadavpur University. His areas of interest are material selection, multi-criteria decision-making, non-traditional optimization, fatigue and fracture.

Arkadeb Mukhopadhyay received Bachelor of Technology in Mechanical Engineering from West Bengal University of Technology and Master of Mechanical Engineering from Jadavpur University. Currently he is pursuing his doctoral work at Jadavpur University under DST PURSE II scheme of Govt of India. His area of research interest is tribology of coatings.

Raghu Raja Pandiyan earned his Bachelor of Technology in Chemical Engineering at Bharathiar University, India (2003) and obtained his Master of Technology in Chemical Engineering at Anna University, India (2005). Then, he carried out his doctoral program at Composite Applications Laboratory, Department of Chemical Engineering, IIT Kharagpur, India, where he received his Ph. D. in composite product development. Dr. Raghu Raja Pandiyan won prestigious Marie Curie - Amarout Europe Programme Fellowship and joined the Materials Department of Madrid's Institute for Advanced Studies (IMDEA-Materials) as Research Associate in 2011. In IMDEA, his Post Doctoral work was on the title "Development of Integrated Composites for Applications in Aerospace Industries", Sponsored by Airbus Operations S.L., Getafe, Madrid, Spain. Dr. Raghu Raja is currently working as Assistant Professor, Department of Chemical Engineering, Birla Institute of Technology (BIT), Mesra: Ranchi and actively involved in the research and development of new innovative composite material and technology for high performance applications.

Noureddine Ramdani received a Bachelor of Engineering at National Preparatory School of Engineers, Ruiba, Alger in 2007 and joined the Military School polytechnics, Alger, where he was an engineer in propellants and energetic materials. He transferred into the College of Chemical Engineering, Harbin Engineering University in 2011, and received a Ph.D. degree in Materials Science at Harbin Engineering University in 2016, under the supervision of Prof. Dr.Wen-bin Liu. He is now Ph.D. supervisor in Materials Science, postgraduate's supervisor in Materials Science and Chemical Engineering at the Harbin Engineering University. In recent years he focused his research on synthesis and relationship between structure and properties of high-performance resins based composites and nano-composites such as polybenzoxazines, phthalonitrile based composites application as well as the preparations of novel resins and armor composite.

Prasanta Sahoo is Professor at the Department of Mechanical Engineering, Jadavpur University, Kolkata, India. He received Bachelor of Mechanical Engineering from Jadavpur University and MTech and PhD in Mechanical Engineering from Indian Institute of Technology, Kharagpur, India. His research area is tribology and structural mechanics. He is the supervisor of 15 PhD theses and 22 Master's theses. He has authored a textbook on Engineering Tribology (Prentice Hall of India, 2005) and book-chapters in Tribology Research Advances (Nova Science, 2009) and Tribology for Engineers – a practical guide (Woodhead Publishing, 2010). He has co-authored more than 300 technical papers in international journals and conference proceedings. He is Associate Editor (computational mechanics and tribology) of International Journal of Engineering, Science and Technology (Multicraft). He is on the editorial board of Journal of Tribology Research (Serials), Journal of Tribology and Surface Engineering (Nova Science), International Journal of Manufacturing, Materials and Mechanical Engineering (IGI Global), Materials and Design (Elsevier).

Sarmila Sahoo received Bachelor of Engineeing (Civil Engineering) from Bengal Engineering and Science University, Shibpur, Master of Civil Engineering (Structural Engineering) from Jadavpur University and Doctor of Philosophy (Engineering) from Jadavpur University. She was the recipient of Senior Research Fellowship (Council of Scientific and Industrial Research) of Govt. of India. Presently she is working as Associate Professor of Civil Engineeing at Heritage Institute of Technology, Kolkata. Her research interests are finite elements method, vibration of plates and shells, composite materials. She has published over 60 research articles in international journals and conference proceedings. Dr. Sahoo is on the editorial board of two international journals.

Hande Sezgin graduated from Textile Engineering Department of Istanbul Technical University in 2009. She assumed the title of Master of Textile Engineering in 2011. Now, she is studying on PhD at the same department. She has been working at Istanbul Technical University, Textile Technologies and Design Faculty, Textile Engineering Department as a Research Assistant since 2011. Her research activities are woven fabrics, composite structures and e-textiles.

Neeraj Sharma completed his B.Tech, M.Tech. and Ph.D. in 2006, 2012 and 2017 respectively. His areas of research interest are Material development and its processing, conventional and Non-conventional machining method and optimization techniques.

Rohini Singh is a PhD scholar in the Department of Chemical Engineering, Indian Institute of Technology (ISM), Dhanbad, Jharkhand. Her fields of interest are wastewater treatment, renewable energy, nanotechnology and photocatalysis.

Xun Yu is the Chair and Associate Professor of Mechanical Engineering at New York Institute of Technology.

Divya Zindani (BE, Mechanical Engineering, Rajasthan Technical University, Kota), M.E. (Design of Mechanical Equipment, BIT Mesra), presently pursuing PhD (National Institute of Technology, Silchar). He has over 2 years of Industrial experience. His areas of interests are Optimization, Product and Process Design, CAD/CAM/CAE, Rapid prototyping and Material Selection. He has 1 Patent, 2 Books, 3 Book Chapters, 1 SCI journal, 6 Scopus Indexed international journal and 4 International Conference publications to his credit.

Index

S

T

V

W

www.ingramcontent.com/pod-product-compliance
Lightning Source LLC
LaVergne TN
LVHW080608300926
849162LV00005B/1082